Corals - Catalphyllia Jardinei
Elegant coral, meatcoral
- Acropora / staghorn

Volume One

THE REEF AQUARIUM

A Comprehensive Guide to the Identification and Care of Tropical Marine Invertebrates
by J. Charles Delbeek and Julian Sprung

Book Design by Daniel N. Ramirez

Published by Ricordea Publishing
Coconut Grove, Florida 33133

"Whereas today the average aquarium uses white sterile corals, it is entirely possible that the aquarium of tomorrow will contain nothing but living corals and fish..."

—Robert P.L. Straughan, 1973
The Marine Collector's Guide

First Edition, 1994.

Published by Ricordea Publishing,
4016 El Prado Blvd.,
Coconut Grove,
Florida, USA, 33133.

Printed and bound by Arnoldo Mondadori, Verona, Italy.

Design and production by Daniel N. Ramirez Design, Inc.
Cover photo: Top view of Julian's reef tank. Photographer: Léon Corry.
Page ii photo: John Haydock's reef tank. Photographer: J. C. Delbeek.

ISBN 1-883693-12-8
Library of Congress Catalog Card No: 94-67338

Delbeek, J. Charles, 1958 and Julian Sprung, 1966 - The Reef Aquarium

Table of Contents

This book is dedicated to our parents:
Jan Carel and Ruby Delbeek
and
Doris and Stewart Sprung
for their many years of encouragement, love and support.

Letters from the Authors:

When I first became interested in keeping marine invertebrates, especially corals, I was very frustrated by the lack of information available about how to recognize them and how to keep them healthy in a marine aquarium. Although I was lucky enough to have access to large amounts of scientific literature that was helpful in identification and ecology, there was little available on how to keep invertebrates alive in closed systems. For this information I had to turn to Dutch, German and other European hobbyist literature. Again, I considered myself fortunate to have had access to these, and to have been able to read and understand them! It soon became obvious that very few North Americans had similar opportunities. That was when the idea for this book began to coalesce in my brain. What if I could bring together the scientific literature, the European literature and the experiences of myself and others into a form that could be made available to all? With this idea in mind I approached Julian Sprung, with whom I had become acquainted earlier, with a proposal to co-author a comprehensive guide to keeping marine invertebrates in home aquariums.

The aim of this book is to help the hobbyist and professional aquarist better understand the organisms that we are attempting to keep, and to provide information that has proven successful in maintaining them in closed systems for long periods of time. We offer this information in the hope that it will help avoid the needless loss of animals or their mistreatment in captivity. Not all of you may agree with what you see in this book, and that is fine. We are presenting you with guidelines that have worked for us and others. They may work for you, or you may develop other methods. The important issue is not who is right or wrong but the degree of success and the welfare of the animals.

Although I am sensitive to the ecological problems facing coral reefs around the world, I am still in favour of maintaining corals in closed systems for several reasons. To begin with, a miniature reef aquarium is a tremendous educational tool. For many people, such an aquarium may be their first and only encounter with living corals. It is extremely difficult to arouse the public's awareness about the plight of something they have never seen.

The simple act of keeping such organisms raises the awareness of the aquarist. Not only do aquarists become more aware of the variety of marine life, but they also quickly come to appreciate the extremely narrow environmental conditions that these organisms require to exist, affording greater appreciation of the natural environment. These individuals spread this awareness and concern for the environment to others.

In addition, very little scientific work has been done on keeping and growing corals in captivity. Most of what we know today about keeping corals comes from the marine aquarium hobby. This information will become increasingly valuable as marine invertebrates, especially corals, become recognized in medical circles as sources of new products in the treatment of human diseases. The techniques for captive propagation pioneered by hobbyists are already beginning to have an impact. As you read this, numerous aquaculture projects are already underway in the Caribbean, Europe, the Indo-Pacific, North America and the Red Sea, to propagate corals for medical research, public aquariums and home aquariums.

There are many greater threats to the world's coral reefs than the collection of live specimens for the aquarium trade. Among these are the collection of coral skeletons for the curio trade, which is minor but still many times larger than the hobby trade, siltation of inshore reefs due to deforestation and dredging, dynamiting of reefs for food fish, the use of coral for construction material on many tropical islands such as the Maldives, oil spills, industrial and residential sewage run-off, eutrophication, and ship groundings on reefs, to name just a few.

Many of the live corals collected for the aquarium trade are soft corals, Octocorallia, which can reproduce asexually from small fragments and quickly regrow. The majority of the live stony corals collected for the aquarium trade come from inshore areas, and many are not true reef-building species. Collectors remove branches off of large colonies, leaving them otherwise intact. When whole colonies are taken, they are mostly fist-sized. The impact is highly selective. This cannot be said of any of the other human impacts, which are indiscriminant by nature and tend to affect whole reefs and the major reef-building species of stony corals. Unlike other human influences, only the keeping of live corals offers a return benefit to the reef through the raising of public awareness and education.

Coral reefs are a renewable natural resource that, if managed properly and wisely, should be able to withstand the demands of the marine hobby. Obviously indiscriminant collecting should be discouraged and studies should be funded to determine what the exact level of sustainable yield can be. The aim of the whole hobby should be to preserve and propagate marine life, not destroy it.

I hope this book will improve the success in keeping invertebrates, and encourage their propagation in closed systems, thereby lowering the demand for wild-caught specimens.

Finally, I would like to dedicate this book to the memory of my late father, Jan Carel Delbeek, who introduced me to the sea and its creatures and passed on his love of them to me. I only wish he could have seen the end result of the seed he planted; I miss you Pop.

J. Charles Delbeek
Hon. B.Sc., B.Ed., M.Sc.
Toronto, Canada
May 1, 1993

As I write this letter I am nearly finished working on my sections for this wonderful book that was the dream of my co-author, J. Charles Delbeek. It has been a long project for both of us. Though we were quite capable of writing this book when we started, we have both learned from our aquariums, our travels, the scientific literature, and communication with other reef-keeping enthusiasts and coral reef researchers.

Our combined experience and the invaluable input of other, generous and talented hobbyists and researchers makes this book shine. We thank these people for their generosity and acknowledge that this book exists in such a complete form because of their input.

We did not want to simply re-state what has been written before about coral reef animals. This is a book about aquariums as whole ecosystems, how to create one, and how to care for some of the special living gems that are a part of the whole reef environment. With this book we lay a new foundation for research, and fortify existing ground-breaking work on the care and propagation of corals and tridacnid clams in aquaria. This book, and a subsequent text we are preparing about soft corals and anemones, follows other texts on the subject of reef keeping that have particular emphasis on the identification and biology of the animals and plants of the reef environment. Our books naturally follow Peter Wilkens' and Johannes Birkholz's first-of-a-kind texts on invertebrates and marine aquariums, originally published in German but recently available in English. Other original work includes Rene Catala's *Carnival Sous la Mer,* Riseley's *Tropical Marine Aquaria: The Natural System,* Robert Straughan's *Keeping Live Corals and Invertebrates,* and Dr. Walter Adey's *Dynamic Aquaria.* Alf J. Nilsen's and Svein Fossa's series of books in Swedish and German also offer a comprehensive view of modern approaches to keeping marine aquariums, and extensive information about the biology of the reef and its inhabitants. These will be available in English in the near future. We are additionally indebted to Dr. J.E.N. Veron for his masterpiece, *Corals of Australia and The Indo-Pacific,* which is the best identification guide to stony corals from that region, with magnificent photographic records.

With our books we wish to promote the understanding and appreciation of both captive ecosystems and the natural environment. To that end we emphasize that this is a book about loving coral reef organisms, their natural habitat, and the planet. Charles and I are opposed to any activities that harm the marine environment. We know that this hobby can proceed with harmless impact, in a responsible manner, and we encourage and promote with this book the propagation of specimens to further reduce the small impact from the collection of wild specimens. Our descriptions of methods of care, transportation and handling are intended for the improvement of aquarist success, and to set a standard for proper industry technique.

Furthermore, we have tried to ensure that the identifications in this book are accurate, not merely the passed-down pseudonyms of the trade. Along with the correct scientific name for each specimen we have included the common trade names familiar to hobbyists, dealers, and importers. The common names are for reference, but truly mean nothing since they are often shared by different species. We believe that it is essential for hobbyists to use the correct scientific names for their specimens, so that exchanges of information about them are accurate, and published observations can have greater scientific value and credibility. It is possible that we have made errors in the identification of some organisms, or that revision of taxonomy will make some changes. That is inevitable. Fortunately the photographs and our descriptions of the care of these species will remain accurate!

Charles and I have dedicated this book to our parents. I also wish to acknowledge my brothers Eric, Elliot, and Brad, who have each played a role in shaping my appreciation of science and the marine environment, and Sam Solomon, my late grandfather, who effused a child-like sense of wonder and awe at all things from nature, especially things from the sea.

Watching a section of reef grow is both a learning experience and a source of immense satisfaction. I sincerely hope that you gain as much pleasure and continued fascination with your aquarium as I do with mine, and that you turn to this book for reference and guidance in the study and care of your captive ecosystem.

Julian Sprung B.Sc.
Miami Beach, FL
April, 1993

Acknowledgments:

We would like to extend our special thanks to the following people: Dr. Walter Adey, Marj Awai, Karin Beitinger, Stanley Brown, Dr. Robert W. Buddemeier, Roger Bull, John Burleson, Sheila Byers, Royal Ontario Museum, Dr. Stephen Cairns, Dr. Bruce Carlson, Merrill Cohen, Dr. Phillip Dustan, Svein Fosså, Thomas A. Frakes, Steven Gill, Tim Goertemiller, Rick Graff, Santiago Guetierrez, Beth Hayden, Gerald Heslinga, Jill Johnson, Jens Krarup, Brian LaPointe, Karen Loveland, Dr. John Lucas, Dr. Frank Maturo, Scott W. Michael, Martin A. Moe, Jr. and Barbara Moe, Alf Jacob Nilsen, Dr. James Norris, Mike Paletta, Dr. Gerhard Pohl, Huntsman Marine Laboratory, Carmen Ramirez, Eugene Shinn, Eric Sisitsky, Dietrich Stüber and the members of the Berlin Association for Marine Aquaristics, Steve Tyree, Sonja and Jan VanBuuren, Dr. John Veron, Peter Wilkens, and Dr. Gary Williams. We are also grateful to Daniel Ramirez for putting our words, the photographs, and our ideas together in a beautiful format, and for his patience with us!

We thank Dr. Craig Bingman, Dr. Bruce Carlson, Anne Folan, Thomas A. Frakes, Gerald Heslinga, and Martin A. Moe, Jr. who read parts of the manuscript and gave us invaluable suggestions. They are not responsible for any errors that may still remain. We also wish to thank the various tank owners, retailers and wholesalers who generously allowed us into their homes and businesses to photograph their corals and aquariums.

Foreward

We live in the middle of a very great explosion. I am, of course, referring to the cultural, technological, and scientific explosion that surrounds us. It began only a few hundred years ago. Up till that time one's world was never much different from the world that one's great grandfather occupied. The same tools, the same culture, the same earth—the life of one generation was much the same as those of the recent past. Technology and development crawled at a snail's pace for over two million years, almost the entire time line of humanity on the Earth. Suddenly, within the lifetimes of many of you who read these words, we have electricity, we have the internal combustion engine, we have nuclear power, we have four billion people, we have had world wide wars, we have a global economy, we are actually changing the planet's ecosystem, we have personal computers, we have reached the moon, we break the bonds of gravity and go into space, we have supersonic jumbo jets and air conditioning, and we have reef tanks. We who love aquatic life and work to bend the realm of Neptune to our will are in the middle of our own little technological and scientific explosion.

Just as the great masses of humanity have accepted our times as normal and do not face each new day trembling with awe and wonder at the explosive rate of change in our societies, we marine aquarists are so caught up with our new capabilities that we do not realize just how quickly our hobby is changing and growing. Although freshwater fish have been kept and bred in captivity for many thousands of years, efforts to breed marine fish date back to just the 1850's, and commercial success in breeding marine fish extends back to the 1950's. A few species of tropical marine fish have been bred in quantity since the early 1970's. And, with few exceptions, soft and stony corals have been kept and bred in captivity only since the mid-1980's. Thus, many aquarists alive today have seen the inception of almost every modern innovation in the science and art of keeping aquatic organisms.

Not so long ago, when freshwater fish were kept in glass jars and pitch sealed tanks, servants were instructed to often stir the water with their hands to keep the fish alive and well by preventing accumulation of stagnant vapors. And if you can look back to the 1950's and early 60's you can remember slate bottomed tanks, and, for marine aquarists, cast iron frames and rusty "stainless" steel framed tanks with bubble-up filters and incandescent tank lights. The first steps in the modern marine aquarium revolution—all glass tanks, artificial sea salts, hang-on-the-back power filters, fluorescent light fixtures designed for marine aquariums, undergravel filters, and a knowledge of the nitrogen cycle—appeared in the mid 60's and were basically the state of the art until the mid 80's. In the 50's and early 60's, despite the publications of pioneers like Robert P. L. Straughan and Helen Simkatis and the limited success of many actual marine aquarists, the popular perception was that marine fish were impossible to keep. In the 60's and 70's it was generally conceded that a dedicated aquarist could keep marine fish and some hardy invertebrates alive for a long time, even years, but breeding marine fish at home and keeping corals and other photosynthetic invertebrates was an impossible task. Many marine aquarists proved otherwise as the 80's flew by and now in the 90's, the actual creation of captive reefs with soft and hard corals in healthy and breeding condition in a 10, 30, 50, 100, or even larger gallon tanks is the accepted epitome of the marine aquarist's art.

It took a technological explosion to make this possible. Hobbyists applied the essentials of many new concepts and techniques in filtration, lighting, and water quality management and carved new frontiers in the creation of marine aquarium systems. Many advanced hobbyists created new business to provide this expertise and equipment to other hobbyists, and the availability of this equipment and technology changed the hobby with an explosive force. And, as in any time of rapid technological expansion, conflicting ideas, opinions, and results foment controversy. The use of copper, use and composition of artificial sea salts, and the effect and even the existence of the nitrogen cycle in new tanks were controversies of the 60's that still echo today. Trickle filters, wet/dry filters, lighting, and use of algae scrubbers are subjects of some current controversies.

Aquarists are pragmatic if nothing else, and systems and techniques that prove to be the most effective, despite contradicting popular and scientific views, almost always become the prevailing practice over time. Home aquarists have no economic or philosophic need to follow a "company line" or stay within the boundaries of any particular technology or methodology. We do what is necessary to maintain the animals and plants we admire in our own little "captive oceans" and we are constantly pushing back the boundaries of knowledge and technology in our quest. The cutting edge of aquarium technology is often found in the home tanks of aquarists rather than in scientific laboratories or huge public aquariums.

And this brings me to this book by Charles Delbeek and Julian Sprung. Both of these young men are dedicated marine aquarists and scientists. I first met Julian when he was 12 years old and came for a tour of our clownfish hatchery in the Florida Keys and I've known Charles since I visited the Toronto marine society in 1988, a year before the first Marine Aquarium Conference of North America. They have both kept reef tanks, almost since reef tanks were first possible to keep, and they have traveled extensively all over the US, Canada and Europe talking with aquarists of all means and philosophies and all levels of success—always looking for what works and analyzing why it works. They are also keenly aware that correct identification of a species is of utmost importance not only when reporting on the occurrence, ecology and distribution of that species, but also when describing the water quality and physical parameters required for its maintenance and propagation in a captive marine aquarium system.

From this background, they have written a book on the art and science of creating and maintaining a captive reef tank ecosystem. They provide detailed information on the water quality and physical parameters necessary to maintain stony corals and tridacnid clams in reef aquaria and describe the techniques used by the most successful marine reef aquarists to achieve a good culture environment. They also provide the information that allows an aquarist to identify the specimens in their care, and detail current knowledge on the culture and maintenance requirements for each species commonly found in reef tanks today. Most books written about corals and other invertebrate builders of tropical reefs are concerned with the natural history, taxonomy, distribution and ecology of these fascinating organisms. Delbeek and Sprung, however, use these basics as the foundation for a text that brings the capability to maintain and propagate stony corals into the realm of any advanced aquarist. This book will play an important role in the continuing explosion of the marine reef tank technology.

Martin A. Moe, Jr.

Preface

The marine aquarium hobby, if we disregard the millenium-old cultivation and propagation of goldfish, is older than the freshwater hobby. Approximately in the mid-nineteenth century the marine hobby began in Great Britain, from whence it rapidly expanded throughout central Europe. Before me lies the oldest reference book in the German languge, which appeared in 1872 and is, for me, a bibliophilic treasure.

What is, then, this curious phenomenon which lures people of the most diverse education, origins, and traditions to engross themselves in the ever mysterious world of marinelife? All life began in the water, as Thales of Miletus so nobly proclaimed in the seventh century of the common era. Who would gainsay him? Even as the salt in our own blood testifies, our origin too lies somewhere in the oceans, which were and still are sacred to many cultures and nature-worshipping religions. It often seems to me a miracle that our hobby unites me in the most heartfelt friendship with countless people throughout the world.

Particularly in our own time, attention to the ocean and its abundant manifestations of life should become the chief concern of mankind, for without these vast water-expanses of our blue planet, we ourselves cannot survive. The marine aquarium hobby pursued with patience, dedication and a sense of responsibility, can contribute something to this end. By setting up even a very small aquarium, we all assume an obligation to provide the best possible living conditions for the animals and plants in our care, and never to forget the wonder of the greater whole. With this in mind, I wish the authors, both of whom possess a rich store of practical experience, a wide dissemination of their book, which was, I know, written with great idealism.

Peter Wilkens,
Winterthur, Switzerland
June 1993

Introduction

This book is the fruit of over 3 years of writing and library research, 6 years of photographing reef aquarium organisms across two continents, and decades of combined experience with reef aquarium systems and coral reefs in the natural habitat. Our overall aim is to educate marine aquarists and scientists about the techniques that provide successful maintenance of stony corals and tridacnid clams in closed systems, and to give the proper identification of the species most commonly maintained in aquariums. The long range goal of this activity is to foster a better understanding of reef ecosystems and the biology of life associated with them, and to promote the propagation and breeding of corals, clams, and other reef life.

Chapter 1 deals with the reef in nature; the types of reefs, the physical and biological factors that shape reefs and their distribution, and reef zonation.

Chapter 2 builds on the framework of chapter 1, by delving into the regulation of nutrients in natural coral reefs. Topics covered include: nitrogen, phosphorus and carbon cycling; sources and sinks, import and export of nutrients; and the effects of eutrophic conditions on coral reefs.

Chapter 3 reviews the biology of stony corals. Topics include classification, polyp anatomy, the relationship between corals and zooxanthellae, calcification, coral nutrition, depth zonation of corals, ultraviolet light and corals, reproduction strategies of stony corals, and the competitive and defense systems of corals.

Chapter 4 deals with the biology of tridacnid clams. This chapter covers the distribution, morphology and anatomy, nutritional requirements, reproduction and growth of tridacnid clams.

Longnose Hawkfish, *Cirrhitichthys oxycephalus*. J.C. Delbeek.

Chapter 5 describes the techniques required to maintain stony corals and tridacnid clams in captivity. Of all the chapters in this book, this is the one that may create the greatest controversy, mainly because there are many different techniques that can be employed for running reef aquariums. However, we feel confident that the methodology we encourage here gives the best results in the long-term. Our review of alternative techniques is based on our personal experience with them and our opinion of the results. Topics covered include, tank selection and location, filtration, methods for establishing reef aquariums, alternative filtration systems, and methods for achieving water movement.

Chapter 6 deals with the topic of lighting, both on the reef and in the aquarium.

Chapter 7 covers aquascaping materials, construction techniques and aquascape designs.

Chapter 8 deals primarily with maintaining water quality and the parameters that should be measured and controlled. This chapter also includes discussions of element additions and instructions for preparing your own trace element solutions.

Chapter 9 is devoted to a discussion of nutrient regulation in closed systems and offers techniques for controlling nutrient levels. There is also a thorough discussion of controlling undesirable algae, along with suggestions for eradicating them.

Chapter 10 covers the major diseases, parasites and pests that can be present in reef aquariums containing corals and tridacnid clams. These include bacteria, protozoans, anthozoans, flatworms, polychaete worms, snails, nudibranchs, crabs and shrimp. We have also written a trouble-shooting section that includes the majority of problems one may encounter with organism health.

Although this book deals mainly with stony corals and tridacnid clams, many pests and parasites affect soft corals also, so we have included these for the sake of completeness, and for the majority of aquarists who also keep soft corals in the same aquarium as stony corals. We will cover in a separate book the identification and care of the soft coral genera commonly kept in aquariums.

Chapter 11 provides the basic information required for the collection and shipment of corals and tridacnid clams. We provide this information primarily for the importers and their collectors, in the hope that this will help reduce mortality in coral shipments. We do not condone illegal collection and/or shipment of corals and clams. It is also hoped that the majority of coral keepers will use this information to help them ship propagated specimens among themselves.

Chapter 12 contains information on the purchase and proper placement of tridacnid clams in the aquarium. This is then followed by individual species descriptions with notes on their colouration, distinguishing features, similar species, natural habitat and aquarium care.

Chapter 13 covers the stony corals most commonly kept in the marine aquarium trade. The chapter begins with a discussion of propagation techniques and guidelines for purchasing healthy stony corals. For each coral genus we give the scientific and common name, colouration and distinguishing features with a discussion of similar species, the natural habitat, and care in the aquarium. Where it is known, information is also provided on propagation and reproduction in the aquarium and in nature.

Chapter 14 offers a view of some of the successful reef aquariums that we have had the pleasure of seeing over the last few years. These aquariums serve as an indication that the techniques we describe do work. We hope that these aquariums will serve as sources of inspiration as well.

Appendix A contains a table outlining the light and water movement requirements for all of the stony corals and tridacnid clams mentioned in this book. This offers a quick reference for the aquarist and provides information that we know we have both been asked many times!

Appendix B offers a brief listing of some of the daily, monthly and yearly duties that should be performed to maintain your ecosystem in peak condition.

Appendix C provides the aquarist with a listing of suggested readings.

Appendix D provides a listing of aquarium societies that are strictly marine or have an active marine segment. We have spoken at many of these societies, and we strongly urge you to join them. Many of these societies produce highly informative newsletters and this alone is often a good reason to join.

Ethical and Ecological Concerns About Our Hobby

Throughout the history of marine aquarium keeping, some people have questioned whether confining wild marinelife in a captive environment is morally wrong, and if the methods of capture and handling are cruel and careless. Even now, as we aquarists are able to duplicate a reef environment in a closed system, affording the fantastic opportunity to study a whole ecosystem in captivity, we are especially aware that the stuff of our hobby still is taken from the wild. We must question whether we are doing any harm to the natural environment in supplying the large demand by aquarists around the world.

Live Rock

The "backbone" of a successful reef ecosystem is live rock. Live rock is described in detail in Aquascaping, chapter 7. The center of the controversy concerning the impact of reef tanks on the environment, is the collection of this live rock, and it's reputation for involving habitat destruction. It is easy to understand how such harvesting from the reef can be perceived as destructive. In the 1970's Jacques Y. Cousteau showed pictures of Maldives Islands' residents carrying huge blocks of limestone taken from the reef for use in constructing roads and houses. Cousteau warned that this activity threatened their very livelihood since the reef they were removing was the only buffer to protect them against the fury of the sea in the event of powerful storms. This image sticks in our minds. When the figures are given of how many hundreds of tons of live rock are taken for our hobby from rubble areas surrounding reefs, it is easy to imagine ruthless individuals using dynamite to blast away huge chunks of reef, thus destroying what took eons to build, simply to earn a few dollars and please some rich hobbyist in a distant land. This image is a myth in the minds of emotionally charged environmentalists, who simply are unaware of the facts involving live rock collection and coral reef ecology.

While we estimate that one thousand tons of live rock (a figure including all pieces attached to anemones and soft corals) are removed each year from the vicinity of coral reefs and hardbottoms around the globe for the marine aquarium hobby, none of these rocks are blasted or dredged from the reef. It is far more practical to harvest rock that has already broken loose from the reef during storms, and accumulated in rubble or shingle piles. The constant wave-driven tumbling of the rocks, in combination with the action of boring and grazing organisms, ultimately reduces much of the loose rock to sand. Rocks are selected for their shape, size and types of growths occurring on them. Therefore, only some of the rock in a given area is suitable, and any method other than picking them up one by one would be counter-productive, as it would lack selectivity, and would break the rocks. These rocks, by the way, do not take long to grow. Most are chiefly composed of the old skeletons of *Acropora* and *Porites* stony corals, which are among the fastest growing types (up to 15 cm (6 in.) per year; Veron, 1986). An average rock might only represent two to three years of original coral growth, and a year or more subsequent growth after the coral died. Also included in the rocks taken are the large shells of conchs, clams, and the aggregates of coralline algae, sand, sponges, and calcifying tube worms, any of which can produce substantial solid mass in a few years. These rocks are not produced on a geologic time scale, as is commonly mis-perceived because of the confusion arising from the name "rock", and confusion over the difference between the slow net growth of a coral reef and a reef's fast gross production of calcium carbonate. The reefs are more akin to giant slow growing trees, and live rocks are like leaves, seeds and twigs that the trees shed on a regular basis.

The other element in the name "live rock", adds even more confusion. The description "live" suggests that the rock taken from reef areas is live coral. In fact, in Indonesia where it is legal to do so, live corals are taken for the aquarium hobby, but they are not confused with live rocks. In Indonesia, and elsewhere, soft corals and anemones are often taken with a bit of the substrate that they are growing on, a procedure which sometimes involves chipping or chiseling the substrate off the reef. Though most chipped pieces are insignificantly small bases for the targeted specimen, here we have potential room for abuse, for if the pieces being chipped off are large, then the practice actually does involve breaking up the surface complexity of the reef, and removal of stable habitat.

Opponents of the collection of live rock also know that by virtue of their origin, some live rocks may have small button-sized colonies of living coral on them. This is a problem that cannot easily be resolved. On the one hand, these buttons of living coral are growths only a few months old, which have the misfortune of having landed on a substrate in an unstable area. Their days are numbered until the next storm that will bury them or grind them away. On the other hand, where the collection of live coral is illegal, there is supposedly "zero tolerance" for removal even of these little buttons. This, in combination with the perception that taking live rock is a means of habitat destruction, has been the fuel that has ended this practice commercially in Hawaii. Furthermore, a moratorium on live rock collection in Florida state waters was implemented in 1989, and a ban on landings (in Florida) of live rock collected in the federal waters of the United States was scheduled to come into effect by 1995. Florida set yearly collection quotas, with reductions of the take each year until the complete ban on landings in the state would take effect. With a harvest season beginning July 1992, by February 1993 the state stopped the landing of live rock from federal waters, since the allotted yearly quota had already been reached. It was expected that no live rock would be harvested until the beginning of the next season in July 1993. However, a group of live rock collectors called the Florida Live Rock Alliance filed a case against the state in federal court that won them an injunction, re-opening the collection of live rock immediately. Their case claimed that the state does not have the power nor authority over federal waters to prohibit the taking of live rock there, and cannot prohibit the landing of federal live rock. Furthermore, their case included mandamus action for the state to put into effect procedures for obtaining aquacultural bottom leases. The state had already promised to make provisions for bottom leases as an alternative to the harvest of wild live rock, but bureaucratic red tape stalled the efforts of those who sought to begin live rock aquaculture. As we complete this book, the federal government has decided on a harvest ban to take effect January 1, 1996, with an annual harvest quota until then of not more than 485,000 lbs. The state has also begun issuing bottom leases for aquaculture.

The progress made by the Florida Live Rock Alliance is unprecedented in the aquarium industry, though similar cases exist in the food fish and sponging industries. Nevertheless, it seems likely that time is running out for the collection of marinelife, live rock and corals in particular. Environmental groups lobby governments to ban live rock collection and the importation of wild-caught marinelife for aquariums. Such decisions are critical to the survival of our hobby, though their benefit to the natural reef is questionable. We also support protection of the marine environment and coral reefs. We believe that environmental activists are over-reacting when they concentrate their efforts on stopping the activities of marinelife fishermen. We can understand, however, that the marinelife fisherman is an easy target. They are few and have little money to spare to fight the forces that try to put them out of business.

It doesn't seem fair that while there is zero tolerance for an incidental by-catch of button-sized live coral by marinelife fisherman, outside the aquarium industry those who fish for food in the shellfish industry suffer no such restrictions nor harassment by enforcement agencies and environmental groups. Here coral buttons *are* a tolerated by-catch. Furthermore, trawlers disturb huge tracts of bottom in the same area where collection of ornamental marinelife is prohibited. The ornamental marinelife, which is a non-targeted by-catch of the trawlers, is allowed to die on deck and is shoveled back to sea, a dead, wasted resource. In the same location where marinelife fishermen are not allowed to use their handnets to catch some species of juvenile fish that appear abundantly in the spring and summer months, the sports fisherman may spear, or catch on a hook and line, the adults that produce these offspring. Finally, the worst insult of all is the permission to dredge boat channels and renourish eroding beaches. This is done mainly to preserve property value at the expense of nearshore reefs and live bottom areas. It seems that when it comes to a question of money and power, massive habitat destruction is allowed. The aquarium industry has neither the money nor the power, and is an easy scapegoat to blame for the destruction caused by too many people in an environmentally sensitive area. Surely a double standard is at work here.

It is the destruction of habitat and disturbance to the environment that has a real impact on an ecosystem's health. The aquarium hobby does not employ destructive methods in the supply of living substrates and invertebrates for reef aquariums. In fact, in Indonesia, coral collection is done in areas already severely damaged by human industry and pollution (Wilkens, 1990). These areas offer proximity to the airports from which the specimens are shipped. They also contain large populations of popular aquarium species, such as mushroom anemones, that often proliferate in such nutrient enriched nearshore waters. These facts aside, the progress in the marine aquarium hobby owing to the availability of living substrates, has allowed it the opportunity to follow in the footsteps of its sister, the freshwater hobby. Captive propagation is now a realistic alternative to supply the living substrates and many of the invertebrates that currently are being harvested. Captive propagation also presents the opportunity to grow corals to repair damaged reefs, and to provide materials to the medical and pharmaceutical industries.

When this book is published there will be several companies working on cultivation of live rock, corals and other items, and within a few years it will not be so unusual to have a successful reef aquarium containing rock, corals, plants, and anemones all grown in captivity. This is not a dream, it is happening right now! On the hobbyist level alone, there are thousands of pieces of tank-raised soft and stony corals that have been exchanged among fellow aquarists.

It is our opinion that it's a small sacrifice when we take a piece of the ocean for our aquariums. We gain knowledge from exposure to the mystery of the sea in our home, and our aquaria foster increased public awareness of the frailty of the marine environment. Our experience can be used for understanding the reefs and managing human behavior to be in harmony with them and our planet.

We aquarists are well aware that not all activities related to the aquarium industry are without reproach. It is an amazing fact that for over twenty years the hobby has allowed the use of sodium cyanide in the capture of fish. Not only has this practice survived but it has also flourished and spread beyond its core of operation in the Philippines to other collecting locales such as Indonesia.

This fishing practice produces a high mortality rate during and immediately following capture, and in the home aquarium. It is also believed that cyanide fishing damages habitat and decimates fish populations to such an extent that collectors must move on to new areas. This results is a spread of this destructive practice. The fishermen even collect their food fish with cyanide, so they are unwittingly poisoning themselves, their families, and others. Much has been written in popular aquarium literature about the situation, and environmental organizations such as the International Marinelife Alliance (IMA), aquarium societies, businesses, and the Canadian Government have played a role in reversing the cyanide practice, by funding net training programs and educating the collectors about proper collection and handling techniques. Still, the success has been small, and the use of cyanide has continued to grow and spread to new areas. It will take much planning, negotiation, effort, diplomacy and organization to end this practice. It is by no means a simple matter. For example, stopping the use of cyanide in the collection of aquarium fish would not necessarily stop its use in collecting food fish, a larger industry with greater impact on the environment. Furthermore, stopping the cyanide fishing altogether could encourage an escalation of the likewise illegal and much more destructive use of dynamite for (food) fish collecting. One encouraging project involves the implementation of tests using ion specific probes for the detection of cyanide. Under this proposal, selected fish will be tested before shipment out of the Philippines. If cyanide is detected in the test fish, then the entire shipment is confiscated. This approach is being pursued by the International Marinelife Alliance (IMA) USA and Philippines, and once the testing procedure is determined reliable, its implementation will have an impact on cyanide fishing.

Another complicated project whose time has come is the implementation of standards and regulations concerning the handling and holding practices involved in the tropical marine fish and invertebrate industry. It is entirely possible for the aquarium industry to police itself, but there has been no reason or pressure for it to do so. This situation cannot last for long as the loss of life that occurs from poor handling or inadequate holding facilities will not always be accepted or tolerated. In the long run it behooves everyone to do the best job possible. Perhaps through the efforts of national or international aquarium societies, or organizations such as the American Association of Zoological Parks and Aquariums (AAZPA), it will be possible to create standards that may be implemented for certification of people involved in the industry, and inspection of their facilities to ensure that they meet all criteria for the animals' care. If certification can be perceived as a desirable attribute for attracting customers, rather than just excessive regulation and a nuisance, it might be possible to win acceptance of this idea.

In the reference section of this book, please find a list of aquarium societies and other organizations with their addresses. By joining and becoming involved in these you can become a part of the movement to end the cyanide fishing practice, and to make self-regulation of the aquarium industry a reality. Furthermore you will make new friendships and promote sharing of information that makes beginning hobbyists achieve success more readily.

This hobby you are involved in, while it is primarily for your visual and intellectual satisfaction, is also one that involves the husbandry of living organisms. Therefore, it is your responsibility to involve yourself beyond the scope of your aquarium to insure that your investment is not encouraging practices that are harmful to the environment or careless with the very creatures you keep.

Chapter One

The Reef in Nature

One of the main goals in keeping coral reef aquariums should be to recreate the environment as closely as possible. Not only the design of the aquarium, but also the selection of the organisms should be based upon knowledge of the natural environment. Yet, there are almost as many different reef environments as there are marine aquarium hobbyists. This fact makes it almost impossible to generalize when it comes to recommending strategies for recreating reef environments.

Too often aquarists try to combine organisms from different areas of the reef into one system, expecting them to survive together. When they don't survive, the system or the quality of the organism is blamed. Often, the lack of knowledge of the conditions that occur on the reef where the organism originated is the problem.

Before one can start to duplicate the various physical and chemical conditions that occur on and near coral reefs, knowledge of these parameters in nature is essential. There are a few key factors that are important in shaping the composition of flora and fauna on a reef. These are temperature, depth, light, water motion and nutrient levels. In addition, the ranges of these factors and the way they interact with each other are often characteristic of certain zones within a coral reef.

Distribution of Coral Reefs

One imagines coral reefs surrounding palm covered islands with year round perfect weather. In reality, coral reefs are found over a much wider range of environments, within certain limits. Tropical coral reefs are usually restricted to areas where warm water and bright light prevail, though reefs do exist in extreme localities such as in the Persian Gulf or the Gulf of Mexico, where the water temperature may fall to below 16 °C (60 °F) in winter. Favourable conditions of light and temperature are generally found between the latitudes of 30° N and 30° S.

A school of *Anthias* feeds on plankton in the water column on a Red Sea reef full of colourful corals. Allan Storace.

In the Pacific region, coral diversity is greatest in the region surrounding the Philippines and Indonesia, with more than 65 genera of stony corals found there. As one moves further away from this region, the number of stony coral genera decreases so that only 13 exist in Hawaii and 7 off the western coast of Panama.

Within the Indian Ocean, the Chagos archipelago and the Maldive Islands are high diversity regions with greater than 60 genera of stony corals represented (Wood, 1983). The Caribbean contains many different stony corals but the number of genera is much lower than in the Pacific. The highest diversity in the Caribbean occurs off the coast of Belize, with only some 35 genera known (Wood, 1983).

Within these regions, extensive coral growths are found in waters that have surface temperatures that range from 25 °C to 30 °C (77-86 °F) throughout the year and rarely fall below 20 °C (71°F) (McConnaughey, 1978; Randall and Myers, 1983). Yet even in the tropics, there are areas where cooler waters can prevent coral reef formation. In certain regions such as off parts of the west coast of Australia and off tropical western South America, winds and currents can cause upwelling of cooler, deeper water, which can limit coral reef formation (McConnaughey, 1978).

Besides temperature, the other main requirement for corals is bright light. Due to the light absorptive qualities of water, reef-building corals are restricted to certain depths. Therefore, the number of species of reef-building corals decreases at depths beyond 30 m (60 ft.), so that by 100 m (300 ft.), most reef-building corals have disappeared altogether (Randall and Myers, 1983). Since different species of coral have different tolerances to both maximum and minimum light levels, light intensity greatly affects both coral distribution and variations in coral reef community structure (Veron, 1986). The reason for this depth restriction is that many corals form a symbiotic relationship with marine algae called zooxanthellae. They depend on these algae to provide them with energy. The algae, in turn, depend on light to manufacture this energy. We will go into more detail on the relationship between zooxanthellae, light and corals in Chapters 3 and 6.

Based on the presence or absence of zooxanthellae, stony corals are divided into two distinct groups. Those that have zooxanth-ellae are termed hermatypic while those that do not are called ahermatypic. Hermatypic corals can deposit calcium at a much greater rate due to the presence of zooxanthellae. This gives them a competitive advantage over ahermatypic corals. Therefore, they are the major contributors in the building of coral reefs. The bright orange Flower Coral, *Tubastrea aurea*, so familiar to hobbyists, is an ahermatypic coral. These corals are commonly found growing under overhangs or in crevices. This is not because they can't withstand light, but is due to the simple fact that they cannot compete successfully for space with the faster growing hermatypic corals.

Of course bright light is of little use unless the water is clear enough to allow the light to penetrate to support coral growth at depth. In areas where weather or water conditions keep particulates in suspension, the light is greatly reduced and so is coral growth. For example, in places where rivers empty into coastal areas (e.g. off the coast of north eastern Brazil), the degree of siltation and turbidity can limit coral growth and can greatly influence coral diversity.

Table 1.1
Light levels (lux) measured at 12:00 noon, in tropical waters.
After van Ommen, 1992

Depth	Minimum	Maximum	Yearly Average/whole day (including clouds)
0 m	114543	126520	77420
5 m	28636	31630	19355
10 m	16039	17713	10839
20 m	9136	10122	6194
100 m	46	51	31

Water motion is another influence on the formation of coral reefs and on the type of community that develops. In areas of extreme wave action, either constantly or periodically, very few species of coral can gain a foothold. Those that do, often take on stunted growth forms and may not resemble specimens of the same species from sheltered areas (Veron, 1986). Also, those specimens that are found in areas with strong currents often have different growth forms than those from calmer areas. Water movement is an important factor in respiration, feeding, photosynthesis and calcification in hermatypic corals, with higher flows increasing productivity (Sebens and Johnson, 1991; Patterson et al., 1991). Please refer to chapter 5, subheading water movement, for additional information about water movement and its affect on different marine organisms.

The level of nutrients (predominantly nitrates and phosphates) in the water can also have an effect on coral reef distribution and the type of corals found in an area. In regions where nutrient levels are high, reef building stony corals tend to be less common while leather corals *(e.g. Sarcophyton* spp.*)*, marine algae, zoanthid anemones and mushroom anemones (Corallimorpharia) can dominate. Such areas are usually found along coastlines, near river outlets and in areas where man-made pollution predominates (Wilkens, 1990). Coral reefs can occur in a wide range of oceanic and coastal environments, from water very low in plant nutrients to water that is quite enriched. In the enriched environments, the

Sargassum algae overgrowing stony corals, Whitsunday Islands, Australia. J. Sprung.

predominance of nutrients can promote macroalgae, which are easily able to out-grow corals.

Types of Coral Reefs

Coral reefs can be divided into two basic classes, shelf reefs and oceanic reefs (Wood, 1983). Shelf reefs grow on the continental shelf that surround the major continents, while oceanic reefs develop in deeper waters, beyond the continental shelf. Oceanic reefs generally develop along the slopes of submarine mountains, often of volcanic origin, that reach within a few hundred metres of the surface. In some cases, such as Hawaii and Tahiti, these volcanic mountains will extend above the water and reefs develop around them.

There are various reef types that can occur in both regions. Fringing reefs, barrier reefs and atolls are believed to correspond with stages of reef development, with fringing reefs being the earliest (see Myers, 1989). As the island sinks over time, the coral reef grows upwards to keep pace. Platform and bank reefs are found mainly on continental shelves, although some bank reefs are oceanic (Wood, 1983).

Fringing reefs are found close to shore and have poorly developed back regions while their seaward sides exhibit strong coral growth (Myers, 1989; Wood, 1983). These reefs are the most common shelf reefs (Wood, 1983). The reef surrounding Pago Pago Harbour in American Samoa, is an excellent example of a fringing reef.

As the reef grows upwards or outwards and the sea floor subsides, the inner-most corals die-off, leaving a shallow lagoon between the outer reef and the shore (Myers, 1989). Such reefs are known as barrier reefs. Barrier reefs form along the edge of a continental shelf, such as the Great Barrier Reef in Australia, or around islands such as Belau (Palau), and Bora Bora in French Polynesia. If the sea level continues to rise or the island continues to sink, the central land mass will disappear, leaving a ring-shaped reef surrounding a central lagoon. In some cases, small islands will develop on the exposed reef. This type of reef is known as an atoll, an example of which is Enewetak Atoll in the Marshall Islands. This sequence of reef development, from fringing reef to atoll, was first proposed by Charles Darwin, and has since been substantiated by drill bores made on existing atolls.

Fringing Reef in Pago Pago harbour, American Samoa. B. Carlson.

A barrier reef surrounding Bora Bora in French polynesia. S. W. Michael.

Ngemelis Island and drop-off, Palau. B. Carlson.

Along some regions of a continental shelf, raised sections will reach close enough to the surface to allow corals to develop. If the reef is shallow enough to be exposed at low tides it is known as a platform reef. If it is not shallow enough to be exposed it is known as a bank reef (Wood, 1983).

Reef Zonation

Within each reef there are a series of regions or zones that contain characteristic organisms. The types of zones are determined by their location and physiographic conditions. There are various zones found on reefs, and these can vary depending on geographic location, the type of coral reef, the degree of turbidity and the type of wave action. There are also distinct differences between Caribbean and Pacific reefs with respect to zonation and species composition.

We present here some general descriptions of use to the aquarist. Since the vast majority of our aquarium specimens are from the Indo-Pacific, we will concentrate on the zonation of reefs in this region. Kaplan (1982) offers an excellent overview of Caribbean reef types and zonation. Refer to the references mentioned in this chapter for other examples of reef zonation.

greater nutrient inputs and lower wave energy (Veron, 1986). Likewise, windward reefs tend to show greater zonation in both morphology and animal life than leeward reefs (Stoddart, 1973). Oceanic reefs encounter the cleanest of open ocean water, poor in nutrients. Reefs in these areas exhibit pronounced zonation due primarily to strong wave action. In a study of soft coral distribution across a section of the Great Barrier Reef in Australia, Dinesen (1983) found that branching soft corals of the family Nephtheidae (e.g. *Litophyton, Nephthea, Lemnalia,* etc.) are more common on outer- and mid-shelf reefs than on inner-shelf reefs. In contrast, *Briareum, Cespitularia, Clavularia, Pachyclavularia* and *Xenia* are more common on inner-shelf reefs. The Leather Coral genera *Lobophytum, Sarcophyton* and *Sinularia* are common in all areas. Inner-shelf reefs exhibit a lower number of soft coral genera than outer- and mid-shelf reefs. Therefore, the high water clarity, good circulation and associated conditions are apparently favourable to growth of a wider range of soft coral genera than turbid, nutrient enriched, inner-shelf reefs (Dinesen, 1983).

Outer Oceanic Barrier Reefs and Atolls

Deep Forereef

The deep forereef extends below 20 m (66 ft.). In this region the coral colonies are composed of broad, flattened plates, adapted to catch as much light as possible. *Echinopora, Porites, Turbinaria* and some species of *Acropora* are usually the dominant stony corals (Veron,1986). Soft corals such as gorgonians and *Dendronephthya* are usually abundant at these depths.

Upper Reef Slope

This zone extends from the surface down to 20 m. This is the most densely populated zone, and at 20 m has the greatest diversity of stony and soft corals (Veron, 1986). In the shallower regions stony coral cover generally exceeds soft coral cover (Dinesen, 1983). Soft coral genera found in this region include *Lemnalia, Lobophytum, Nephthea, Sarcophyton, Sinularia* and *Xenia*; the zoanthid *Palythoa* is also common (Dinesen, 1983). This region is often composed of steep terraced or spur and groove formations. Commonly encountered stony corals include various species of *Acropora, Favia, Favites, Leptoseris, Lobophyllia, Plerogyra, Pocillopora, Porites* and *Stylophora,* as well as *Millepora* (Stoddart, 1973).

Reef Front

This region is exposed during low tide and receives the greatest wave action. The corals tend to be short and robust in structure, in response to growing in such a high energy area. Stony coral populations tend to be dominated by pocilloporids and acroporids (Veron, 1986). *Lobophytum, Sarcophyton* and *Sinularia* are the most abundant soft corals in these higher wave energy areas (Dinesen,1983).

The outer reef flat lies above the mean low-water mark of spring tides. Seaward regions contain spur and groove zones in which fleshy green algae, sponges, encrusting stony and soft corals can grow in protected grooves and tunnels formed by the algae growth (Stoddart, 1973). These include *Acropora, Goniastrea, Favia, Favites*, fungiids, the soft coral *Lobophytum* and the zoanthid *Palythoa*. The algae are constantly being removed by wave action and are carried towards the leeward side of the reef, providing an input of nutrients. Some areas are devoid of coral growth and are dominated by coralline algae encrusted surfaces. In areas of lower wave energy or in leeward areas, extensive coral growth may be present that is contiguous with the coral flat. On some Indian Ocean reefs, seagrass beds are found in this area, composed of *Thalassia, Cymodocea* and *Syringodium* (Stoddart, 1973). In the Caribbean, *Thalassia* and *Syringodium* also may occur here, though they are more typical of back reef margins and bays.

Coral Reef Flat

Behind the outer reef flat lies an area of aligned corals, where tongues of sand and rubble extend leeward giving a pronounced spur and groove appearance. Stony corals such as *Acropora* and *Pocillopora* are dominant in these areas, along with *Goniastrea, Favia, Favites* and *Porites* (Veron, 1986; Davies et al.,1984). Waves carry sand and rubble from these areas to the sand flat and lagoon areas behind them (Davies et al., 1984).

The coral reef flat gives way to a sandy area in some reefs. These areas are formed by the erosion and destruction of reef flats. Stony corals such as *Acropora* and *Pocillopora* gradually colonize this area, followed by more massive corals such as *Favia*, extending the coral flat leeward. Similarly the sand flats also move leeward, into the lagoon, burying patch reefs (Davies et al.,1984).

A fringing reef surrounding a small island in Fiji. Sonja VanBuuren.

Reef front in Rarotonga, Cook Islands. S. W. Michael.

Acropora sp. growing at 1.2 m (4 ft.), fore-reef zone of Augulpelu reef, Palau. B. Carlson.

Reef flat of Agincourt Reef, GBR, July 1991. A.J. Nilsen.

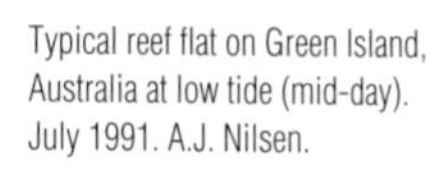

Typical reef flat on Green Island, Australia at low tide (mid-day). July 1991. A.J. Nilsen.

Backreef edge of Agincourt Reef, GBR. July 1991. A.J. Nilsen.

Lagoon

Lagoon bottoms are composed of sands eroded from the windward sections of the reef. As a result they are depositories for sediment, organic materials and nutrients, and act as major nutrient sinks for the reef (Davies et al., 1984). The leeward end of the lagoon is often the deepest and sediments tend to be muddy. Lagoons may be sparsely populated by corals or they can contain numerous patch reefs. Deeper lagoons (>10 m; 33 ft.) tend to be more muddy and can have extensive stands of sediment resistant stony corals like *Catalaphyllia, Euphyllia, Goniopora, Leptoseris, Pachyseris* and *Montipora* (Veron, 1986). The numerous patch reefs that usually rise vertically from the lagoon floor are composed of many stony coral genera. Various species of *Acropora, Favia, Favites, Galaxea, Pavona, Pocillopora, Porites, Seriatopora* and *Stylophora* can be found on these patch reefs. In calmer areas of the lagoon, branching stony corals can form beautiful, delicate growths. On the sandy bottom one can find numerous free-living stony corals such as *Fungia, Heliofungia* and *Herpolitha*, as well as *Euphyllia. The soft corals Heliopora, Sarcophyton, Lobophytum* and *Sinularia* occur frequently, as do zoanthid anemones and corallimorpharians. Many of the corals collected for aquariums come from this region. In the case of coral atolls such as the Maldive Islands, the lagoon backs onto a beach or rocky shore.

Back Reef Margins and Slopes

Towards the leeward edge of oceanic reefs, the lagoon may reach another algal flat and then the rear of the reef forms a back reef margin, followed by a descending slope. Back reef areas tend to be quite luxuriant, with impressive stands of stony corals separated by sandy channels. *Acropora* spp. tend to be the most common in shallow waters, but as depth increases more massive corals tend to dominate such as *Lobophyllia, Porites, Turbinaria* and faviids (Stoddart, 1973; Veron, 1986). Deeper slopes contain large soft coral populations such as *Dendronephthya* and gorgonians.

Inter-reef Sea Floor

The areas between outer reefs tend to be mainly sandy flats. In these areas, free-living fungiid corals such as *Cycloseris* and *Diaseris* are the most commonly encountered stony corals (Veron, 1986). Soft corals such as *Clavularia* and *Dendronephthya* are common here where there are pieces of rubble for attachment.

Inner Fringing and Barrier Reefs

Inner fringing and barrier reefs tend to be influenced by turbidity from terrestrial sources such as freshwater run-off, resulting in elevated nutrient levels. This is particularly pronounced when the land is heavily forested. Islands or coastlines that are drier tend to show reef formations more like those found on oceanic reefs. Both stony and soft coral diversity tends to be lower on near-shore reefs (Dinesen, 1983).

Intertidal mudflats

Along the coastline, one can often find extensive mangrove beds. These areas are home to many of the juvenile forms of creatures found on the nearby reef. It is not unusual to find some stony corals in this area as well, particularly *Pocillopora* spp., growing on the roots of the mangroves (Veron, 1986). As one moves seaward the water increases in depth, and coral cover begins to increase. Seagrasses may be encountered growing with the corals. Corals found in this area include *Catalaphyllia, Goniopora, Euphyllia, Montipora* and certain *Acropora spp.*

Sand Flats

Between the seagrass bed and the outer slope a broad sandy area may exist. In this area several mixed coral communities may occur composed of *Acropora, Favia, Favites, Galaxea, Goniastrea, Platygyra, Plerogyra, Pocillopora,* and *Porites.* Soft corals include *Heliopora, Lobophytum, Sarcophyton* and *Sinularia,* but *Xenia* and *Cespitularia* tend to be the most common soft coral genera (Dinesen, 1983).

Outer Slope

Near the upper edge of the reef one finds low encrusting corals, mostly *Acropora, Pocillopora,* and *Millepora. Tubipora, Heliopora, Lobophytum, Sarcophyton, Sinularia,* and *Xenia* are common soft corals in this zone, and the zoanthid *Palythoa* is also abundant here (Stoddart, 1973).

As the water deepens, the reef slopes down towards the sea floor, which may be only 10 m (33 ft.) deep in some cases (Veron, 1986). These slopes usually have a wide variety of stony and soft corals, but few *Acropora* species are present. Most other stony coral genera are found, with *Goniopora, Pavona, Porites, and Turbinaria* being the most dominant (Veron, 1986). Free-living fungiids are not uncommon in clearer waters and many shelf-forming species occur i.e. *Turbinaria, Echinopora,* and *Montipora* (Veron, 1986). Many of the corals collected for the aquarium trade occur in this region.

Acropora sp. with Green Chromis and Butterfly fish on back-reef flat, Fiji. B. Carlson.

Lagoon in New Guinea. S. W. Michael.

Green water turbid lagoon at 12 m (40 ft.) in Palau. *Pavona, Plerogyra,* and *Lobophyllia* visible. B. Carlson.

Chapter Two

Nutrient Regulation on Coral Reefs

Nutrients dissolved in the aquatic environment make life possible since aquatic organisms have to directly or indirectly extract most of their required elements from the water that surrounds them. The availability of organic and inorganic nutrients limits biological productivity. Coral reefs develop best in "nutrient poor" water, but reef communities may also exist in more nutrient rich coastal waters.

In order to appreciate some strategies employed in maintaining corals, one needs to understand the nutrient regulating regimes that occur on actual coral reefs; the sources, sinks, cycles and the influences the major nutrients can have on a reef community.

Nutrient regulation is a major area of study in coral reef ecosystems. The principle nutrients studied are nitrogen (as ammonium, nitrite and nitrate), phosphorous (as both inorganic and organic phosphate compounds) and carbon. Although coral reefs have been shown to be areas of high gross primary productivity, their net primary productivity is quite low. The oceanic waters surrounding most coral reefs are characteristically nutrient poor, and the majority of the primary productivity comes mainly from the bottom (benthic) community not the water column (pelagic) community (D'Elia, 1988). The low net primary productivity of the pelagic ecosystem is reflected in low phytoplankton standing stocks and in high water clarity. The general explanation for this high gross, but low net primary productivity is that tight cycling of essential nutrients must occur within the reef and it's organisms. The problem is that the principle pathways of cycling essential nutrients are neither well defined nor well understood (D'Elia, 1988).

Three species of *Acropora* on a back-reef flat, Fiji. B. Carlson.

Other problems include the fact that the majority of studies have focused on Pacific coral reefs, more specifically coral atolls, away from large land masses (D'Elia, 1988). These studies have usually concentrated on sections of reef that included "reef flats", areas of intensive macro- and microalgae growth. Kinsey (1983) has pointed out that different zones of a coral reef function differently, and to categorize sections of reef dominated by macro- and microalgae as "coral" reefs may not be appropriate. Therefore the experiences gained from these studies is somewhat restricted by geography and reef type.

Presumably studies of reefs near large islands, river outlets, areas of deep-water upwelling, groundwater intrusion, and even inhabited areas, also could reveal different nutrient regimes (D'Elia, 1988). Seasonal variations in light, rainfall, wind and temperature do exist in tropical regions and can have an effect on nutrient export from, and import to, a reef. Due to time and funding constraints, however, many of these studies occurred only during certain times of the year. Therefore, seasonal fluctuations in nutrient availability are often neglected. Seasonal differences in algal composition and cover, as well as seasonal abundance of some fish species, can also affect nutrient availability (Webb et al., 1975).

Despite the problems in studying coral reef nutrient regulation, it is obvious that such regulation does occur and that it may be tightly controlled within the ecosystem of the reef. Although internal cycling definitely plays a role in nutrient regulation there is some evidence that the importance of this cycling may not be as significant as previously thought. In fact, D'Elia (1988) suggests that coral reefs may be more of an open system with a greater throughput of nutrients such as nitrogen occurring than once believed.

Just to clarify before we go any further, the levels of ammonium, nitrate and phosphate that occur on coral reefs are very small compared to closed systems. Although there is some variation between reefs, ammonium, nitrate and phosphate values are all generally well below 0.05 mg/L.

Table 2.1

Examples of surface oceanic nutrient levels found in seawater impinging on different coral reef sites around the world.

Site	Season	NH_4^+	NO_2^-	NO_3^-	PO_4^{-3}
Canton Atoll,Phoenix Is.	Summer	0.03	0.05	0.16	0.06
Discovery Bay, Jamaica	Mean, All Seasons	0.004	0.006	0.02	0.02
Enewetak, Marshall Is.	Spring	0.004	NA	0.007	0.02
Great Barrier Reef	Summer	< 0.004	< 0.0009	0.03	0.03
	Winter	0.01	0.007	0.03	0.03
Great Sound, Bermuda	Mean, All Seasons	0.02	NA	< 0.03	0.002
Guam	Mean, All Seasons	NA	NA	0.05	0.02
Houtman-Abrolhos Is.	Winter	0.01	0.002	0.073	0.04
	Summer	0.0004	0.002	0.06	0.02
Salomon Atolls	Summer	NA	NA	0.061	0.04
Sesoko Island, Okinawa	Spring	NA	< 0.006	< 0.02	< 0.07

* All levels in mg/L, converted from µmol/L.
NA = not available
Modified from D'Elia 1988

Sources, Storage and Cycling of Nutrients in Coral Reefs

Although nutrients are tightly cycled within a reef, there must be some nutrient input or else no growth would occur (Szmant-Froehlich, 1983). There must be mechanisms by which reefs collect and store nutrients.

The importation of nutrient poor oceanic waters contributes little in the way of nitrogen or phosphorus. Generally this is not

Back reef zone in Bora Bora, French Polynesia. S.W. Michael.

regarded as significant. Yet, if one considers the sheer volume of water transported, multiplied by the concentration, the potential is there for some input (D'Elia, 1988). For autotrophic (self-feeding) organisms such as corals that can use nitrogen and phosphorus in extremely low quantities, this may be enough.

Upwelling of nutrient rich, deep oceanic waters is another possible source (D'Elia, 1988). Significant amounts of nitrogen and phosphorus can be imported to reefs in the vicinity of such areas of upwelling. Upwelling occurs where major currents encounter islands, and the cold, nutrient-rich deep water flows up the slope of the island. Upwelling also occurs along the equator and western coasts of continents, where prevailing winds drive surface water away, and the deep water rises to replace it.

A recent study suggests that the hydrodynamic geothermal convection process of "endo-upwelling" is linked with reef formation, and provides deep ocean water rich in nitrate and inorganic phosphate to reefs via seepage through the permeable limestone (Rougerie and Wauthy, 1993). The study proposes that an endless supply of nitrogen and phosphorous are continuously provided to reefs by this mechanism, and compares oceanic coral reefs to giant pelagic trees, with their roots in the nutrient-rich deep ocean, and their leaves in the clear, brightly illuminated nutrient-poor shallows (see Isaacs, 1977).

Reefs situated near areas of freshwater run-off can receive significant nutrient inputs, especially when near agricultural areas. Ground-water intrusion also can carry nutrients to a reef (D'Elia, 1988).

Reefs around islands where birds have rookeries receive a high input of nitrogen and phosphate in the form of bird guano. The birds gather fish from a wide range in the surrounding sea and deposit their concentrated waste in a localized area. This input can significantly affect the near-shore fauna.

Herbivorous fish and invertebrates such as tangs and sea urchins may offer another nutrient source. Their constant grazing of algal turfs can liberate vast quantities of detrital-sized algal fragments, resulting in nutrient transport from one area of the reef to another (D'Elia, 1988). Tangs and parrotfish crop these algal turf areas extensively. Since these fish have a low assimilatory rate, much of the fixed carbon, nitrogen, and most likely phosphorus is released back into the water in the form of feces (Weibe et al., 1975). These feces are transported around the reef by prevailing currents, supplying "packets" of nutrients to various organisms, including corals. Schools of planktivorous fish stationed over coral heads also provide nutritious feces. Studies have shown that fish schools provide significant amounts of ammonium to reef corals, resulting in increased growth (Meyer and Schultz, 1985).

Fragmentation of benthic algae by wave action is also an important method of transport of carbon, phosphorus and nitrogen around a reef, and is quite important in supplying fixed carbon to balance the respiratory energy requirements of corals (Wiebe et al., 1975). Algal fragments collect in quiet zones where bacteria and other microorganisms break them down into lighter-weight detrital particles. These become water-borne and may be directly consumed by corals and other benthic invertebrates, or by worms, crustaceans, and planktonic organisms that may also be eaten.

Phosphorus Cycling

Reef sediments are one of the main sinks for nutrients, especially phosphorus and nitrogen. Levels of phosphorus in reef sediments tend to range between 210 and 520 mg/L (Entsch et al., 1983). Approximately 80-83% of this phosphorus is in the form of inorganic phosphate (orthophosphate). Phosphorus in sediments does not appear to diffuse readily into the overlying phosphorus poor reef water. This can occur, however, in confined bodies of water such as lagoons with low flushing rates (Andrews and Miller, 1983). The main way this bound phosphorus is liberated is through the action of photosynthetic organisms such as algae. There is no net liberation of phosphorus into overlying waters from undisturbed sediments covered by algae, implying that the algae are liberating the phosphorus and using it before it can enter the water column. As the inorganic phosphate is used in this fashion, dissolved organic phosphorus (DOP) can be broken down biologically to feed the inorganic phosphorus pool (D'Elia, 1988). The phosphorus that is still bound to the carbonate sediments can be liberated by

Figure 2.1
Phosphorus Cycle
After Wheaton 1977

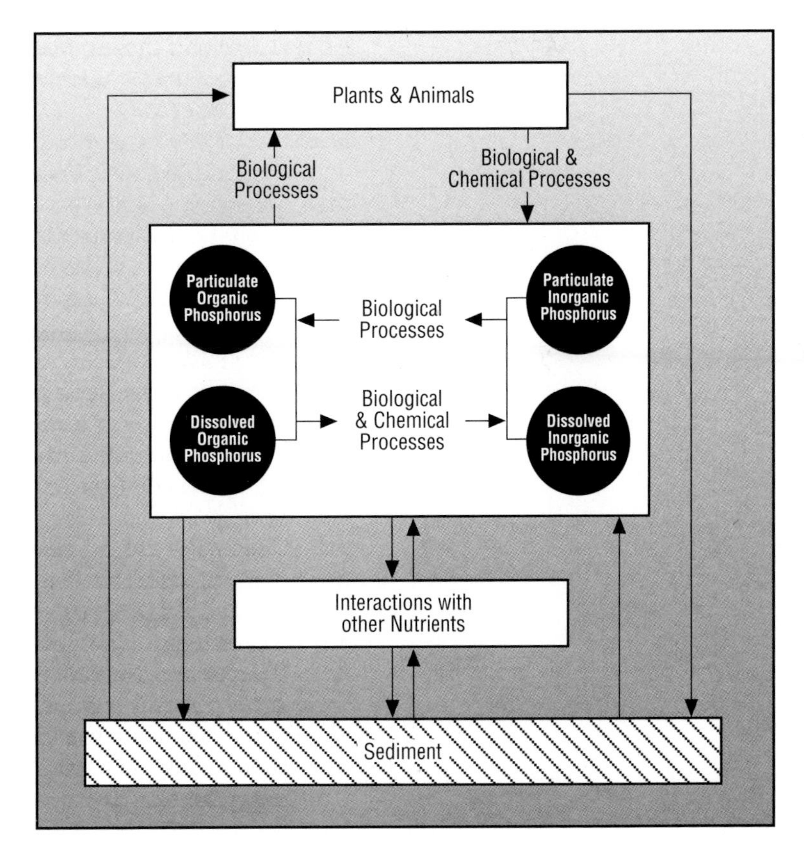

organisms that can dissolve carbonates with the aid of enzymes such as alkaline phosphatase; for example benthic algae (Lobban et al., 1985). This phosphorus can then be transported throughout the reef by fragmentation of the algae caused by wave action and grazing organisms. This form of phosphorus is known as particulate organic phosphorus (POP). The POP is, in turn, acted upon by bacteria to form dissolved inorganic phosphorus (DIP) again. It has also been shown that below a pH of 8.0 and in the presence of carbon dioxide, phosphate can be deadsorbed from calcite (Kanel and Morse, 1978; Dejonge and Villerius, 1989). Therefore, bacterial production of acids and carbon dioxide in the sediments could liberate minute amounts of phosphate. These mechanisms apply not only to soft carbonate sediments but also to consolidated carbonates such as live rock and stony coral skeletons (Webb et al.,1975).

Measurements of interstitial water from surface sediments have shown micromolar concentrations of inorganic phosphorus and even higher concentrations of inorganic nitrogen, primarily as ammonium. Presumably these occur through the actions of certain bacteria. Both of these nutrient concentrations should allow for high uptake rates by benthic algae (Entsch et al.,1983).

Excretion of DIP by reef organisms such as corals and fish is not considered significant. Corals excrete very little DIP as most of their phosphorus is slowly turned over. Recent work, however, has shown that inorganic phosphate release by the reef bottom (i.e. substrate, hermatypic corals, periphyton) is significant, but the subsequent uptake of phosphate from the water column balances this loss (Sorokin, 1992). Ahermatypic corals also show a loss of phosphate to the water column, and rely on their planktonic food supply for phosphate replenishment (Sorokin, 1992). Fish and molluscs make no special effort to retain phosphorus (D'Elia, 1988). Macroalgae can liberate large quantities of DIP and DOP, and can absorb DIP (Spotte, 1979).

Any phosphorus that is liberated, usually as DIP, is quickly used by various organisms on the reef. As mentioned above, algae will readily assimilate DIP, yet, other organisms such as corals, bacteria and foraminiferans are important consumers too. Corals containing zooxanthellae have been found to readily absorb DIP even at low, natural concentrations (D'Elia, 1988). The algae within the coral tissue use the DIP from the coral for biological processes. As this DIP is depleted more diffuses into the coral tissue from the surrounding environment (D'elia, 1988).

DIP may also be removed from solution by precipitation, absorption and chemisorption onto various substrates such as calcium carbonate. The relative importance of these processes compared to removal by biologic processes has not been investigated (D'Elia, 1988).

Nitrogen Cycling

Studies of dissolved nitrogen production show a net export of nitrogen from the reef to the water column, in the form of nitrate and dissolved organic nitrogen (DON) (Webb et al., 1975). Nitrogen fixation by blue-green algae (cyanobacteria) in shallow 1-2 m (3-6 ft.) algal turf zones contribute to this production. These algae can fix atmospheric nitrogen then liberate up to 85% of this fixed nitrogen into the water column as DON (namely peptides and amino acids) (Wiebe et al., 1975). Some of this DON is further mineralized by heterotrophic aerobic and anaerobic bacteria into ammonium, which can be oxidized into nitrate by nitrifying bacteria.

The process of nitrification is well known to most aquarists. In this method ammonia is oxidized by *Nitrosomonas* spp. bacteria into

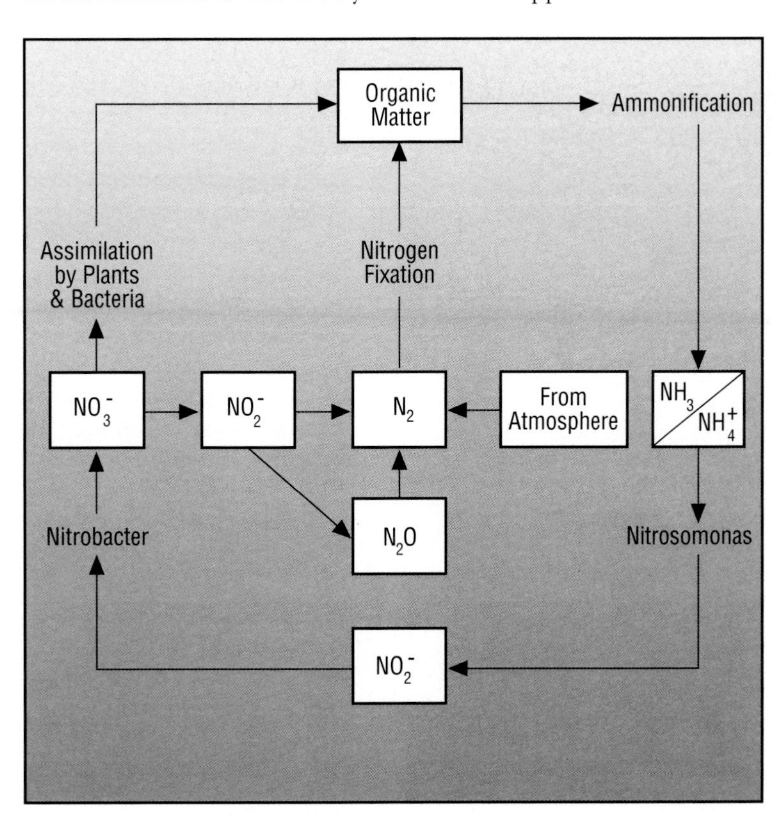

Figure 2.2
Nitrogen Cycle
After Wheaton 1977

nitrite. This nitrite is in turn, oxidized by *Nitrobacter* spp. bacteria into nitrate. Nitrification occurs on living reefs in a wide variety of habitats including algal flats, tidal pools, reef sediments, reef sponges, the interior of coral heads and on coral tissue (Wafar et al., 1990).

Although some ammonium for nitrification comes from nitrogen-fixing cyanobacteria, ammonium excretion by reef grazers, sediment ammonium fluxes and ammonium in interstitial waters of sediments are other important sources (D'Elia, 1988). Ammonium also can be generated by bacterial action on decaying organic matter. Some of this ammonium enters the nitrification cycle and is oxidized to nitrite and then to nitrate. Ammonium also can be directly absorbed and used by algae and zooxanthellae-bearing organisms such as hermatypic corals, tridacnid clams and anemones (D'Elia, 1988).

The nitrate that is produced by nitrification or released by nitrogen-fixing cyanobacteria has a variety of fates. It can be used in the biochemical pathway of respiration by either obligate or facultative anaerobic bacteria. These bacteria reduce nitrate to nitrite, and eventually to ammonium, producing nitrogen and nitrous oxide gases as waste. These gases are then released into the water and are usually either lost to the atmosphere or absorbed by cyanobacteria. This process is generally called denitrification. Since denitrification occurs mainly in the absence of oxygen, it is limited to sediments and anaerobic microhabitats such as coral heads. It has been shown that nitrification and denitrification can also occur in aerobic layers. Here anoxic microsites within fecal pellets and detritus provide a habitat for anaerobic bacteria, while being surrounded by aerobic pore waters. Such a system would allow nitrification and denitrification processes to occur in close proximity (Jenkins and Kemp, 1984). Most of the nitrate supplied for denitrification comes from nitrification occurring within the same sediments. Although a close coupling between nitrification and denitrification has been demonstrated (see Jenkins and Kemp, 1984), the importance of denitrification in coral reef nutrient budgets is not clear, and more information is required on the rates of denitrification in different habitats of coral reefs (D'Elia, 1988).

Nitrate also can be absorbed and assimilated by zooxanthellae bearing corals, tridacnid clams, bacteria, foraminiferans and macroalgae (D'Elia, 1988). Here the nitrate is reduced to ammonium within the organism, where it is then used as a nitrogen source. When the organism dies, this nitrogen is then returned to the system, as ammonium, by decomposing organisms such as mineralizing bacteria.

Carbon Cycling

Carbon, the basic element of life, is the building block of all organic compounds. After hydrogen and oxygen, which form the water that accounts for at least 70% of the composition of living tissue, carbon is the third most abundant element found in most organisms. Carbon is especially important to the coral reef since the reef structure is built from its combination with calcium, strontium and magnesium.

Carbon dioxide from the atmosphere readily dissolves into water, and combines with hydrogen and oxygen atoms to form bicarbonate (HCO_3^-), carbonate (CO_3^{-2}) and carbonic acid (H_2CO_3). Aside from these inorganic forms of carbon in the water, organic forms also exist. These come primarily from the activity of plants, which absorb inorganic carbon (as CO_2) from the water and, through photosynthesis, create organic compounds. Photosynthesis by hermatypic corals also produces organic carbon in such forms as glucose and glycerol, and many varieties of lipids. Coral mucus is rich in carbon and DOC-lipids, and by shedding this mucus a coral may be releasing excess respiratory carbon fixed through photosynthesis (Davies, 1984; Crossland, 1987). This mucus, and the detritus formed by decomposing algae, forms a nutritious source of organic carbon for numerous crustaceans, worms, filter feeders, and microorganisms that respire CO_2 back into the water.

The exudates from algae and seaweeds is rich in simple and complex polysaccharides, organic forms of carbon that are broken down by heterotrophic organisms and bacteria, which respire CO_2 back into the water (Spotte, 1979). The respiration of animals day and night, and plants at night, also adds carbon dioxide to the water. The water is in a state of equilibrium with atmospheric CO_2, so that excess CO_2 accumulating in the water from respiration is readily liberated to the atmosphere. Much of the CO_2, however, does not escape, and becomes part of the carbonate buffer system. The carbonate buffer system receives CO_2 from many sources and acts as a carbon "sink" (Spotte, 1979).

Not all of the exudates from algae are easily broken down, and some end up as refractory substances that colour the water yellow. Phenols released into the water by the plants become poly-phenols and react with carbohydrates as well as nitrogenous compounds from both animals and plants to form this yellow "gelbstoff" (Spotte, 1979). Please refer to the section on chemical filtration in chapter 5 for methods of removing these compounds from the water.

Figure 2.3
Carbon Cycle
After Barnes 1980

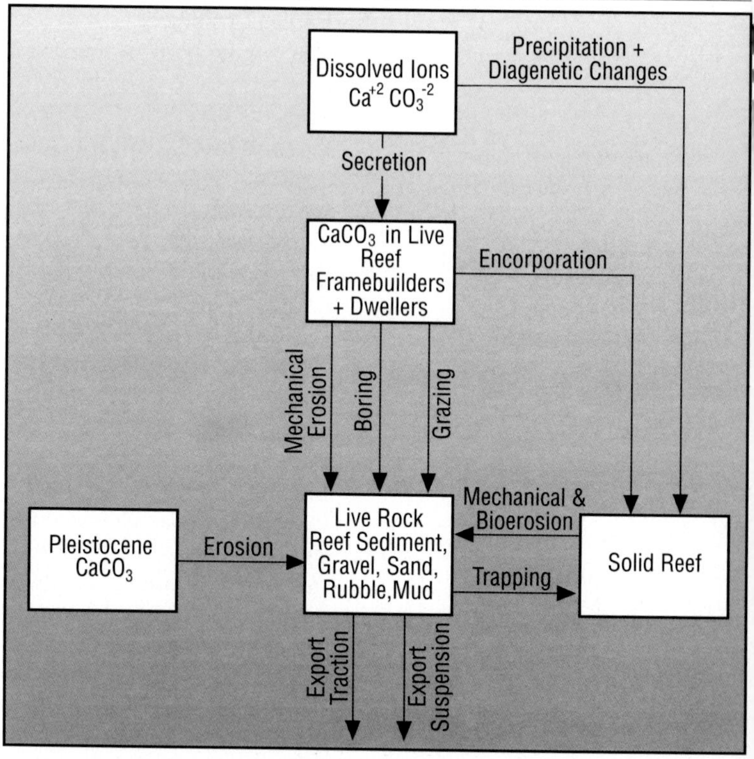

Dissolved Ions
Ca^{+2} CO_3^{-2}
Precipitation +
Diagenetic Changes
Secretion
$CaCO_3$ in Live Reef Framebuilders + Dwellers
Encorporation
Mechanical Erosion
Boring
Grazing
Mechanical &
Bioerosion
Pleistocene
$CaCO_3$
Erosion
Live Rock Reef Sediment, Gravel, Sand, Rubble,Mud
Trapping
Solid Reef
Export Traction
Export Suspension

Figure 2.4
Calcium Cycle
After Wheaton 1977

Ca^{2+}
Chemical and biological action
$CaCO_3$
Precipitation
Sedimentary Rock
Solution
Erosion
Uplift

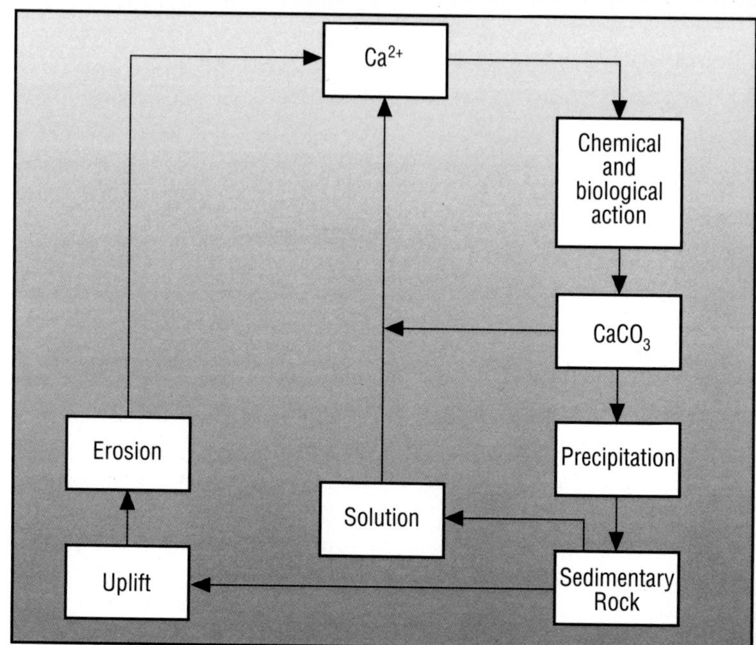

There is also exchange of carbon between the water and sediments. Reef rocks and sediments are composed of calcium carbonate, strontium carbonate and magnesium carbonate. This is deposited by the many algae and photosynthetic animals on the reef that remove calcium, strontium, and magnesium from the water and form a hard skeleton. Thus some of the carbon ends up as limestone or oolitic sand that can later fuse and harden into limestone. The formation of limestone is a global sink for carbon. Some refractory forms of carbon also accumulate in the sediments. In our aquaria this detritus is of concern since it is unsightly and contributes to the growth of undesirable algae because it also contains some organic forms of phosphorous.

In seawater, the buffering system which stabilizes the pH is a dynamic system of carbon compounds. The carbonate buffer system and pH will be described in chapter 8.

To summarize, a variety of mechanisms have been proposed to explain the high gross primary productivity but low net primary productivity of coral reefs. The commonly held view is that tight cycling occurs, both within organisms i.e. zooxanthellae-coral associations, and between organisms. However, there is evidence that suggests that coral reef nutrient budgets may be more open-ended than previously thought. In this scenario, there is a higher throughput of nutrients. Thus the external gains and losses of nutrients to internal nutrient recycling is higher than is generally assumed (D'Elia, 1988). It has also been proposed that the nutrient requirements of autotrophs such as zooxanthellae-coral associations, is much lower than previously believed based on planktonic studies, thus they are not nutrient limited (D'Elia, 1988). Although studies have shown that nutrient input from the ocean is low, it is not limiting (Wiebe, 1988). Indeed, it has been postulated that coral reefs require low nutrient levels to persist, otherwise macroalgae can out-compete corals for space at higher nutrient concentrations (Wiebe, 1988). Furthermore, elevated levels of ammonia, nitrates and phosphates may actually hinder their uptake by corals.

The effects of increased nutrient levels on coral reefs has been studied in only a few cases. It is interesting to note the parallels that can be drawn between keeping corals in closed systems and the importance of nutrient control. For instance increased levels of nitrogen and phosphorus on one Pacific patch reef resulted in increased levels of photosynthesis and a decrease in calcification rates (Kinsey and Davies, 1979). This is not surprising given that phosphate acts as a "poison" in calcification (Simkiss, 1964). Nutrient enrichment of Kaneohe Bay, Hawaii due to sewage outfall, resulted in dramatic changes in the reef community. In areas of the bay where nutrient enrichment was slight, sheets of the Bubble Alga, *Dictyosphaeria cavernosa*, and other macrophytes overgrew and excluded reef corals. In areas closer to

Nutrient enrichment from a palm oil plantation has resulted in the increased growth of macroalgae on this reef in New Guinea. This photo shows *Padina* overgrowing a thicket of *Acropora*. S.W. Michael.

the sewage outlet, calcification decreased, deposition of organic material increased, and deposit and filter feeders dominated the benthic community (Smith et al., 1981). When the sewage outlet was moved further out into the ocean, recovery became equally dramatic. Filter and deposit feeders died off, macroalgae began to decrease and corals began to recover.

Much of the information contained above can be extrapolated to the creation and maintenance of closed coral reef ecosystems. We will delve more deeply into this in chapters 8 and 9. There we will explore the importance of nutrient control and loading on coral reef aquaria, and their effects on system and organism health.

An additional note, minor and trace elements such as iron, iodine, strontium, and molybdenum added to closed system aquariums dramatically affect the growth of corals and algae. Silicate and sulfur compounds also affect growth of algae and invertebrates. Although these elements and compounds are not nutrients in the sense that nitrogen, phosphorous, and carbon are, their effect on the growth of corals and plants can be limiting. We have seen this is so in our aquaria. Little research has been done on natural coral reefs to determine whether these elements can be limiting or are even limited there. We suspect that the study of coral diseases and mysterious blooms of algae in the natural environment should include an investigation into the possible role of trace element depletion or excess respectively. Refer to chapter 8 for information about trace elements.

Chapter Three: The Biology of Corals

Scientific Nomenclature

The naming of living things follows a classification scheme known as scientific nomenclature that groups living organisms on the basis of their shared characteristics. Students remember the hierarchy of classification by means of a phrase like, "keep party clean or father gets sore", each word beginning with the first letter corresponding to the terms kingdom, phylum, class, order, family, genus, species.

Coral taxonomists examine coral skeletal structure to group corals by family, genus and species, using material both from living corals and from fossils of extinct ones. Skeletal characteristics of stony corals vary among individuals of the same species from different regions of the world, or from different locations on the reef.

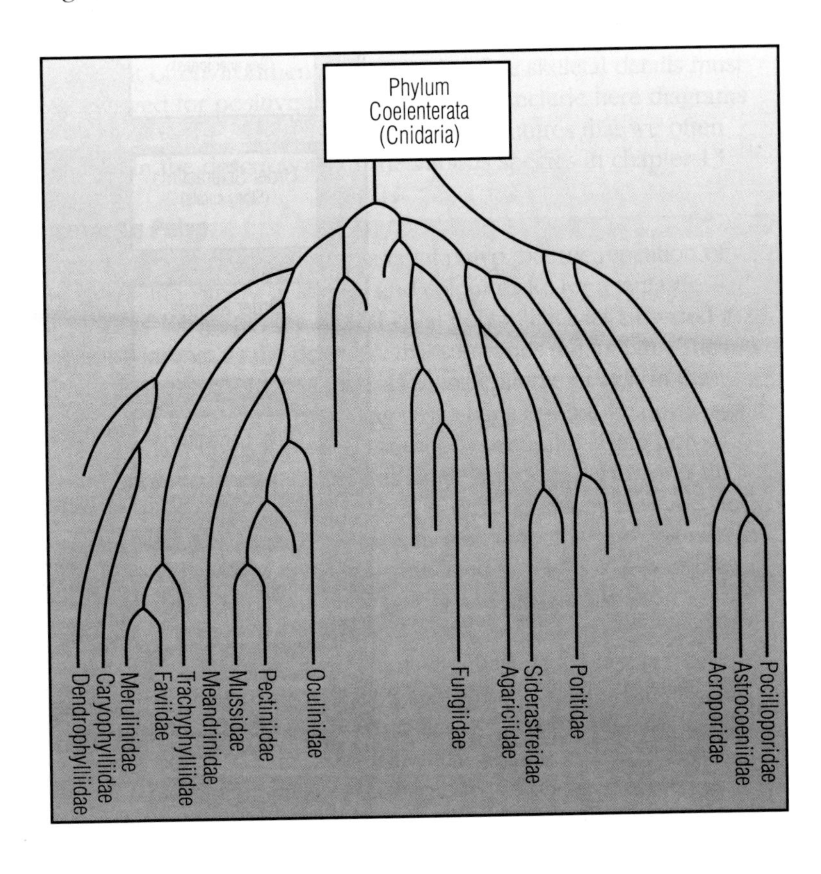

Figure 3.1a
Scleractinian Family Tree
Showing the relationships between hermatypic families.
After Veron 1986

Montastrea cavernosa from the Caribbean, photographed at The New York Aquarium for Conservation. J. Yaiullo and F. Greco.

Table 3.5a
Colony and Polyp Formation

Cerioid - Corallites share common walls but do not form valleys, pressed so close together that the calyces often take on a polygonal shape. eg. *Goniastrea* spp.

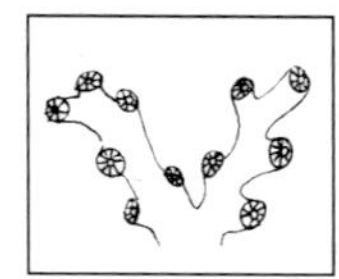

Dendroid - Corallites are tubular & colonies formed of branches with zig-zag alternation of corallites. eg. *Duncanopsammia axifuga* and *Tubastrea micrantha.*

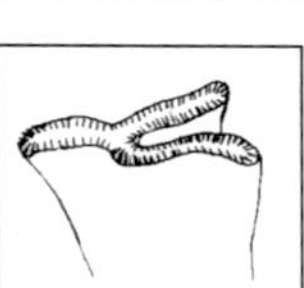

Flabellomeandroid - Forming valleys which do not share common walls. eg. *Euphyllia ancora* and *Catalaphyllia jardinei.*

Hydnophoroid - Formation in which the corallites are grouped around conical hillocks. eg. *Hydnophora* spp.

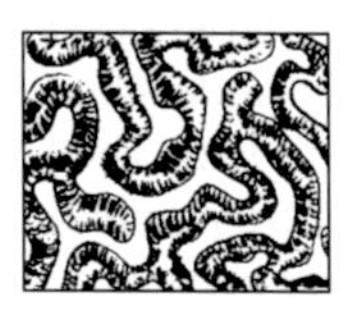

Meandroid - Formation in which groups or series of corallites in valleys share common walls. eg. *Leptoria phrygia.*

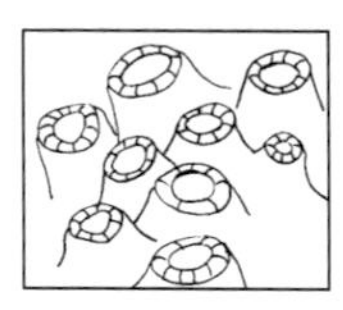

Plocoid - Distinctly separated polyps, not united by their walls, but connected only by epitheca. eg. *Barabattoia* spp.

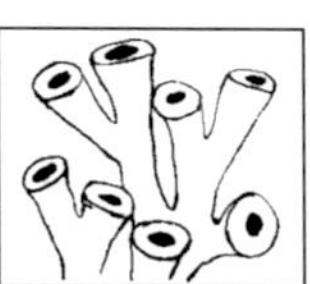

Phaceloid - More dramatically separated polyps than plocoid. Polyps become like columns. Plocoid polyps may become phaceloid as they grow upward. eg. *Caulastrea* spp.

Intratentacular budding - New polyps bud from the oral disc, within the ring of tentacles. eg. *Goniastrea* spp.

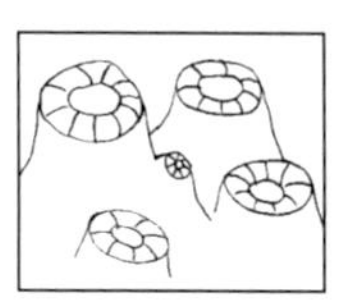

Extratentacular budding - New polyps bud between polyps, from the coenosarc outside of the ring of tentacles of any polyp. eg. *Montastrea* spp.

Table 3.5b
Colony forms
The growth form, referring to colony shape, not structure.

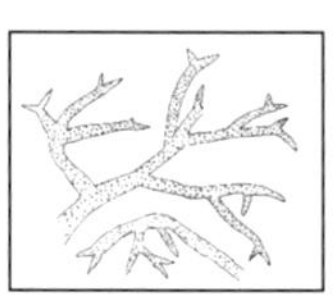

Branching or Ramose - Forming finger-like growths (digitate) or branches like trees. eg. *Acropora, Porites, Seriatopora, Stylophora, Pocillopora* spp.

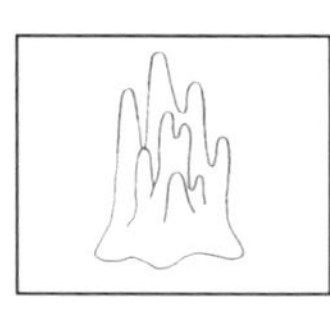

Columnar - Forming prominent upright columns. eg. some *Porites* spp.

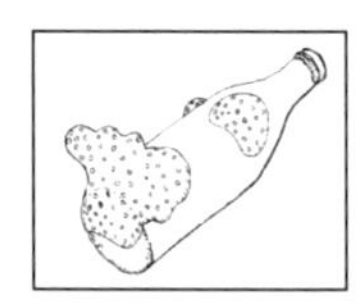

Encrusting - Sheet-like growth form tightly adhering to the substrate and conforming to its shape. May also send growths upward. eg. *Porites* spp.

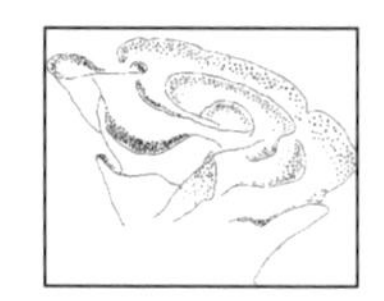

Foliaceous - Leaf-like or forming thin, expanded sheets and whorls. eg. *Turbinaria mesenterina.*

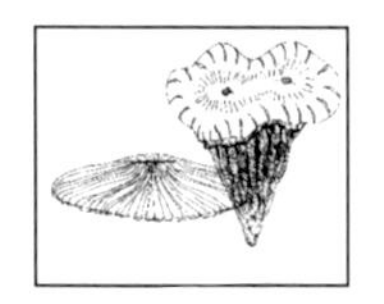

Free Living - Colonies or single polyps typically unattached on sand, mud, or hardbottom. eg. *Fungia* spp. and *Trachyphyllia geoffroyi.*

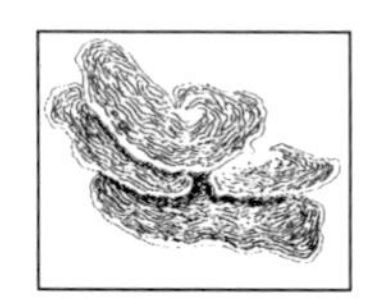

Laminar - Growth like flat plates or tiers. eg. *Pachyseris speciosa.*

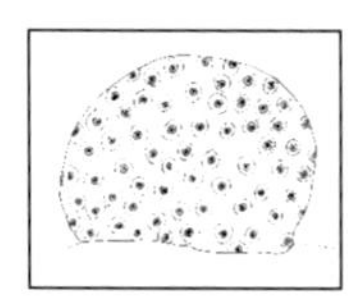

Massive - Colonies are thick, and similar in all directions. eg. *Favia* spp. and *Favites* spp.

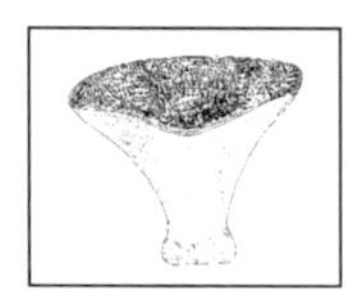

Turbinate - Forming the shape of an inverted cone. eg. *Turbinaria* spp.

Relationship Between Corals and Zooxanthellae

Coral reefs are relatively nutrient poor areas (see Chapter 2). Phosphate, nitrogen, iron and other essential nutrients are barely detectable in the clear water; clear because so little is growing in it (Benson, 1984). To compensate for this lack of nutrients, many invertebrates have developed symbiotic relationships with algae that they hold in their tissues. The larvae of hermatypic corals may already be "infected" with zooxanthellae from the parent polyp or colony (Veron, 1986). These algal cells produce energy that is used by the host. The host produces ammonia as a by-product of metabolism, and the algae use it as an energy source. Freshwater animals, such as *Hydra, Spongilla* and *Paramecium,* contain a green alga known as *Zoochlorella,* while marine organisms generally contain brownish-gold dinoflagellates of the genus *Symbiodinium,* commonly termed zooxanthellae (Greek "xanthos" = yellowish-brown) (Gordon, 1977). Therefore, contrary to popular belief, zooxanthellae are not green algae but gold-coloured dinoflagellates.

At one point it was believed that there was only a single species of zooxanthellae, *Symbiodinium microadriaticum,* but it has been demonstrated that there are in fact several strains of zooxanthellae; some fast growing, some slow (Blank and Trench, 1985; Trench 1979). Using gel electrophoresis, Trench and Blank (1987) found that of four photosynthetic organisms studied, each had its own distinctive species of *Symbiodinium.* The mangrove jellyfish, *Cassiopeia*, incorporates *S. microadriaticum.* The Pacific hard coral *Montipora* has *S. kawagutii. Symbiodinium goreauii* was found in the Atlantic anemone, *"Heteractis" lucida,* and *S. pilosum* in the common Caribbean zoanthid, *Zoanthus sociatus.* Therefore it may no longer be appropriate to say that photosynthetic invertebrates contain *S. microadriaticum,* only that they contain *Symbiodinium* spp.

Zooxanthellae are found in most reef building stony corals, many Octocorallia (soft corals) including some gorgonians, sea anemones, zoanthids, Corallimorpharians (mushroom anemones), and tridacnid clams. Sponges and some ascidians (Sea Squirts) use different types of symbiotic algae, but their role is essentially the same as zooxanthellae (Gordon, 1977). Zooxanthellae are found in the second layer of cells below the outer layer of coral tissue, the epidermis; one algal cell per animal cell. They are important components of reef building corals because they provide them with nutrition, remove metabolic waste, and contribute to the production of calcium carbonate skeletons. Corals with zooxanthellae grow rapidly because they can deposit calcium carbonate 2 to 3 times faster than those that do not have zooxanthellae.

Figure 3.6
Carbon Pathways in Corals
After Barnes 1980

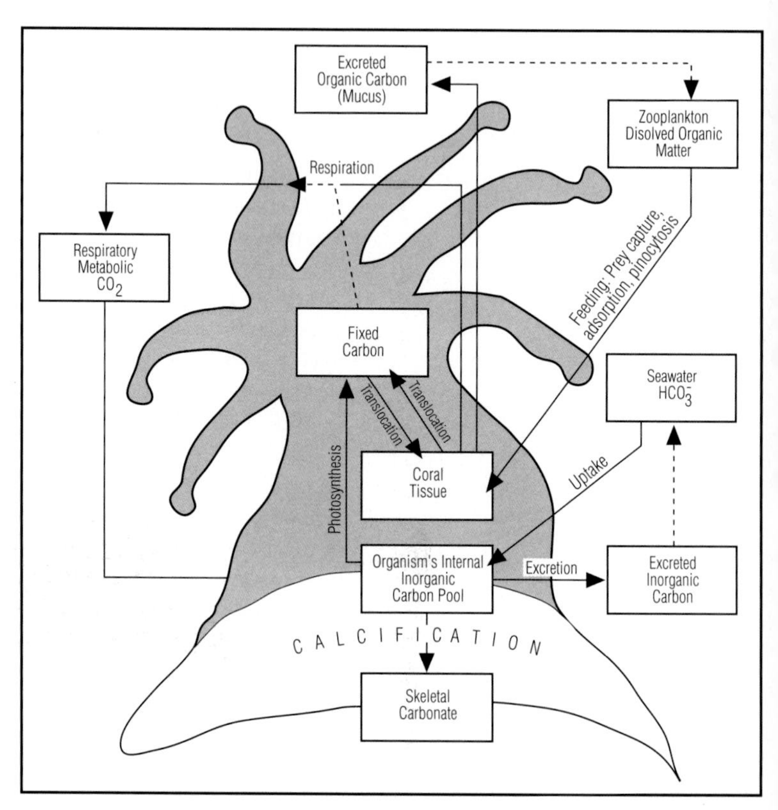

The zooxanthellae of corals are brown in colour, which is the best colour for absorbing blue light (Benson, 1984). If you have ever been SCUBA diving in a tropical ocean, or seen pictures from these areas, you may have noticed that the water is very blue. This is due to the absorption of the longer wavelengths of light (red and yellow) within the first few metres of water. Therefore it is the blue light that extends furthest into the sea. Zooxanthellae have adapted to make the most use of blue light. The accessory pigments isolated from zooxanthellae such as carotenoids and several xanthophylls, all exhibit peak absorptions between 408 and 475 nm; the blue end of the spectrum (Jeffrey and Haxo, 1968). Furthermore, chlorophyll c isolated from tridacnid clam zooxanthellae, has been shown to consist of one form, chlorophyll c2, which exhibits peak absorption in the blue end of the spectrum

(Jeffrey and Shibata, 1969). Just as in terrestrial plants, marine algae adapt to decreasing levels of light (such as encountered at greater depths or under overhangs) by increasing the amount of chlorophyll and accessory pigments in their chloroplasts, which further enhances their ability to use blue light (Benson, 1984).

Zooxanthellae use light energy to fixate bicarbonate, a form of carbon dioxide, into carbohydrates through the process of photosynthesis. The carbohydrates released by the zooxanthellae are in the form of glycerol and glucose; fatty acids and the amino acids alanine and leucine are produced also. This process requires certain nutrients, mainly nitrogen and phosphorus. As a source of nitrogen, zooxanthellae use the ammonia produced by the coral (Barnes, 1974; Gordon, 1977). It has been suggested that the nitrogen and phosphorus produced by the coral is a result of the metabolism of the tiny amounts of zooplankton on which the coral feeds (Barnes, 1974; Johannes et al., 1970). However, as mentioned in chapter 2, both ammonium, nitrate and phosphorus (as DIP) are readily absorbed by corals, and they are then used by the zooxanthellae (D'Elia, 1977; Muscatine and D'Elia, 1978). Internal cycling of phosphorus and nitrogen in a variety of compounds also occurs between the zooxanthellae and the coral (Johannes et al., 1970; Muscatine and Porter, 1977). The carbon used by the zooxanthellae comes mostly from carbon dioxide released by the coral, not from any external food source. Finally, the coral also releases acetate to the zooxanthellae, which utilize it to form fatty acids to help stabilize their chloroplasts (Benson, 1984).

Zooxanthellae can transport up to 98% of their photosynthetic products to the coral. This is assisted by digestive enzymes produced by the coral that act on the cell walls of the algae. These enzymes cause the cell walls of the zooxanthellae to become "leaky", allowing them to pass their photosynthetic products to the coral. Amino acids produced by the zooxanthellae are used by the coral to make proteins, fatty acids are used to produce waxes and lipids, while the carbohydrates provide energy for work and tissue growth (Benson, 1984). The importance of zooxanthellae in coral nutrition will be discussed later in this chapter.

One recently described phenomenon in coral/zooxanthellae symbiosis occurs in a Red Sea species of deepwater hard coral, *Leptoseris fragilis*. This coral contains zooxanthellae, but is most common at depths between 110 -120 m (365-400 ft.); unheard of depths for a photosynthetic coral. At this depth the quality and intensity of light is unsuitable for zooxanthellae. According to

Schlichter and Fricke (1986), *L. fragilis* contains pigments that alter the wavelength of the light, making it useful for the zooxanthellae.

In summary, using light energy, zooxanthellae convert carbon dioxide (from bicarbonate taken from seawater and the carbon dioxide produced by cellular respiration of the coral tissue) into carbohydrates (glycerol and glucose), lipids and amino acids. These products are then passed on to the host animal tissue, which subsequently provides a source of nitrogen (ammonia primarily) and phosphate to the algae.

Calcification of Corals

As mentioned earlier, zooxanthellae also assist in the production of coral skeletons. Corals that are deprived of their zooxanthellae, or are kept in the dark, deposit calcium at a much slower rate than normal, and hermatypic corals with their symbionts calcify faster than ahermatypic corals. It is this ability to rapidly deposit calcium carbonate which has helped the corals to become the dominant animal constructors of the reef, allowing them to grow at a rate which can exceed the rate of destruction by biological and mechanical erosion and storms.

Some stony corals do not have symbiotic algae (ahermatypic). These corals are either deep water species or found in caves and grottoes. The most commonly encountered ahermatypic hard coral in the aquarium is the Orange Flower Coral (*Tubastrea* spp.). Their location on the reef is not an indication of their intolerance of light, rather it is a result of their inability to compete with faster growing hermatypic species and algae. In locations where the water is very rich in the plankton upon which these corals feed, they may compete with hermatypes and grow on upward reef surfaces.

That light enhances calcification in corals has been known for a long time, but it has only recently been demonstrated through scientific analysis. Chalker (1983) gives an overview of the history of scientific investigation into the calcification of corals. Photosynthesis, not any other biological affect of the light, is directly related to this increased calcification rate.

Exactly how photosynthesis enhances calcification is a subject full of controversy, but it seems that the different hypotheses all relate to the benefits that corals derive by having photosynthetic partners. Some hypotheses about symbiont-linked light enhanced calcification are reviewed by Chalker (1983).

Figure 3.7
Calcification Diagram
After Schuhmacher, 1991

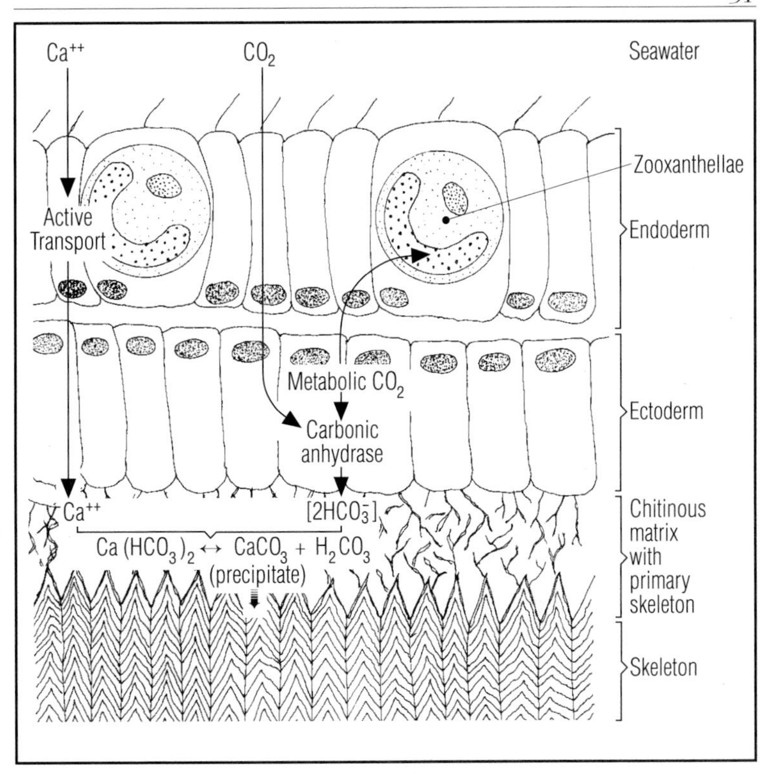

It is believed that algal photosynthesis increases the calcium carbonate deposition by removing carbon dioxide and driving the following reaction to the right:

After: Barnes, 1980

$$Ca(HCO_3)_2 \leftrightarrow CaCO_3 \downarrow + H_2CO_3 \leftrightarrow H_2O + CO_2$$

Calcium bicarbonate	Calcium carbonate	Carbonic acid	Water	Carbon dioxide

Additionally, the zooxanthellae may remove from the site of calcification phosphate produced as a waste product of metabolism by the coral. Phosphate acts as a crystal poison (Simkiss, 1964), and its removal from the site of calcification could enhance the rate of crystal formation while feeding the zooxanthellae and increasing their metabolism.

In aquariums, the calcification process is inhibited when the amount of calcium bicarbonate in the water is low, or when the pH is too high or too low. See the topics calcium additions, alkalinity, and pH in chapter 8 for further information.

Corals do not merely deposit calcium as a solid mass. The intricate design of their skeletons has a framework composed of an organic matrix of filaments. The calcium, magnesium, and strontium carbonate crystals form on this matrix, which is deposited by the coral. The exact composition of the matrix and the mechanisms of its deposition are not completely understood, and they may vary among different species.

Nutrition in Corals

The nutritional requirements of the various organisms that occur in reef systems are extremely varied and/or difficult to ascertain. We are now no longer dealing only with the different feeding habits of the fish that we are keeping but also various orders of invertebrates, each with their own peculiarities. Add to this mass of confusion the fact that very little is known about the nutritional requirements of these organisms, and one can quickly see that the topic of nutrition in reef systems is a most intimidating one.

The general approach taken in feeding organisms in captivity is to closely study the diets of these organisms in the wild and to duplicate this as much as possible. This approach works quite well with most fish and some invertebrates. However, many fishes' natural diets are almost impossible to duplicate (e.g. those that feed on coral polyps and sponges, or bryozoans for certain fish and nudibranchs). Another problem comes with corals. There are few studies published on the diets of corals, and much debate about the amount corals feed or whether corals need to be fed in captivity.

There are various feeding mechanisms used by the inhabitants of our reef aquariums. In some instances the same organism may use more than one feeding strategy, which is probably an adaptation to ensure that as much nutrition as possible can be extracted from the nutrient poor environment of the reef. This is the case with many coral species, and for this reason, we cover coral feeding strategies individually.

Zooxanthellae bearing organisms can use a wide variety of feeding techniques. Not only can they utilize the photosynthetic products of their algal symbionts but they can also feed directly on plankton, bacteria, detritus and fish feces (Sorokin, 1973; Schiller and Herndl, 1989). Some corals even have the ability to directly absorb carbohydrates from the water (Stephens, 1962).

As mentioned previously, most corals contain symbiotic algae in their tissues that can supply some of their nutritive needs. We say some because the degree to which zooxanthellae contribute to a coral's nutrition has been the subject of much research over the past 40 years. It seems that the amount varies between species. In some species of zoanthids over 90% of their nutrition can be met by the zooxanthellae while in others this figure is much lower (60%) (Steen and Muscatine, 1984). However, the general

Large-polyped hermatypic corals such as this *Turbinaria peltata* (left) benefit from the occasional meal of shrimp or other small solid food. The smaller-polyped *Porites* on the right is more dependent on photosynthesis, though it will also feed on smaller particles and dissolved organic matter. J.C. Delbeek.

consensus is that zooplankton do not contribute a major portion of the caloric or carbon requirements of hermatypic corals as the amount of zooplankton available on a coral reef is simply too little to satisfy their energy requirements (Johannes et al., 1970; Muscatine and Porter, 1977). The amount of plankton generated on a coral reef can be significant, while the imports of plankton from the surrounding sea are small.

Corals can feed in a variety of ways. The larger-polyped forms (e.g. *Euphyllia*) can actually feed on shrimp-sized prey that they capture with their tentacles. Other corals such as *Heliofungia* may collect the slime that forms on their large polyps and swallow the microorganisms and detritus trapped in it (Kuhlmann, 1985). Those that feed on detritus either digest the bacteria living on them or the particulate organics that coat such particles (Wotton, 1988). It has even been proposed that some species of coral such as *Acropora* and *Psammocora* may actually "farm" bacteria growing among their coral branches. The coral's mucus acts as a carbon and nitrogen source for the bacteria. These bacteria are either directly consumed by the coral, or the nanoplankton that feed on the bacteria are eaten by the coral (Schiller and Herndl, 1989). Still other corals can directly absorb nutrients used by the zooxanthellae (ammonium, nitrate and phosphate, as well as various amino acids) from the water (Franzisket, 1974; Muscatine and Porter, 1977; D'Elia, 1977; Muscatine and D'Elia, 1978).

Although one can certainly feed the larger polyped corals, in our opinion many coral species do not need direct feeding. Many get more than enough from natural sources in the tank. The live rock and associated algae and bacteria produce copious amounts of nutrients, vitamins and other products through their metabolic processes. In addition, worms and microcrustaceans in the live rock produce larvae and gametes that are food sources. Every time you feed your fish, particles of food and dissolved nutrients are added to the water. Wilkens (1990) found that even in the presence of an efficient skimmer the levels of amino acids in the aquarium were many times higher than on the reef. It is safe to assume that many other "nutrients" are just as abundant, despite our best efforts.

Those polyps that are large enough to be fed small pieces of shrimp can be fed once a week or so by directly placing pieces of food on some of the polyps. Zooxanthellae require phosphate, and although they may be able to absorb this from the water, it is generally felt that the main source is from the tiny amount of prey captured by the polyps (Johannes et al., 1970). However, some large polyped soft corals (e.g. *Xenia* spp.) have never been observed feeding. Lacking stinging cells in their tentacles, *Xenia* probably absorb nutrients directly from the water. Judging from the large number of successful aquariums that we have seen in which the corals are never directly fed, most zooxanthellae bearing corals do not require direct feeding to survive, grow and multiply. We will provide more information on feeding individual coral species in Chapter 13.

If you do decide to try and feed your corals be very careful about over-feeding i.e. feed SPARINGLY. An occasional feeding of live baby brine shrimp or liquid foods may be appropriate for some specimens but not others. Pay careful attention when feeding, if it looks like the coral is not ingesting any food then perhaps it does not require additional feeding.

When one is dealing with corals that do not contain zooxanthellae (ahermatypic), feeding takes on extreme importance. Examples of such organisms include certain gorgonian species, *Dendronephthya* spp. soft corals, and Orange Cup Coral (*Tubastrea* spp.). For these corals live or prepared foods should be given often. Live foods such as baby brine

Ahermatypic soft corals such as the *Scleronepthya* sp. shown here, require frequent feeding to survive in the aquarium. A. Storace.

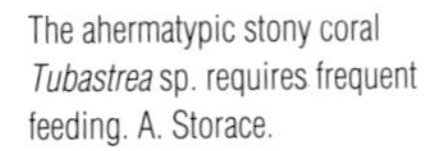

The ahermatypic stony coral *Tubastrea* sp. requires frequent feeding. A. Storace.

shrimp, daphnia and rotifers are excellent for most gorgonians and *Dendronephthya,* whereas *Tubastrea* should be fed larger items such as live adult brine shrimp, small pieces of shrimp, scallop or fish. Prepared foods can be used as well. Dried or freeze-dried foods can be finely ground and soaked in a vitamin preparation. This sludge is then fed directly to the coral through a pipette or baster. You should not feed such food by simply placing it into the water. This only results in added pollution as most of it ends up in the filter, in the gravel or under rocks.

The final method of feeding that might occur in reef tanks is through the direct uptake of organic compounds through the body walls of various sponges, marine worms, ascidians (i.e. tunicates), bryozoans, etc. (Sepers, 1977). The mechanisms, importance and role of such feeding in marine ecosystems and our aquariums is not

well understood and certainly bears more extensive research.
In conclusion, ecological, biochemical and physiological data indicate that symbiotic algae are of major importance in the nutrition and growth of coral reefs. They are important not only to the reef building corals, but also to other reef dwelling animals such as sea anemones, giant clams and sponges. Symbiosis between animals and algae appears to be a highly successful adaptation for solving nutritional problems in nutrient-poor areas (Gordon, 1977).

As mentioned at the beginning of this section, the topic of nutrition in aquariums is poorly understood at best. This is an area where the experiences of hobbyists can be of value to scientific researchers. There are more hobbyists out there than there are people actively researching this area. It would be a shame if this tremendous pool of information and experience went unused. Share your information with others, write articles for club or national magazines, keep detailed notes on each of your specimens, spread your knowledge and experience.

Depth Zonation of Corals

The diversity and abundance of corals on a reef, their distribution with depth, and the shapes and colours of coral colonies are affected by numerous environmental factors. We offer here a brief summary of the types of factors that may control coral abundance, distribution, colour and shape.

Light

Light is one of the most important physical parameters that controls the distribution, morphology, and colour of corals. For more detail on the importance of light to the coral reefs and to photosynthetic corals and clams, see chapters 1 and 6.

We know that corals of the same species from different depths or zones can have different growth forms (see Falkowski and Dubinsky, 1981), and that light plays an important role in coral morphology. Light also affects coral pigmentation. Corals from shallow, brightly illuminated water manufacture special pigments to absorb UV light (see next section). Corals in deep water are generally darkly pigmented, to absorb more of the available light, while their counterparts in shallow water tend to be paler. Furthermore, the symbiotic zooxanthellae have photoadaptive states, depending on the location of their host and the species of zooxanthellae. In transplant experiments, zooxanthellae in corals from shallow water that are adapted to high intensity light, function poorly when their host is moved to a deeper, less illuminated environment. Likewise, zooxanthellae adapted to the deep environment are damaged by high intensity light when moved to shallow water (Dustan, 1982). Therefore, different species of zooxanthellae may function best within different ranges of light levels. Adaptation by the coral via bleaching or shedding of excess zooxanthellae, or modifications in pigment density, affords some flexibility for location with respect to light. It has been proposed that changes in the lighting (or other parameters such as temperature) beyond the range of flexibility for a particular species of zooxanthellae necessitates that the coral bleach (shed all of its zooxanthellae) and adapt by re-populating its tissues with a different species of zooxanthellae compatible with the new range of light (Buddemeier and Fautin, 1993). The growth of corals may also be controlled in part by the species of zooxanthellae, and its ability to function optimally with respect to the light field in the location of its host coral.

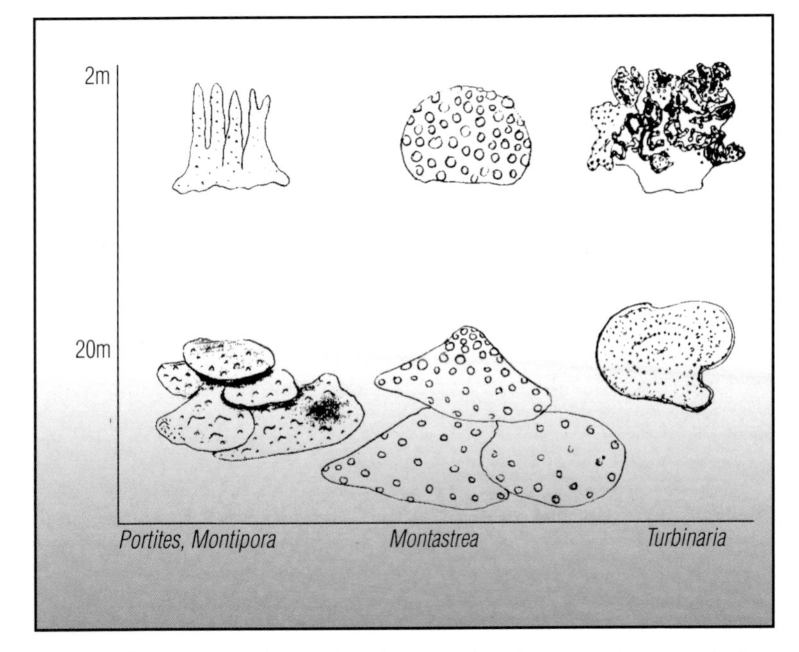

Figure 3.7
Change in Morphology
In four genera of stony corals from 2 and 20 m.

Growth form in corals is related to calcification rate. In general, the side or portion receiving the most light grows fastest. In shallow water, the light is not only intense, it comes from many directions because of surface wave refraction and reflection of light off the sandy bottom. Corals in shallow water typically form massive domes or, if they are branched species, they form heads of fingers or branches. The growth form of these corals shows that they are utilizing light from many directions. In deep water, the light is very directional, mostly from above. When the light is coming uniformly from above, all upward facing surfaces on the coral grow at the same rate. Corals therefore tend to be plate-like in deep water. This phenomenon has also been demonstrated in an experimental aquarium under bright illumination, because of the directional nature of the light (Jaubert and Gattuso, 1989). Corals that normally form thin sheets, crusts, or scrolls (i.e. agariciidae and *Turbinaria* spp.) only grow rapidly at the outer edge. In deep water they form nearly horizontal, shingle-like plates (an angle off of horizontal may facilitate the removal of settling detritus), while in shallow water they can grow in vertical plates or scrolls.

Many species show a slight increase in growth rate in a gradient from the surface down to about 5 meters (16 ft.) depth, because of the effects of photo-inhibition by the intense light at the surface (Huston, 1985). In general, however, the coral growth rate within a species decreases with depth (Huston, 1985). Some species do not show this trend, or exhibit just the opposite response.

Factors Other Than Light

Though some corals grow faster in deeper water, other factors may affect their distribution or growth. For example, in transplant experiments, caged colonies of *Pocillopora damicornis* grew faster at 15 meters (50 ft.) than at 2.4 meters (8 ft.), even though *P. damicornis* is seldom found at this depth. The cage protected the specimens from coral-eating fish (Huston, 1985).

Corallivores

As we just explained, in deep water, corals may be subject to predation by fishes and invertebrates (i.e. butterflyfish, parrotfish, Crown-of-Thorns starfish, predatory snails, etc.) that may not be able to reach them easily in the most shallow environment because of wave action (Huston, 1985). Therefore certain species that easily fall prey to corallivores may be more abundant in the shallowest water, though they may grow well (or better) under the environmental conditions found in deep water.

Sedimentation and Turbidity

The average turbidity on a given reef as a result of land run-off, tidal currents or wind generated waves affects the distribution of corals and their growth forms. Turbidity blocks the light, and the reduction in intensity affords ideal conditions for growth of both shallow and deep water species in a mixed zone, in relatively shallow water. Those species that typically occur in the brightest illumination will be restricted to very shallow water in a turbid environment, whereas they might occur over a broader depth range in clear, sediment free water. Sedimentation can also inhibit the settlement of coral larvae and is a major factor in the prevention of recolonization on disturbed reefs (R. Richmond, pers. comm.).

Growth of Filamentous Algae

Filamentous algae and algal turfs, most abundant in shallow water, can limit the settlement of coral larvae, and thus affect the zonation of species. Herbivores that clear the algae away allow some settlement and counter the effect of rapid algae growth in shallow water, but the reduced growth of algae in deep water makes for better coral larvae survival there. Therefore the rapid growth of algae in shallow water is a factor that tends to increase coral abundance in water deeper than a few meters (Huston, 1985). In areas where the loss of herbivores has resulted in uncontrolled growth of algae, resident corals may be smothered, and there is a lack of recruitment of new stony corals. This has become a problem in some localities in the Caribbean (e.g. Jamaica) due to the harvest of parrotfish via fish traps, and the sudden loss of most *Diadema antillarum* sea urchins in the early 1980's, as a result of a mysterious illness. Where this has happened, subsequent damage from hurricanes can eliminate the remaining live corals, and the algae prevent the reef from recovering.

Fast Growing, Over Shading Species

Rapid growing species such as table acroporids create large shade areas below their branches. This shading affords them a competitive advantage in the fight for space and use of light. Therefore they can easily dominate the tops of reefs. Shade-loving species may settle below their "umbrellas" (see coral aggression at the end of this chapter).

Water Motion

The motion of water has a strong influence on both the zonation and shape of corals. The design of coral skeletons makes them either suited or unsuited to the harsh surge and currents found in shallow water fore-reef zones. Some corals adapt to different water flow by altering their shape, but some corals cannot adapt to certain flow regimes (see chapter 1).

Although it may seem a paradox, one often finds the more delicate, branched forms in the shallows where water motion can be most severe, while in deep, calmer water the more massive and robust forms are common (Vine, 1986). In areas of high water motion, the current can be so great that polyp extension would be inhibited. Therefore, a hydronamic shape that offers high resistance to the water flow (i.e. ramose), is an asset, as it dampens the flow enough to allow the polyps to open (Vine 1986). Likewise, less resistive hemispherical shapes are favoured in low-flow environments, since the shape allows better gas exchange across the coral surface (Vine, 1986).

Temperature

High water temperatures that occur in calm, shallow water near shore, prevent most species of coral from growing there. Cold temperatures in the same environment likewise limit coral growth (see chapters 1 and 8).

Disturbances

Natural disturbances such as hurricanes, if they are frequent and not too severe, tend to increase the diversity of coral species in shallow water because they prevent dominant species from achieving stable growth for too long. Unnatural disturbances, such as oil spills or ship groundings, and severe disturbances, such as powerful hurricanes, can impact coral diversity and abundance on a reef for many years. Chronic disturbance, such as pollution and constant damage from daily visitation by careless divers can also impact coral abundance.

Interactive Competition Between Coral Species

Because of competitive interaction between corals, the presence of a particular species can affect the presence of other, less competitive species in localized zones. Soft corals are especially capable of dominating areas of reef (see coral aggression at the end of this chapter).

Ultraviolet Light and Corals

One of the common misconceptions in the marine reef hobby has been that ultraviolet (UV) light does not significantly penetrate through seawater. In fact, as long ago as 1950 it has been known that UV can penetrate as far as 20 m (66 ft.) in clear seawater (Jerlov, 1950). Yet, it has only recently been appreciated that UV light can be a significant factor in shaping shallow water coral communities (Chalker et al., 1986).

Ultraviolet light extends below the range of visible light (400-700 nm) and can be divided into three classifications: UV-A, UV-B and UV- C. The wavelength of UV-C extends from 200-280 nanometres and is not considered a factor in marine aquariums since only germicidal lamps produce UV-C (Mohan, 1990). UV-C is also not a factor in nature since light below 286 nm does not penetrate the Earth's atmosphere. In contrast to UV-C, both UV-A (320-400 nm) and UV-B (280-320 nm) penetrate the atmosphere and can be physiologically and photosynthetically damaging to many forms of reef life (Chalker et al., 1986). UV-B has been shown to cause photo-oxidation in corals, to destroy DNA and RNA, and to inhibit the formation of chloroplasts (Halldal, 1968; Mohan, 1990). Under

Acropora sp. from shallow water exhibit UV protection pigments. S.W. Michael.

artificial sources of UV-B, corals have shown withdrawal of polyps, discharge of mucus, swelling of tissue, ejection of mesenterial filaments and eventually death (Mohan, 1990). UV-A is somewhat less damaging then UV-B but excessive levels can inhibit calcification in corals and can cause damage to DNA and RNA at shorter (320-350 nm) wavelengths (Mohan, 1990).

It would therefore appear that many shallow water organisms are at risk from exposure to UV light. In most cases, UV sensitive organisms such as algae, sponges and bryozoans exist in shallow areas by growing between coral crevices or underneath overhangs, thereby avoiding direct UV exposure. When such organisms are placed in full sunlight they quickly succumb to UV light but if these same organisms are placed under a UV absorbing shield, they do just fine (Jokiel, 1980). Therefore UV light can be shown to be a major factor in organism distribution on a coral reef. However, there are still many organisms that exist quite well in shallow waters, exposed to large amounts of UV light. These include, stony and soft corals, anemones, giant clams, zoanthids, some sponges and algae (Chalker et al., 1986). Many of these invertebrates contain zooxanthellae, which require light for photosynthesis. Therefore the tissues of these organisms must be transparent to allow for the transmission of light. Jokiel and York (1984) showed that isolated zooxanthellae quickly die when exposed to UV-A and B at levels above 20% of incident surface radiation. It has been shown, however, that oxygen production does occur when zooxanthellae are exposed to UV-A, indicating that it can be used for photosynthesis (Halldal, 1968). Still, corals and clams are quite common in shallow waters, suggesting that they must have some mechanism for protection from UV light. In most cases these organisms have developed UV absorbing compounds in the zooxanthellae and tissue cells. One class of compounds is called S-320, named after its absorption spectrum, which peaks at 320 nm. Currently S-320 is known to consist of three separate mycosporine-like amino acids; mycosporine-Gly, palythine and palythinol (Dunlap and Chalker, 1986). These compounds were originally isolated from the colonial anemone *Palythoa tuberculosa* and have since been found in sponges, algae, molluscs, echinoderms and tunicates (Dunlap and Chalker, 1986). Other pigments act by absorbing UV light and re-emitting the energy as fluorescence. These pigments are responsible for the bright greens often seen in corals, anemones and clams (Mohan, 1990). Other pigments that result in violet and bright whites, block UV by being good reflectors of UV light (Mohan, 1990)

Jokiel and York (1982) demonstrated that when placed under near-UV absorbing material, the hard coral *Pocillopora damicornis* grew faster and had lower S-320 levels. This suggests that S-320 concentrations are directly related to UV intensity and that near UV light is an important factor in the growth and physiology of corals. It has also been shown that the concentrations of UV absorbing compounds are lower in organisms found deeper on the reef. Therefore, such organisms are more sensitive to UV light and caution should be exercised when dealing with them in the aquarium. We will go into more detail on UV light and its role in the aquarium in chapters 6 and 10.

Reproduction Strategies of Corals

Sexual Reproduction

Sexual reproduction in corals may occur in two ways: some corals are brooders, with fertilization and embryonic development occurring internally, in their coelenteric cavity; others are broadcasters, releasing their gametes synchronously into the water where fertilization and embryonic development occur (Atkinson and Atkinson, 1992). Depending on the species and its location, the reproductive period may be seasonal, monthly, or continuous (Atkinson and Atkinson, 1992). Synchronization of spawning is critical to the success of broadcast spawners since the tides rapidly wash their gametes away. On the Great Barrier Reef of Australia, a mass spawning event occurs once per year, after sunset about five days after the full moon in late spring (Veron, 1986). During this time it is estimated that 90% of the stony corals on the entire reef release their gametes!

Release of eggs by a female colony of *Sandalolitha robusta* in May, 1992, at the Waikiki Aquarium; a nearby male colony released sperm shortly thereafter. The female specimen has been in captivity since 1983. T. Kelly.

In other areas of the Pacific, and in the Caribbean and Red Sea, corals spawn over a wider range of months, weeks, or days. There are numerous cues that can trigger coral spawnings and maintain synchronicity. Temperature, photoperiod, and nocturnal illumination all appear to be important. However, even within a species there can be a great deal of geographic variation in the spawning season (Richmond and Hunter, 1990).

Figure 3.8
Reproduction Cycle
After Sammarco 1986

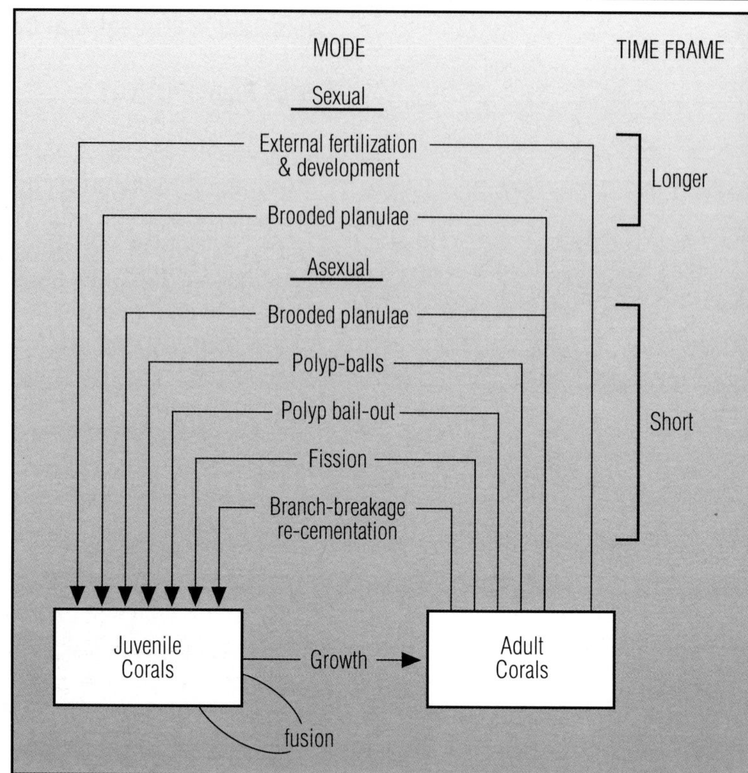

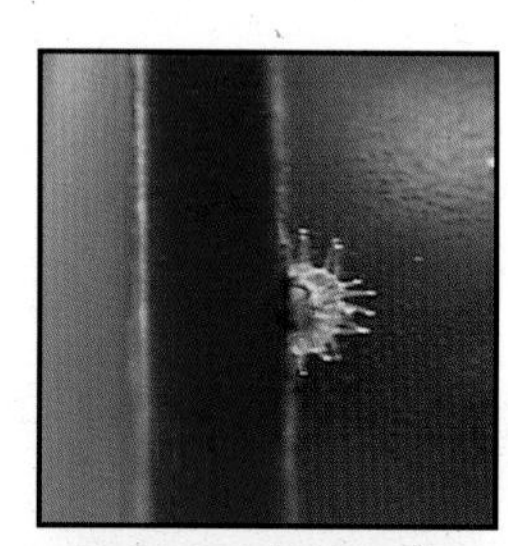

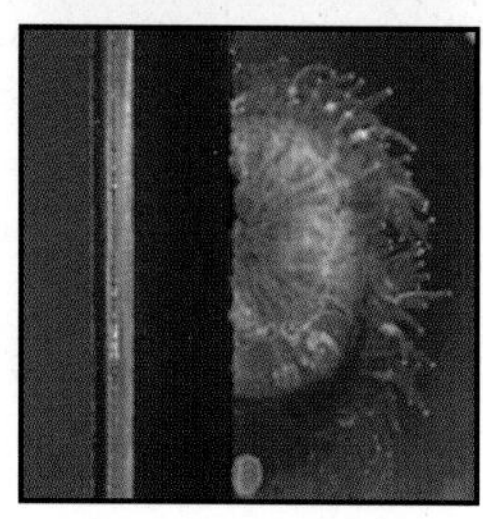

Growth series of a *Favia fragum* that arose from a planula larva in the aquarium and settled on the glass. October, 1992; December, 1992; and February, 1993.
J. Sprung.

The annual temperature range is thought to be directly correlated to the degree of multispecies synchronicity, such that areas that have the greatest annual temperature range exhibit the highest percentage of synchronicity (Richmond and Hunter, 1990). The affect on coral fecundity of the loss of these temperature cues in a closed system aquarium has not been explored.

While temperature is a seasonal cue, nocturnal illumination (lunar phases) can stimulate corals to spawn on a particular night (Richmond and Hunter, 1990). A coral species' spawning response differs according to the location. For example, *Stylophora pistillata* colonies in Palau planulate according to a distinct lunar cycle, while Red Sea specimens do not (Richmond and Hunter, 1990).

Most corals (68% of those studied) are hermaphrodites, having both sexes in one individual (Richmond and Hunter, 1990). Some corals, however, have separate sexes (gonochorism). In hermaphroditic species, the eggs and sperm may be released separately, or together as an egg-sperm bundle, that breaks up rapidly after release before the eggs are ready for fertilization (Veron, 1986).

Hermaphrodites can also be brooders. Sperm enter the gastric cavity and fertilize the eggs within. These develop into planulae that are released when mature. Brooding appears to be the dominant form of sexual reproduction in Caribbean corals, while broadcast spawning seems to be more common in Pacific and Red Sea corals (Richmond and Hunter, 1990). Self-fertilization is suspected to occur, but has only been shown in *Acropora tenuis* (Richmond and Hunter, 1990). Barriers to self-fertilization appear to break down some time after spawning for some species (Richmond and Hunter, 1990).

The typical method for ahermatypic corals is internal fertilization and brooding of larvae (Veron, 1986). Those corals that brood their larvae do not need to synchronize spawns since the eggs are always retained in the polyps; only sperm are released (Veron, 1986). As a result, these corals can spawn year-round.

The larvae of corals are called planulae, and they may be produced asexually as well (see Ayre and Resing, 1986). This has been documented in at least one species, *Pocillopora damicornis,* but may be more widespread since it effectively maximizes production of larvae (Veron, 1986). *Pocillopora damicornis,* which is hermaphroditic and also able to produce planulae sexually, has been propagated for years by European aquarists. It has reproduced by means of asexually-produced planula larvae and polyp "bail-out" in many aquaria (Stüber, 1990; J. Sprung, pers. exp.; S. Tyree, pers. comm.).

Asexual Reproduction

While sexual reproduction offers the species-preserving advantages of genetic variability and long distance dispersal (Veron, 1986), it is the least common means of duplication that corals employ. Far more regular is asexual reproduction, which comprises numerous strategies, including the production of planulae, as mentioned earlier. Sexual reproduction can be the primary means of recruitment for some coral populations, with asexual reproduction becoming the dominant or sole form of reproduction at the ecological limits of the coral's distribution (Richmond and Hunter, 1990). Temporal and regional variations in biotic and abiotic factors can cause differences in reproductive and recruitment patterns in coral communities (Richmond and Hunter, 1990).

Asexual reproduction strategies in corals include asexually produced brooded planulae, the formation of polyp "balls" or

Pocillopora damicornis from a planula larva spawned in the aquarium and settled on the glass. J. Sprung.

Budding of polyps along the growing edge of the stony coral, *Duncanopsammia axifuga*. J. Sprung.

Polyps budding along the growing edge of the related coral *Turbinaria*. J. Yaiullo and F. Greco.

Acropora elseyi in an outdoor aquarium in June 1992. B. Carlson.

Same colony 7 months later. Growth rate was 13.2 cm (5 in.) per year, per branch. B. Carlson.

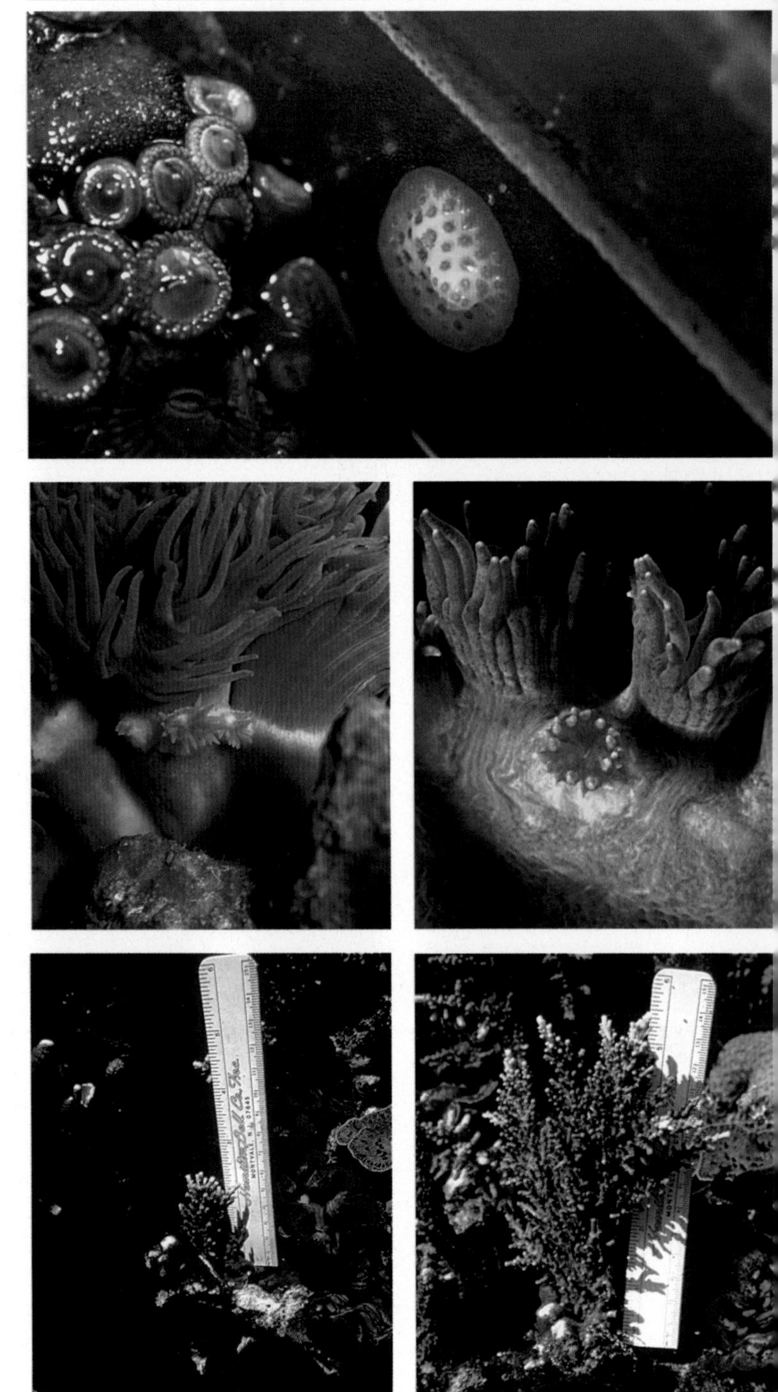

Acropora sp. fragment from D. Stüber in J. Sprung's aquarium, November 1992. J. Sprung.

January, 1993. J. Sprung

February, 1993. The clam had to be moved because of its own increase in size and the encroaching branches. J. Sprung

July, 1993. J. Sprung

February, 1994. The growth of Stüber's *Acropora* and other *Acropora* spp. has injured, shaded, and nearly killed the adjacent colony of *A. cervicornis* that had previously grown well in this aquarium for three years. The front view of this aquarium has been obstructed as the encrusting growth is difficult to remove. J. Sprung.

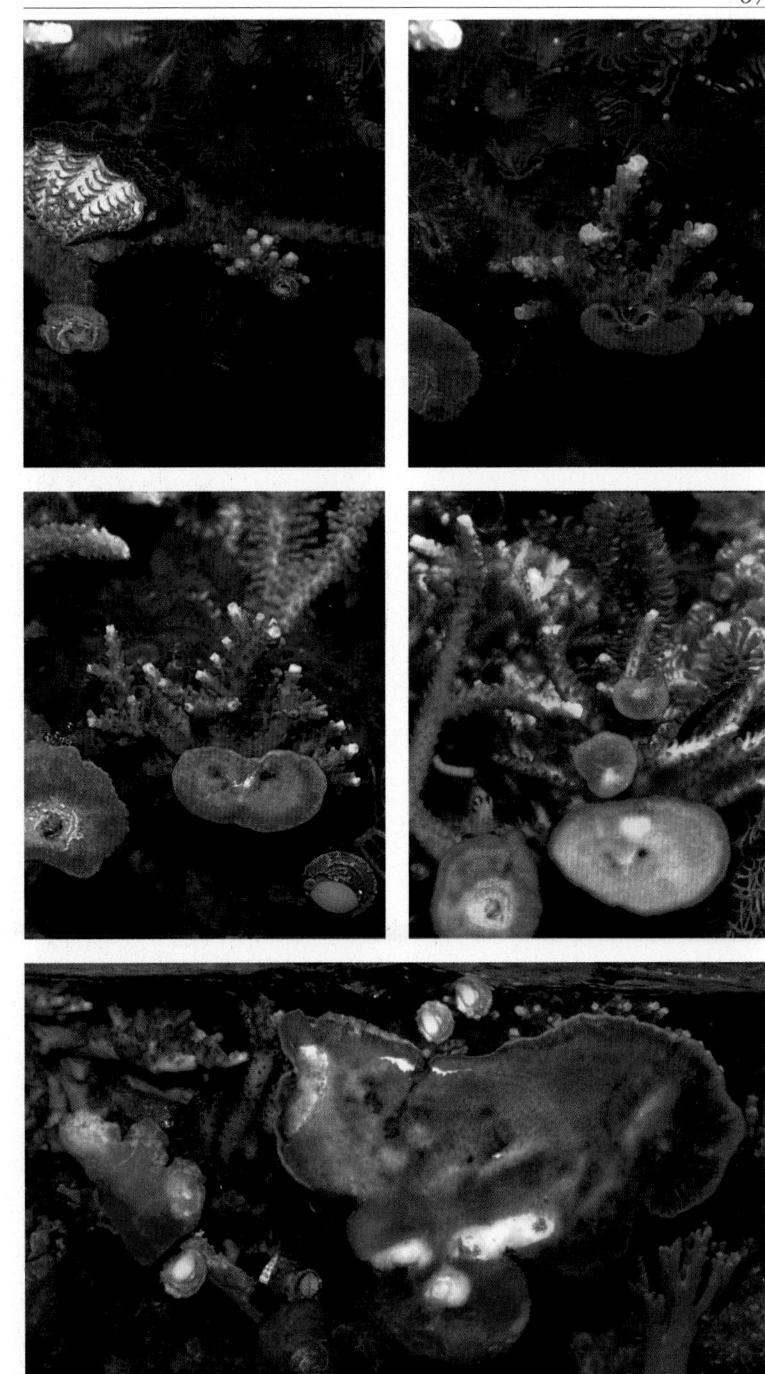

Goniopora stokesi. M. Awai.

Same colony with polyps retracted to show daughter colonies. M. Awai.

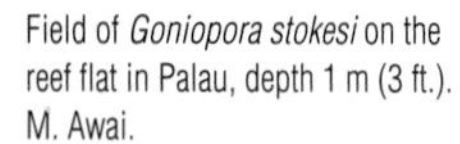

Field of *Goniopora stokesi* on the reef flat in Palau, depth 1 m (3 ft.). M. Awai.

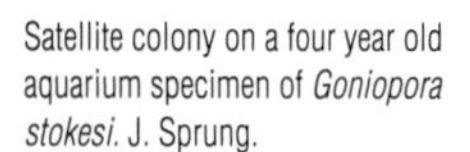

Satellite colony on a four year old aquarium specimen of *Goniopora stokesi.* J. Sprung.

Daughter colony after detaching from mother colony in previous photo. J. Sprung.

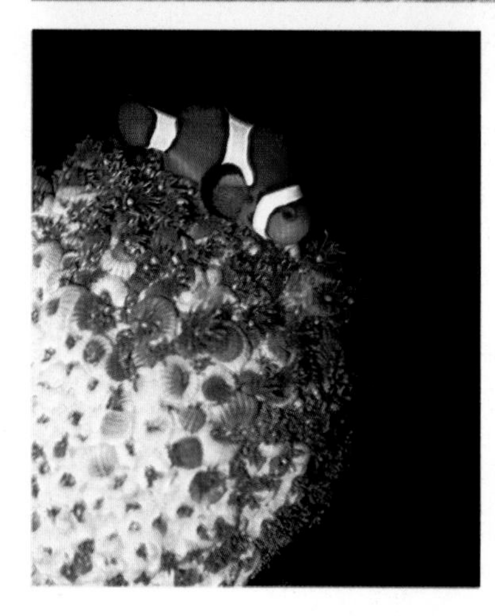

A newly introduced *Platygyra* in June, 1991. J.C. Delbeek.

The same coral in April, 1992. J.C. Delbeek.

March, 1993. J.C. Delbeek.

A polyp "ball" forming in March, 1993. J.C. Delbeek.

The resulting free-living polyp, complete with skeleton. J.C. Delbeek.

"satellites", polyp "bail-out", fission and fragmentation, outgrowths, and various means of "budding".

The process of growth for corals is a means of asexual duplication of polyps. Colonies of polyps that comprise a coral head are all clones from an original polyp that have arisen by budding. Such budding may occur by division of polyps, or by formation of new polyps between others. The distinction of some species of corals is based on the means of polyp duplication. Some corals are really composed of one large polyp that forms additional mouths as it grows.

The coral head need not have been produced by the settlement of an original polyp. It may, instead, have arisen from a fragment containing a few living polyps, broken off of a larger head. Such asexual reproduction is called fragmentation, and while it may be an accidental means of reproduction for some corals, it may be the most important means of dispersal on the reef for some species, *Acropora* being a primary example. Fragmentation increases recruitment in locations where sexual reproduction is also common, and it is an especially important means of forming new colonies during periods of environmental stress, when larval recruitment is low (Richmond and Hunter, 1990). In another example, the term fission refers to non-accidental fragmentation, as in the solitary coral *Diaseris fragilis,* which is closely related to the many *Fungia* species. It always develops weak spots in its round skeleton that cause it to fragment into several daughter colonies (Veron, 1986).

The formation of so-called polyp "balls" or "satellite colonies" among the polyps of the "mother" colony is another form of asexual reproduction that seems to combine clonal duplication of polyps and fragmentation. When the satellite colony of, for instance, *Goniopora stokesi* is heavy enough, its weight causes the tissue connecting it to the parent colony to tear, freeing the new colony which drops down next to its parent. This means of reproduction can result in vast monospecific fields of *Goniopora.*

A similar technique employed by faviid corals is the formation of a single polyp with a bit of skeletal material which, when heavy enough, separates from the parent colony and settles on the adjacent substrate.

Furthermore, large single-polyped corals such as *Trachyphyllia, Euphyllia* and *Catalaphyllia,* may produce unattached septae that may

Asexual reproduction in *Catalaphyllia* (see text for a complete description). J. Sprung.

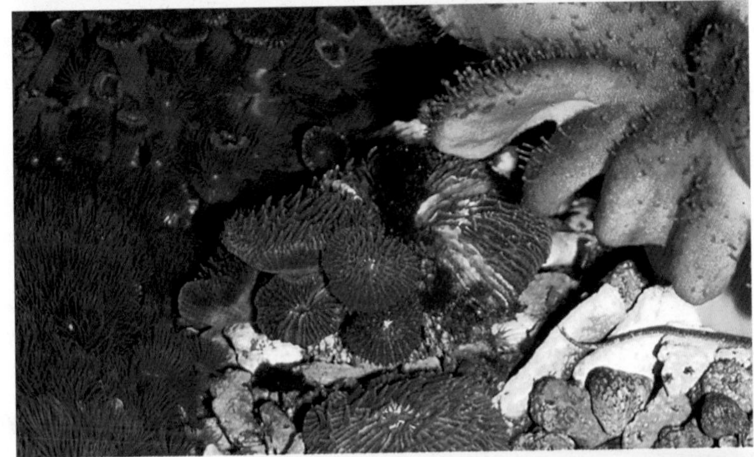

Anthocauli production on *Fungia* sp. J. Sprung.

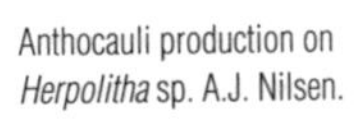

Anthocauli production on *Herpolitha* sp. A.J. Nilsen.

drag a bit of tissue with them as they separate from the parent by their weight, forming a new colony (de Greef, 1990; J. Sprung pers. obs.).

Environmental stress plays a role in the stimulation of other means of asexual reproduction. In the case of polyp "bail-out" water temperature, oxygen concentration, pollution, or other environmental stimuli cause a polyp to separate from its skeleton and drift free (Sammarco, 1982). This affords great opportunity for dispersal away from the site of stress, without formation of planula larvae.

In aquaria, polyp separation from the skeleton can occur slowly as a result of numerous circumstances (see trouble-shooting section, chapter 10). This polyp separation is quite distinct from the bail-out method that occurs in nature. It is much slower, and is more akin to slow death than a quick escape. Such separation can be healed if caught in time, and it is also possible for a separated polyp to form a new skeleton. Portions separated may form new skeleton and drop off as a new colony as in the *Catalaphyllia* pictured.

The other environmentally induced means of asexual reproduction is the formation of anthocauli in fungiid corals. The environmental stressor is typically an injury to the original polyp, such as burial, stinging by an adjacent anemone or coral, or predation by fish or invertebrates. The area of tissue loss soon produces tiny individual polyps called anthocauli. These form a skeleton of their own as they grow, and eventually they separate from the original, injured fungiid, forming a complete new coral. The small point of attachment, once broken, forms a new anthocaulus on the original *Fungia*, so the reproduction is perpetual. Hobbyists have used such

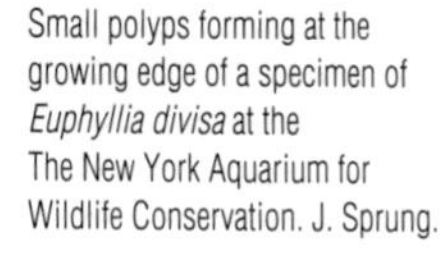

Small polyps forming at the growing edge of a specimen of *Euphyllia divisa* at the The New York Aquarium for Wildlife Conservation. J. Sprung.

injured, reproductive *Fungia* and *Herpolitha* specimens to propagate new colonies for sale and trade.

Finally, *Euphyllia, Catalaphyllia* and *Trachyphyllia* species may also produce new polyps like anthocauli between the septae when injured, and *Euphyllia* spp. regularly bud new polyps along the walls

A well-stocked reef aquarium dominated by soft corals requires frequent pruning. Terpenoids also accumulate unless removed by chemical filtration. J.C. Delbeek..

of the skeleton when they are healthy. These new polyps break off at the constricted point of attachment when they are large enough, or fuse with the main colony to form a new branch.

Competitive and Defensive Mechanisms of Corals

Competition for space is one of the most important factors limiting populations on marine hard substrata. Obviously there must be some sort of controls to allow a high diversity of life to exist in such a limited area. This is why sessile colonial marine organisms such as anemones, sponges, soft and stony corals have developed various mechanisms for defending their space and moving into new ones (Sammarco et al., 1983). Failure to recognize this fact can cause a lot of frustration for the aquarist, and can lead to many expensive losses. Sometimes predation controls growth, but this is usually not a factor in our aquariums, at least it shouldn't be! There are four main mechanisms that marine invertebrates use to establish their territory: rapid growth to "shade-out" competitors; the development of aggressive structures such as mesenterial (gut) filaments, sweeper tentacles and acrorhagi and; the release of toxic compounds into the water. In many cases an organism will use a combination of these tactics.

Rapid Growth

The growth rates of hard corals are species dependent, with certain species growing much quicker than others. This is a definite advantage and allows these corals to quickly colonize new areas. A rapid growth rate also allows these species to achieve dominance over other species by over-topping them, thereby reducing the amount of light and water flow they can receive (Huston, 1985). This mechanism has been proposed as an explanation for the dominance of *Pocillopora* corals in the Pacific and *Acropora* corals in the Atlantic (Huston, 1985). Presumably this mechanism is of limited importance in a closed system due to the paucity of branching hard corals. However, the aquarist should not forget the effects of over-topping and shading caused by large expanding anemones, corals or macroalgae growths. This can occur in an aquarium. One should always ensure that a specimen is receiving adequate lighting and water circulation. Enough space must be allotted for both expansion and growth when one first places a specimen in the aquarium.

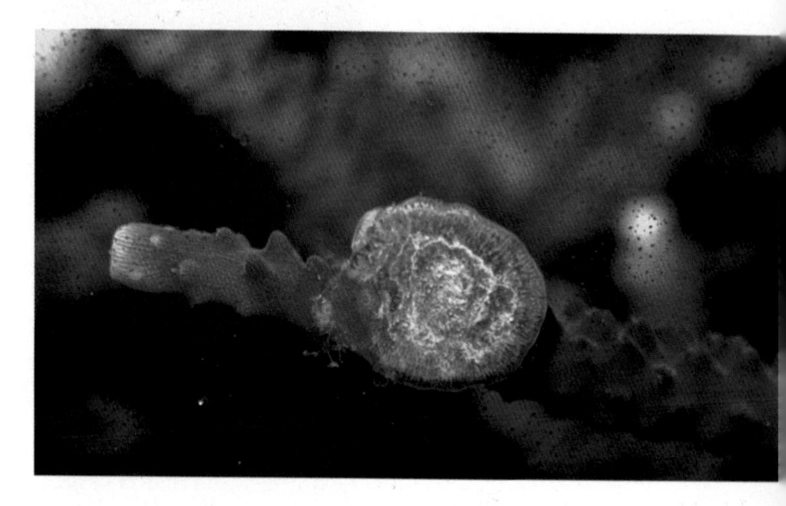

Acontia filaments produced along the growing margin of a section of an *Acropora cervicornis* branch that had grown onto the front glass of an aquarium. These acontia swept the surface of the glass, killing algae there before the tissue growth advanced over the area. J. Sprung.

Aggressive Structures

Acontia (Mesenterial Filaments)

Mesenterial filaments, or acontia, are part of a coral's digestive organs and can be used quite effectively as aggressive structures. When two hard corals come into contact (either different species or the same) one of them, the aggressor, can extrude mesenterial filaments through the mouth cavity or the body wall, onto the surface of the other, literally digesting it's tissue. This results in a zone of naked skeleton that can then be overgrown (Sebens and Miles, 1988). This zone can be overgrown by the attacking coral or it can be colonized by encrusting organisms, thereby creating a "buffer zone" between the two species (Huston, 1985). In the aquarium, such damaged areas can become infected by bacteria, attacked by protozoans or colonized by microalgae. We will deal with these problems in greater detail in chapter 10. In most cases, corals of the same species or genus do not cause damage when within touching distance, for example several *Euphyllia* spp. can usually be placed next to each other without any problems. Smaller polyped corals, such as *Acropora* spp., have also been observed producing acontia on an almost daily basis (J. Sprung; J.C. Delbeek, pers. obs.). Not only can they cause damage to adjacent corals, but the acontia also serve to keep a microalgae-free zone around, not only the base of the colony, but also around any tips that may have grown onto an adjacent substrate, like the glass of the aquarium.

Acrorhagi

Acrorhagi are specialized structures that were first recognized in coldwater species of the anemone family Actiniidae. They consist of inflated sacs that protrude from below the tentacles and are loaded with stinging cells. When they make contact with another anemone they leave behind a layer of tissue that results in localized tissue death of the intruder (Sebens and Miles, 1988). It is not clear whether these structures appear in tropical species, but one should at least be aware of the possibility.

Sweeper Tentacles

Sweeper tentacles are specialized tentacles that appear on polyps after several weeks of contact with other organisms (Hidaka and Yamazato, 1984; Sebens and Miles, 1988). In some cases contact is not necessary and some species always seem to possess sweepers e.g. *Galaxea.* These tentacles are usually much longer and thinner than normal tentacles and have many more stinging cells (nematocysts) than natural. As a result, their function has changed from one of feeding or light gathering, to one of defense or aggression. Elongated polyps of some corals such as *Goniopora,* can also be used as "sweeper polyps" for aggressive purposes (Sheppard, 1979). Although the production of sweeper tentacles are usually associated with stony corals, a recent study has shown that they can also develop in soft corals such as the encrusting Caribbean gorgonian *Erythropodium caribaeorum* (Sebens and Miles, 1988). These sweeper tentacles were found to lack pinnules on the tentacles and had bulbous tips loaded with nematocysts

(Sebens and Miles, 1988). Such specialized tentacles form only along the edge of the colony that is in contact with another coral, encrusting algae, or by nematocyst discharge (Hidaka and Yamazato, 1984). Notorious developers of sweeper tentacles in the aquarium include *Catalaphyllia, Caulastrea, Euphyllia, Favia, Favites, Galaxea, Goniastrea, Leptoria, Platygyra,* and *Plerogyra.*

Ates (1989) provides a listing of aggressive stony corals, some of which are regularly kept in aquaria such as Bubble Coral (*Plerogyra sinuosa*), Anchor Coral (*Euphyllia ancora*), *Favia* spp., *Favites* spp. and *Galaxea fascicularis.* We urge you to read this article, it is an excellent discussion of the phenomenon of coral aggression from a European hobbyist's point of view. Another excellent reference is an article written by Mike Paletta (1990) that deals with aggression in stony and soft corals. This article provides a listing of those species commonly found in home aquaria, arranged in order from most to least aggressive.

Various authors have tried to quantify the aggressive capabilities of corals and to rank them in order of aggression. Thomason and Brown (1986) found that in stony corals there was a direct relationship between aggressive proficiency and the number of nematocysts per polyp and/or mesenterial filament. Interestingly, it is the number of nematocysts per structure that is important, not the size or number of the polyps and mesenterial filaments. Combining their findings with those of Sheppard (1979), we have placed various Indo-Pacific stony corals in the following aggressive categories. Aggressive: *Fungia* spp., *Goniopora* spp. (nematocysts concentrated in the polyps not in the mesenterial filaments), *Galaxea fascicularis* and *Acropora* spp.; Intermediate: *Lobophyllia* spp.; Subordinate: *Montipora* spp. and *Porites* spp. (Thomason and Brown, 1986). Although *Acropora* spp. are classified as being aggressive it is generally believed that they rely more on over-topping and asexual reproduction by fragmentation to compete for space. The small size of *Acropora* spp. nematocysts supports this hypothesis (Thomason and Brown, 1986). However, in our experience, some species of *Acropora* can severely damage other corals if brought into contact with them. In the Caribbean, Lang (1973) classified *Isophyllia sinuosa* as very aggressive, *Montastrea annularis* as moderately aggressive and *Porites* spp. as weakly aggressive. Interestingly, the initial dominance of *Montastrea* is due entirely to the action of its mesenterial filaments. Bak et al. (1982) found that as the length of the aggressive encounter increased, the subordinate coral was able to develop sweeper tentacles and reverse the interaction in its favour.

This photo appears to be a scene from a reef in Fiji, but was actually taken in the basement of a house on an island off the coast of Norway! It shows the top view of Alf Jacob Nilsen's reef aquarium. Just how dense can the corals be allowed to grow? With fast growing branched *Acropora, Pocillopora,* and *Montipora* shown here one must be careful to prune the growth often to prevent overshading and stinging of neighboring colonies. A.J. Nilsen.

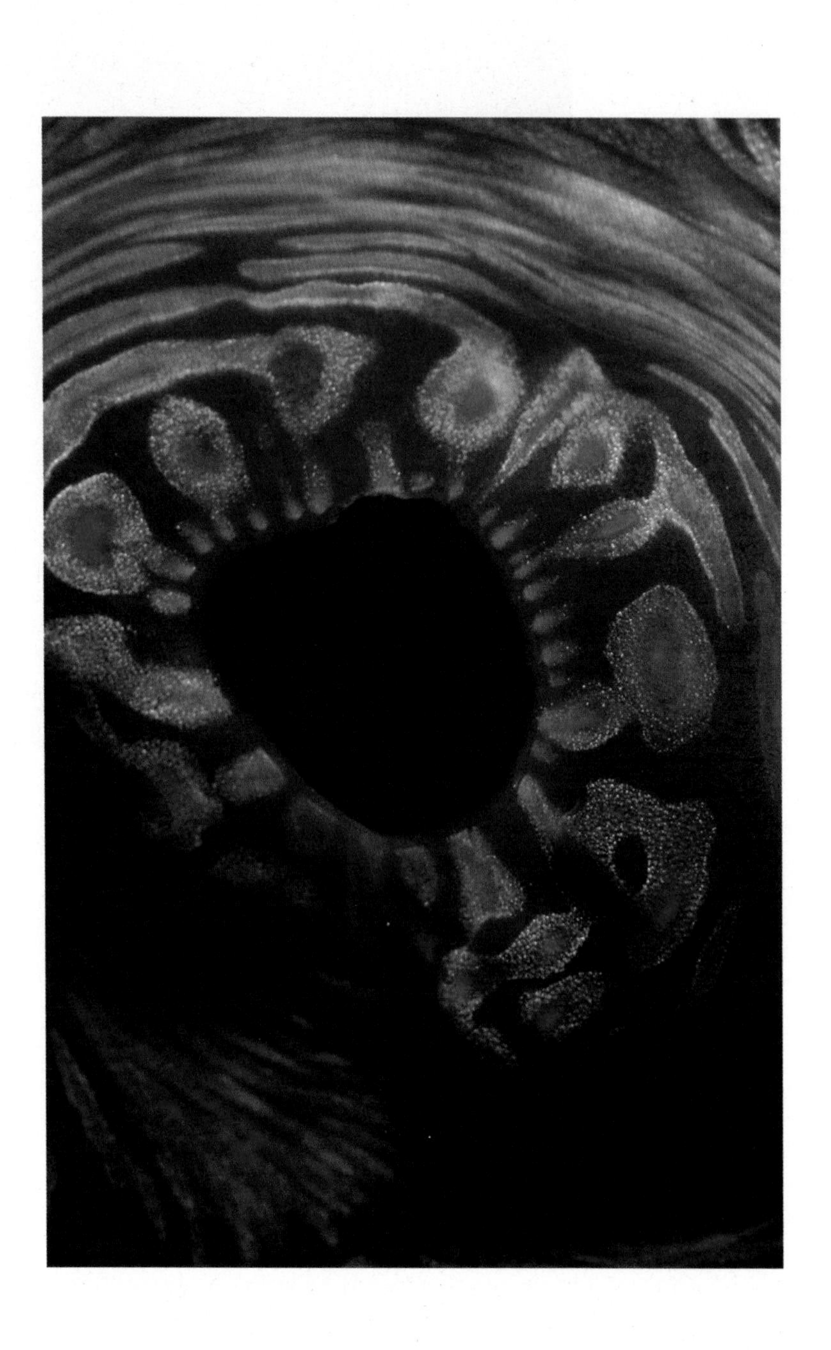

Chapter Four

The Biology of Tridacnid Clams

In popular fiction, no picture of a coral reef environment is complete without a giant clam maliciously capturing an unwary diver in its cavernous maw. This image is based more on romanticism than fact. In truth the giant clam, *Tridacna gigas,* is not dangerous, but it is the largest bivalve in the world, reaching lengths greater than 1 m (3 ft.) and weights up to 400 kg (800 lbs.). It represents only one species and there are other, smaller species, that are much more suitable for the home aquarium. Tridacnid clams belong to the order Bivalvia which includes the various families we commonly refer to as clams. The family Tridacnidae currently contains eight species in two genera, Hippopus and Tridacna: *Hippopus hippopus, H. porcellanus, Tridacna crocea, T. derasa, T. gigas, T. maxima, T. squamosa and T. tevoroa.*

Distribution

Tridacnid clams are found throughout the Indo-Pacific and Red Sea, and are usually associated with coral reefs, either amongst live corals, or on sand and rubble areas adjacent to reefs. *Tridacna squamosa* and *T. maxima* have the widest distributions, being found throughout the Indo-Pacific, from the Red Sea in the west to Tonga and Pitcairn Island in the east, respectively. *Tridacna crocea, T. derasa,* and *T. gigas* are found from the Nicobar Islands in the west to Fiji in the east, while *H. hippopus* is found from the Nicobar Islands to Tonga (Yonge, 1975). *Hippopus porcellanus* has a limited distribution being found only between eastern Indonesia and western Papua-New Guinea (Lucas, 1988), while *T. tevoroa* has so far only been found in eastern Fiji and islands within the Ha'apai and Vava'u groups, Tonga (Lucas et al., 1991).

Morphology and Anatomy

At first glance tridacnids resemble normal clams by having two valves (shells). However, it soon becomes apparent that they are different in a number of important facets. The major factor that has resulted in these differences is the presence of symbiotic zooxanthellae in the mantle tissue. It is thought that it is the presence of these algae that has allowed these clams to do as well as they have in nutrient poor areas. Due to the presence of these zooxanthellae, tridacnids have undergone a number of behavioural and physical changes.

The exhalent siphon of a *Tridacna squamosa.* Sonja VanBuuren.

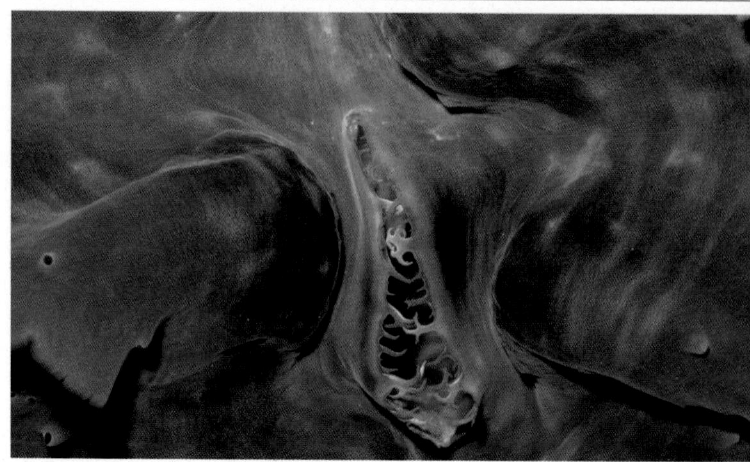

The inhalent siphon of *T. squamosa* has numerous, large fringing tentacles. J. Sprung.

Tridacnids are generally limited to shallow waters where they can receive the maximum amount of light. In fact, some specimens are found in water so shallow that they are exposed to the air during periods of low tide. *Tridacna gigas* can be found as deep as 20 metres (66 ft.), however, and *T. tevoroa* is only found in deep water (Crawford and Nash, 1986; Lewis and Ledua, 1988). Physical adaptations include a large, fleshy mantle that increases the surface area available for exposure to light. Actually, the mantle of a tridacnid is simply an extension of the inhalant and exhalent siphons. The inhalant siphon consists of an elongated opening, often surrounded by fringing tentacles, which act to strain out large particles. The exhalent siphon is located further along the mantle and forms a raised cone, through which water leaves the body cavity after being filtered by the gills. In order for the siphons and mantle to be in an upper position, the internal organs have been twisted 180 degrees such that the heart, inhalant and exhalent siphons, and the stomach lie near the top of the body, just below the mantle. This allows the siphons to be on top, further increasing the available surface area. As a result of this rotation, the muscular foot, so prominent in other clams, has become greatly reduced and is found next to the hinge of the valves. To compensate for the small, functionless foot, tridacnids have a much more prominent byssus gland. The byssus gland produces filaments (byssal threads) that extend through an opening between the two valves and fasten the clam to the substrate. The larger species, *T. gigas, T. derasa, T. tevoroa,* and *Hippopus* spp. lose these glands as they grow larger, relying instead on their size and weight to hold them in place (Lucas, 1988).

Figure 4.1a
Anatomy of a cockle
After Yonge, 1975

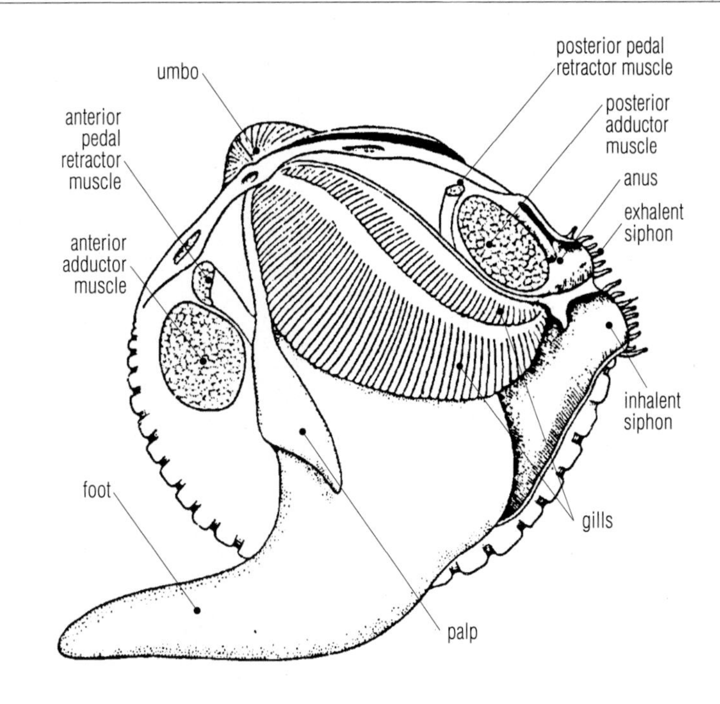

Figure 4.1b
Tridacnid Clam Anatomy
After Yonge, 1975

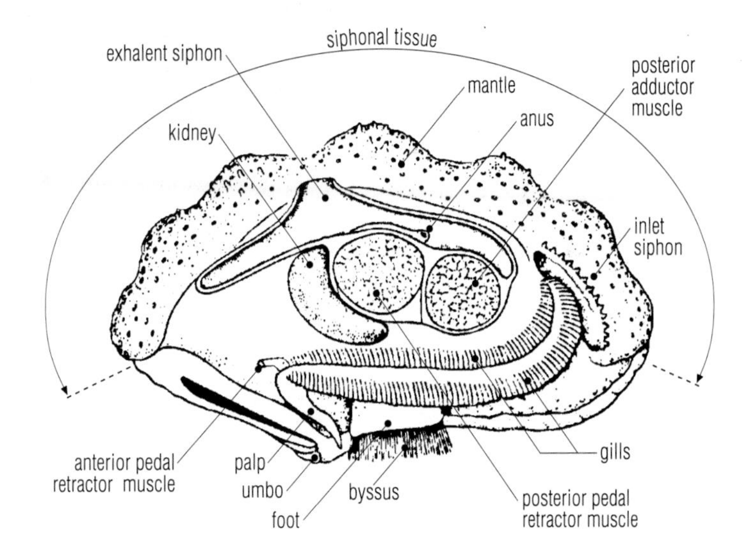

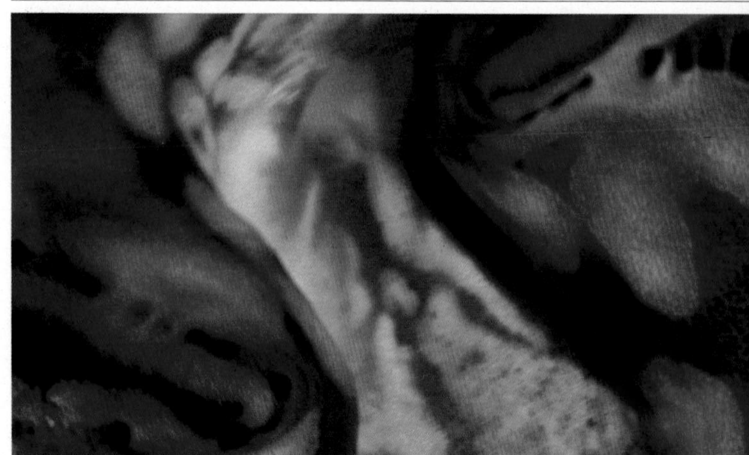

Close-up of the blue UV-absorbing pigment pattern on a *Tridacna maxima*. J.C. Delbeek.

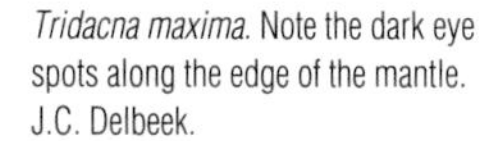

Tridacna maxima. Note the dark eye spots along the edge of the mantle. J.C. Delbeek.

The siphonal tissue (mantle) contains the majority of the zooxanthellae as well as fixed cells called iridophores that contain numerous pigments. Mainly in the colour range of blue to brown, or green to yellow, combinations of these pigments give rise to the wide range of colours and patterns that make these clams so desirable to the marine aquarist. However, the main function of these pigments is to protect the clam against excessive light and UV radiation (Yonge, 1975). If the clams do not receive the proper light intensity and quality, they will quickly lose their bright colours. Loss of bright colour results in the underlying brown colouring of the zooxanthellae becoming visible. Unless conditions are improved, the zooxanthellae may begin to disappear too, and the clam will take on a whitish-brown colour (see topic "bleaching" chapter 10). Once this stage is reached, death shortly follows (Achterkamp, 1987a). We have seen bleached clams that were highly illuminated by intense metal halide lighting. Though this lighting is normally ideal, bleaching may occur when the trace element iodine is depleted. See topic iodine in chapter 8 under trace element additions, and in chapter 6 with respect to lighting.

Tridacnids have hundreds of eyes along the edges of the siphonal tissue and some specimens of *T. crocea* and *T. maxima* can also have eyes on top of raised tubercles scattered over the mantle surface. These eyes are used primarily to detect shadows, warning the clam of the passing of potential predators (Wilkens, 1986). The eyes are also sensitive to green, blue and ultraviolet light (Wilkens, 1984). It is felt that these sensitivities help the clam to orient toward the light, in order to maximally expose the zooxanthellae. Even clams that are lying on their sides will stretch their mantle toward the light (Wilkens, 1986). Given the transparency of reef waters to UV light, the eyes could also function to detect excessive amounts of these potentially harmful wavelengths. Tridacnid clams also have light concentrating organs in their mantles called hyaline organs. These are translucent "windows" that allow more light onto pockets of zooxanthellae, thereby further enhancing their metabolism (Rosewater, 1965).

Nutrition

Most clams can obtain nutrition from a variety of sources such as filter feeding and direct absorption from the surrounding waters. Tridacnid clams have gone one step further by harbouring symbiotic algae (zooxanthellae), that manufacture food for them just as in hermatypic corals. The zooxanthellae of tridacnid clams are located within zooxanthellal tubules that extend from the stomach into the mantle tissue, not within individual cells as in corals (Norton, et al., 1992). Through photosynthesis, the zooxanthellae provide clams with the same products corals receive: carbon, in the form of glycerol, and amino acids such as alanine. Under sufficient light, zooxanthellae can provide 100% of a clam's respiratory carbon requirements (Fisher et al., 1985). In return, the zooxanthellae use the nitrogenous wastes produced by the clam, primarily ammonia, as a nitrogen source.This method of nutrient recycling benefits the tridacnids a great deal, allowing them to utilize a highly efficient, internal food source that minimizes energy loss between trophic levels (Heslinga and Fitt, 1987). It has also been shown that tridacnid kidneys contain large amounts of calcium phosphate (Trench et al., 1981). In low phosphate areas such as reefs, it is tempting to speculate on the

Figure 4.2
Tridacnid Clam Anatomy showing zooxanthellal tubules extending from the stomach into the mantle tissue.
After Norton et al., 1992.

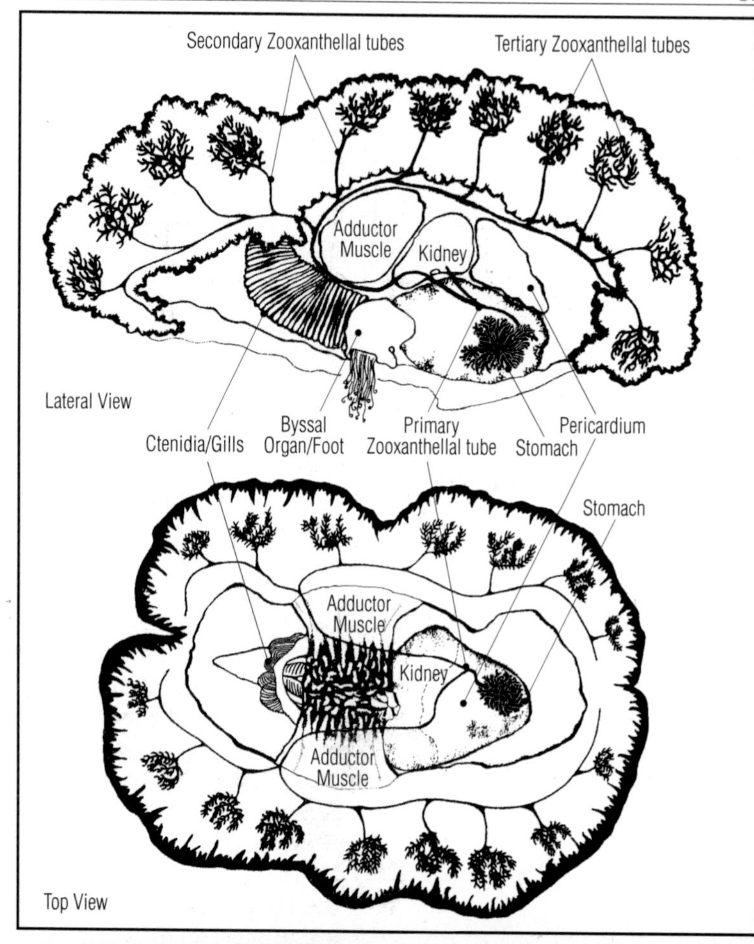

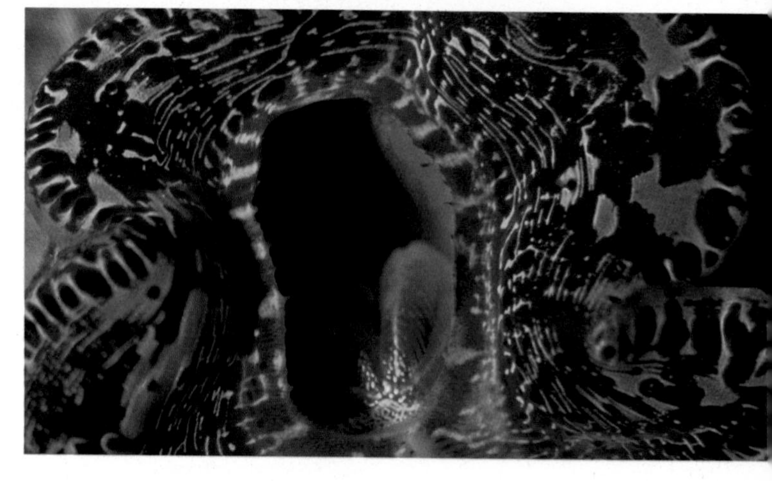

The gills are evident through the inhalent siphon of this *T. maxima*. J.C. Delbeek.

possible role this phosphate could play in zooxanthellae nutrition. However, clams without zooxanthellae also have these deposits.

The large, convoluted mantle is not only efficient in capturing light but also in absorbing dissolved nutrients from seawater. In light, zooxanthellae in the mantle take up ammonia, nitrate, phosphate and sulfate from the surrounding water and use them to make amino acids. This accounts for the ability of tridacnid clams to lower levels of these substances in closed systems (SeaScope, 1991). The clams can also expand and contract the mantle as light intensity changes, depending on their need to eliminate excess ammonia (Benson, 1984). Finally, it has recently been demonstrated that additions of ammonia, nitrate and ammonium, primarily in the form of ammonium nitrate, to culture systems has improved the growth rate of juvenile tridacnid clams (Heslinga, 1989; Hastie, L.C. et al., 1992). The presence of ammonium, however, interfered with the uptake of nitrate. Ammonium is the preferred nitrogen source since it does not require energy to be absorbed or reduced, as nitrate does (Fitt et al., 1993). It should be pointed out that these were not recirculating systems, but open ones that received constant exchange of nutrient poor ocean water. In a closed system like an aquarium, where these nutrients are generally 10 to 100 times higher than in natural seawater, no such additions are necessary.

The role of phytoplankton in tridacnid nutrition is not clearly understood. It is believed that phytoplankton provides the clam with some of its protein, but it is more likely a source of carbohydrates. However, as Yonge (1975) points out, the amount of phytoplankton available on tropical reefs is probably not sufficient to meet the needs of the clam. It has been argued that since clams possess feeding appendages such as gills, palps, and an efficient digestive system, they must be actively feeding. However, the gills are still required for respiration, ammonia expulsion, and possibly nitrate uptake (Fitt et al., 1993), the palps are greatly reduced and the digestive system is used to expel excess zooxanthellae (Norton, et al., 1992). Furthermore, studies of *T. maxima* have shown that zooxanthellae can produce excess oxygen, far above what is required by the clam (Trench et al., 1981). High levels of oxygen are potentially lethal and need to be eliminated, either through the mantle or perhaps via gills. It has been speculated that tridacnids actually digest their zooxanthellae, in effect "harvesting" excess senescent zooxanthellae as a source of protein (Yonge, 1975). Yet, several studies have shown that many

Measuring the growth in captive-bred *Tridacna derasa* in the nursery beds off of the Micronesian Mariculture Demonstration Centre, Palau. G. Heslinga.

of the zooxanthellae in the stomachs, rectum and feces of tridacnids are still viable and fully functional (Trench et al., 1981).

Zooxanthellae are introduced into juvenile clams via their feeding organs and move from the stomach of the clam into the mantle via the tubule system. Zooxanthellae are generally resistant to digestion so it should not be surprising that viable cells can be isolated from the feces (Heslinga and Fitt, 1987). If the zooxanthellae were not resistant, they could not survive long enough to make it into the mantle. This is easily shown in closed systems as clams are often observed to release thin, brown strands from their exhalent siphons, especially after periods of stress. When examined under a microscope these strands can be seen to be composed mainly of viable zooxanthellae (Achterkamp, 1987a; J.C. Delbeek pers. obs.). It is even possible to cultivate these expelled zooxanthellae (Trench et al., 1981); a potentially useful technique for those interested in breeding tridacnids.

In summary, zooxanthellae photosynthetic products appear to be the major source of nutrition in tridacnid clams. However, other sources such as dissolved nutrients in the surrounding water, phytoplankton and senescent zooxanthellae may contribute to a lesser extent.

Reproduction

Tridacnid clams are simultaneous hermaphrodites (i.e. they possess functional sets of both male and female gonads) and reach full sexual maturity after about 5 - 7 years. The age of maturation is different for each species (Heslinga, et al., 1990). Some clams can become sexually mature as males within two years and then gradually acquire female gonads as they mature further (Lucas, 1988). Although clams possess both male and female sex organs at maturity, the release of sperm and eggs are separate events. This tends to prevent self-fertilization. Generally, the sperm are released into the water first, followed shortly thereafter by the eggs. Self-fertilization, and cross-fertilization between different species can occur (Alcazar, 1988). Hybrids can be formed that exhibit characteristics of both species. Known and suspected hybrids include *Hippopus hippopus* x *H. porcellanus* (Alcazar, 1988), *T. maxima* x *T. crocea, T. derasa* x *T. gigas,* and *T. maxima* x *T. squamosa* (J. Sprung pers. obs.; G. Heslinga, pers. comm.)

Breeding may occur throughout the year at lower latitudes, but it appears that each species of tridacnid may have its own breeding season at higher latitudes (Fitt and Trench, 1981; Lucas, 1988). *Tridacna maxima* and *T. squamosa* spawn during the winter months, *T. derasa* in the spring, *T. crocea* and *H. hippopus* in the summer, and *T. gigas* in the fall (Fitt and Trench, 1981) (no data available for *T. tevoroa*).

The release of sperm can be triggered by a number of environmental factors such as temperature, light, salinity changes that might occur during, diurnal, lunar or annual cycles, and the presence of pheromones (Heslinga and Fitt, 1987). It is thought that the release of sperm is a cue for the release of eggs by other clams, and that the presence of eggs in the water is a cue for clams still further downstream, to release sperm (Lucas, 1988). Under hatchery conditions sperm release can be artificially induced by introducing macerated clam gonads (fresh, frozen or freeze-dried), or neurotransmitters such as serotonin (Heslinga and Fitt, 1987). In the home aquarium, tridacnid clams have been known to spawn both spontaneously and after some sort of disturbance in the aquarium i.e. after adding large amounts of freshwater, increasing

Aerial view of the Micronesian Mariculture Demonstration Centre, Palau. G. Heslinga.

lighting, adding large quantities of activated carbon, excessive UV exposure, etc. (J.C. Delbeek, pers. obs.; pers. comms. M. Paletta, J. Burleson, L. Jackson).

In closed system aquariums, sometimes the clams die a few days after spawning. This is most likely a direct result of the severity of the disturbance, and not an end result of spawning. It is also possible that sperm release is toxic, and the concentration of the toxin in a small closed system could kill the clams (B. Carlson, G. Heslinga pers. comms.). After the clams spawn we suggest performing a partial water change.

Strong contractions of the adductor muscles, during spawning close the valves vigorously, sending the sperm or eggs out of the exhalent siphon and into the water column. These contractions and expulsions can continue for over 30 minutes during which time millions, and in the case of the larger species, hundreds of millions of 100 micron diameter eggs are expelled into the water. When the eggs hatch (roughly 12 hours after fertilization), the larvae are called trocho-phores. This stage only exists for 12-24 hours, during which no solid food is ingested. Within two days metamorphosis occurs and they become 160 micron long, bivalved veligers. At this point, the veligers begin to take up dissolved nutrients from the surrounding water, and start to ingest zooxanthellae and other phytoplankton. Symbiosis, however, doesn't occur until after the final metamorphosis. Generally about a week after fertilization, the veligers will transform into pediveligers (pedi = foot), developing a larval foot, and begin to settle. During this period they alternate between swimming and resting on the substrate. Within 9 days, they settle permanently onto the substrate, using byssal threads to attach the 200 micron juveniles. Nevertheless, they can still travel short distances using their foot until a suitable place is found. The factors responsible for triggering metamorphosis and substrate selection are not yet known. The time from fertilization to settlement and establishment of a symbiosis with zooxanthellae usually takes about 1 to 2 weeks, with the larger species having the shorter larval periods (Heslinga and Fitt, 1987).

As mentioned above, zooxanthellae are introduced into the stomachs of the developing clam during the veliger stage. The zooxanthellae may remain in the stomach for as long as a week. A few days after metamorphosis, zooxanthellae are seen in tissues adjacent to the stomach and are subsequently found in rows in the tubules extending into the developing mantle (Heslinga and Fitt, 1987). The zooxanthellae are moved along the tubular system by the beating of cilia that line the tubule (Norton, et al., 1992). The final step in the development of the symbiosis is the growth of the zooxanthellae population within the mantle.

In the last 10 years a great deal of information has been acquired on the artificial propagation of tridacnid clams (see Heslinga et al., 1990). Various commercial breeding programs have arisen in Palau, Australia, Micronesia, the Philippines and Tonga, to name just a few. Currently *T. gigas, T. derasa, T. squamosa* and *H. hippopus* are the main species being propagated for food, restocking programs and the aquarium industry. However, more colourful, commercially raised *T. crocea, T. maxima* and *T. squamosa* are now being produced for the aquarium market, and especially colorful varieties of *T. derasa* may appear in the near future.

The white margin along the edge of the shell of this *Tridacna maxima* is new growth. J.C. Delbeek.

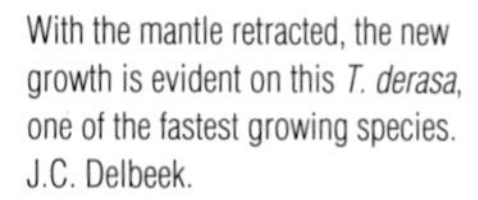

With the mantle retracted, the new growth is evident on this *T. derasa*, one of the fastest growing species. J.C. Delbeek.

The propagation of tridacnid clams in the home aquarium is a very real possibility. In fact, they may prove much easier to breed and raise than clownfish. The main hurdle is to acquire specimens that can produce eggs; sperm production is easily induced. Once the larvae reach the veliger stage, they can be fed unicellular algae such as *Isochrysis galbana* but success can be had without feeding. After metamorphosis, zooxanthellae need to be introduced into the clam. This is another hurdle. Although growing cultures of zooxanthellae is not difficult, acquiring a suitable strain may be. Once symbiosis is established, all that is required is light and the proper nutrients to promote shell and tissue growth i.e. calcium, strontium, iodide, ammonium, sulfate and nitrate. Some initial mortality results from bacteria, but these losses can be curbed with antibiotics (Fitt et al., 1992). For those of you interested in pursuing this topic further, a breeding manual *Giant Clam Farming* was published by the Micronesian Mariculture Demonstration Centre, P.O. Box 359, Koror State, Republic of Palau, 96940, and it covers the subject thoroughly. It was possible to purchase the manual at one time through the Pacific Fisheries Development Foundation, P.O. Box 4526, Honolulu, Hawaii, USA 96812, but the supply has run out. We hope that it will be published again, but for now one may have to find this manual in a library. Another book, *Giant Clams in Asia and The Pacific,* also contains a wealth of information about giant clam farming. It is available from the Australian Center for International Agricultural Research G.P.O. Box 1571, Canberra, A.C.T. 2601 Australia.

Growth

When one looks at the large size attained by some species of Tridacnidae, it is easy to imagine that these individuals became so large because they were very old. Although some species such as *T. gigas* can be over a hundred years old, it is now believed that their large size is more a function of their rapid growth rate than their extreme age. The two largest species, *T. derasa* and *T. gigas,* can grow over 10 cm (4 in.) per year, while the smaller species such as *T. crocea* and *T. maxima* grow much slower, only 2-4 cm (0.8-1.6 in.) per year. On average, *T. gigas* can reach a length of over 60 cm (2 ft.) within 10 years (Crawford and Nash, 1986). Although growth is relatively slow in the first year, it increases rapidly after that for the larger species but slows for the smaller species (Lucas, 1988). As the clams become sexually mature their growth and calcification rates can also slow noticeably (Jones et al., 1986; Lucas, 1988). For example, at an age of approximately ten years, growth of *T. maxima* slows appreciably, which was found to coincide with sexual maturity (Jones et al., 1986).

The life span of these animals can range from 8-200 years depending on the species (Achterkamp, 1987a), however, very little work has been done on age measurements of tridacnid clams. Since clams form seasonal growth bands in their shells it is possible to age sections of dead shells; perhaps in the future, more accurate measurements will be made (Lucas, 1988).

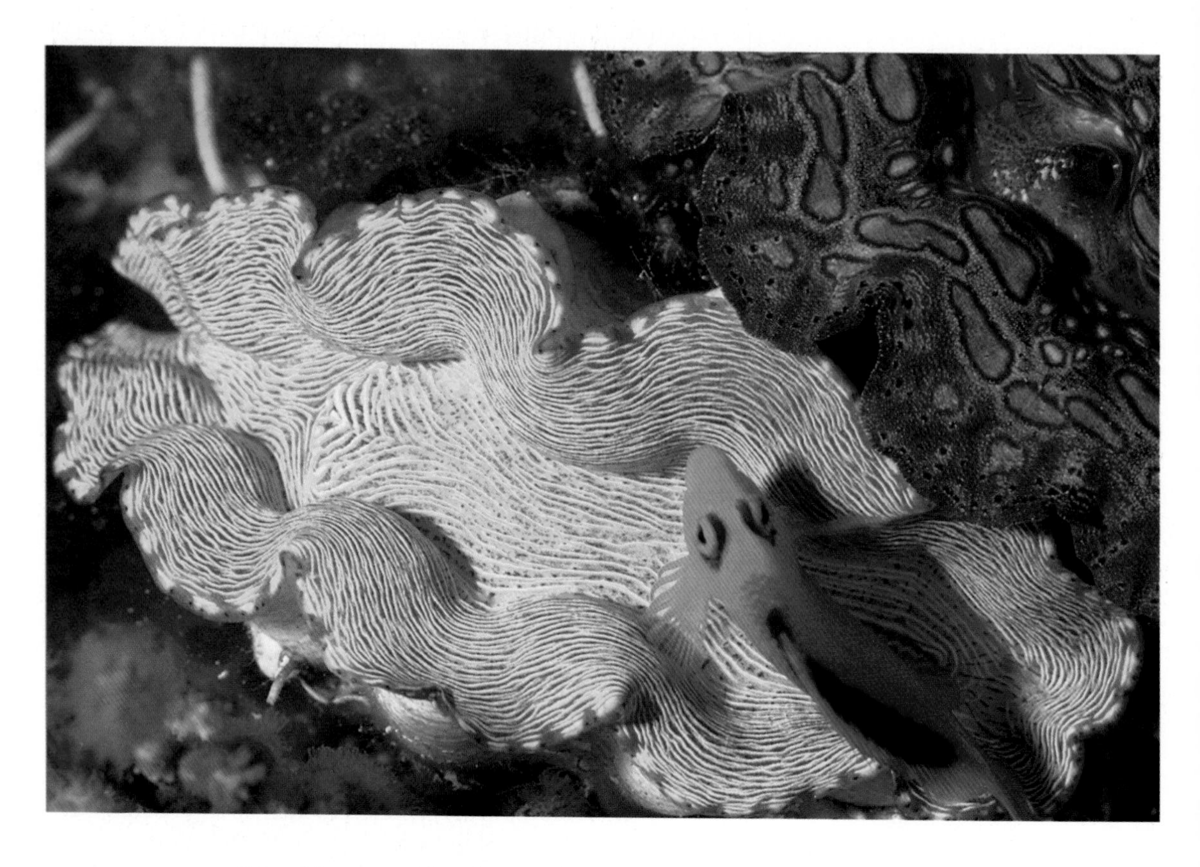

"My object in writing this article is to introduce to you this new system, (no, it is not new, it is as old as when God first created the oceans and their contents) to help marine hobbyists all over the world to enjoy their hobby with the greatest ease and simplicity."

Lee Chin Eng, 1961
Nature's System of Keeping Marine Fishes

brand that is safe for aquariums; if it doesn't say so on the package, don't use it! In addition, brands of silicone do vary with respect to their colour, clarity, and adhesive powers. In North America, clear silicones are the most common type used, but in Europe some tank manufacturers use a special black silicone that affords a really different look. When purchasing pre-cut glass, have the edges ground to reduce their sharpness, primarily to avoid injury when constructing the tank. Have any holes cut before you assemble the tank. This will save headaches later, by avoiding a return trip to the glass cutter with a completely assembled tank for drilling.

Most aquarists simply buy a commercially-made glass aquarium from their pet dealer, since these aquariums are usually the most readily available and least expensive option. There are numerous manufacturers that supply tanks nationally and locally, and some pet shops even build their own aquariums. When buying a glass aquarium, observe the quality of the workmanship and construction. The joints should be clear, with very few bubbles. The pieces of glass should fit evenly at all joints so that no piece sticks out beyond the others. Large aquariums and tall ones have top braces installed to prevent bowing of the glass, and some large custom aquariums also have bottom strips of glass inside, to provide additional bracing and prevent leaks from the bottom joints. Aquariums made from tempered glass resist bowing and can be made of thinner glass, so they are lighter and cheaper than aquariums made from non-tempered, plate glass. Tempered glass cannot be drilled, however, so it is unsuitable for special modifications.

Another option is acrylic. Acrylic is difficult to glue properly, requiring special solvents, a knowledge of the unique characteristics of the particular type (or batch) of acrylic used, and special equipment to cut the material properly. It's best to leave the construction to the manufacturers who have the expertise and proper equipment to make strong joints that will not come apart or leak. Acrylic tanks are more expensive than glass, but many aquarists prefer their appearance and the ease with which drilling and custom modifications can be done. Acrylic acts as a thermal insulator, and maintains temperature about 20% more effectively than glass. Furthermore, while some exotic custom shapes can be made with glass, acrylic offers even more possibilities. Acrylic tanks have a greater clarity than glass tanks (they are colourless, while glass is slightly green), and are lighter, weighing less than half as much as glass, but they do tend to scratch easily and are more difficult to clean properly than glass tanks. The interior surface can become scratched due to the difficulty of removing algae. For this reason it is our opinion that while acrylic tanks are aesthetically nice, they are better for fish-only aquariums than reef aquariums because reef aquariums encourage more growth of algae on the viewing windows. To make matters worse, coralline algae have an affinity for plastics, including acrylic. The coralline algae in particular create a problem because the calcium they deposit will scratch the acrylic when you try to remove them to keep the viewing window clear.

Fiberglass tanks are commonly used in research institutions and commercial aquaculture operations. They are basically fiberglass tubs with one large glass or acrylic window bonded or secured with a seal. One advantage is that plumbing holes can be incorporated with relative ease, and another is that the tank is fairly light-weight, yet strong.

Wooden tanks have been popular among freshwater aquarists for years. They are relatively easy to make and, if well constructed, can last a long time. They are easy to drill should you require drainage or return holes. Drawbacks include their heavy weight and possibly cost. Small wood tanks can cost more than similar sized all-glass aquariums. For really large exhibits though, they are cost effective.

Concrete tanks are used by public aquariums for really big displays, typically with thick acrylic sheets for the viewing windows. Aquaculture facilities also use concrete to make long, raceway style aquariums. Concrete is not typically used for home aquariums, but it is a good material for making really big tanks when weight is of no concern. It is not the material of choice for building an aquarium that you plan to move someday. If you wish to use concrete to build that giant aquarium of your dreams, we suggest that you contact several public aquariums about the best materials and construction techniques.

Filtration

As in conventional marine aquariums, filtration in reef aquariums consists of three main types: mechanical, biological and chemical. However, there are differences in the way they are incorporated and utilized in a reef tank. Mechanical filtration is usually accomplished by an inert filter pad or floss, but may also be accomplished passively through settling. Biological filtration is provided by a trickle filter and/or living rock and substrates, and chemical filtration may consist of various gadgets and substrates used to remove specific dissolved chemicals from the water.

Mechanical Filtration

The main function of a mechanical filter is to remove large particles from the water before they begin to decay. This helps to reduce the amount of substrate or "stuff" available for mineralizing bacteria to convert into ammonia and other substances. The mechanical filter could be considered equivalent to the currents and tidal flushing on a natural reef, as both help to remove particulates. The key to proper mechanical filter design, if one is used at all, lies in regular cleaning. The easier it is to clean, the more likely you are to clean it on a regular basis. Frequent cleaning of the mechanical filter helps to maintain a high redox potential (Paletta, 1989), which is one of the keys in preventing microalgae problems. Mechanical filters come in a variety of designs, though not all of them are well-suited to reef aquarium husbandry. Common materials for reef aquarium mechanical filters include sponge pads, fibrous mats, or polyester floss. Rapid sand, pleated cartridge, and diatomaceous earth filters are best suited to fish systems, or only for temporary use on reef aquariums. Mechanical filtration is not absolutely necessary for reef aquariums, as we will shortly explain

One feature of European aquaristic that has gained prominence in North America is surface skimming. This is the process whereby water is drawn off the surface of the aquarium, resulting in a clean water surface, and enhanced gas exchange and light penetration. Surface skimming is accomplished by the use of an overflow design that may incorporate some form of mechanical filtration. With this design one can efficiently remove floating particulate matter and dust that settles on the surface, and even some amino acids and fatty acids that collect at the surface because of their polar nature. Early designs consisted of an overflow located in one corner of the aquarium. A stand pipe was fitted into a hole drilled into the bottom of the tank behind the overflow partition, and some type of filter padding was wrapped around the pipe. A problem with this design is that the aquarium hood and/or lights have to be removed to get at the pad. Another problem occurs in tall aquariums when animals or pieces of equipment fall into the overflow. They are difficult to remove. Since such overflows are difficult to service, they tend to be neglected and gradually become clogged. One then sees overflows filled with water circulating around a stained filter pad. "Conversion" units, designed to hang on the back side of the aquarium have also come onto the market. Operated by a siphon, they eliminate the need for drilling holes in the aquarium, and allow you to easily convert an aquarium to surface skimming, with a trickle filter or a reservoir below the tank. A mechanical filter can be located either in the overflow box or in the top half of the trickle filter or reservoir, and it can be cleaned without disturbing the main tank. However, siphon boxes are not without their problems. If they become clogged or lose siphon, your tank may overflow onto the floor. If you change water pumps you should check first to see if your siphon can handle the new flow rate. An alternative to the above designs is to use a hole drilled in the side of the aquarium near the upper rim, fitted with a bulkhead fitting. By constructing an overflow partition in front of the hole, you can make an easily

In this simple sump, a plastic basket serves as a diffuser to slow the rushing flow of water draining down from the tank. A larger basket could be employed to hold mechanical filtration media, i.e. floss. Note the activated carbon located in bags outside of this basket. The bags of carbon can be located in a basket as well, but must not be placed where strong water flow will cause the carbon grains to tumble and disintegrate. Also note protein skimmers located in sump. J. Sprung.

accessible chamber. You must ensure that the size of the hole is large enough to handle the anticipated flow rate. We recommend a mechanical filter basket with polyester floss just where the water enters the sump, if you use a mechanical filter at all. This design simplifies filter changing and inspection compared to location in the overflow chamber. We recommend that you avoid the use of pleated cartridges and canister filters for mechanical filtration in reef tanks since these are unlikely to be serviced frequently.

Mechanical filtration can be accomplished more passively, by settlement in the sump or a connected "refugium" aquarium. If the sump has a sloped bottom, then particulate matter will settle at the low end where it can be removed by siphon or by means of a purge spigot installed there. When combined with surface

This leads us to void space, which is basically the amount of empty space available in the dry chamber of a trickle filter after the media has been added. Void space is important for gas exchange, and to prevent clogging of the media by detritus or bacterial slime. If the media is too densely packed, then there will be a low void space and a large surface area. This impedes gas exchange, and clogging of the media can also become a problem. If the media is too loosely packed you will have a large void space and a low surface area. Although this may be ideal for gas exchange, it does not offer the most surface for nitrifying bacteria. The best media offers the greatest amount of surface area without significantly decreasing the amount of available void space.

In reef tanks using sufficient live rock, the issue of whose media has the greatest surface area is unimportant. These aquariums run quite successfully without a trickle filter. They rely instead on the bacterial population in the rock and other substrates, and the associated algal growth and photosynthetic organisms that consume nitrogenous wastes. With regard to gas exchange, the water skimming over the overflow picks up oxygen and gives off carbon dioxide as it tumbles down to the sump below. A protein skimmer further assists with gas exchange, and removes compounds that contribute to biological and chemical oxygen demand (BOD and COD).

Trickle filters are among the most efficient means of biological nitrification available to aquarists, but one does not need a trickle filter, or external biological filter of any kind, in order to maintain a reef aquarium. In fact, the trend of relying on protein skimming combined with natural biological processes within the aquarium for the maintenance of water quality has finally achieved wide acceptance, but there will always be a spectrum of opinion about what pieces of equipment are really useful, and we wish to present all options.

Chemical Filtration

The topic of chemical filtration is described in just about every textbook written on marine aquarium keeping, yet few aquarists really seem to understand it's capabilities, limitations and applications. There are numerous forms of filtration that fall under the category of chemical filtration, depending on their mode of operation. For our purposes we will limit our discussion to the common forms of chemical filtration used in reef systems, namely activated carbon, protein skimming, molecular adsorption media and ozone.

Due to the various biological processes that occur in an aquarium, a build-up of organic substances takes place. They are referred to as organic because they all contain the element carbon in their chemical composition. The list of these things is quite lengthy, and includes such goodies as amino acids, proteins, phenols, creosols, terpenoids, fats, carbohydrates, hydrocarbons, plant hormones, vitamins, carotenoids and various organic acids such as fatty, acetic, lactic, glycolic, malic and citric (deGraaf, 1968; Moe, 1989). Fortunately for us, we generally lump all these things together under the all-encompassing term, dissolved organic carbon (DOC). These organic substances can have various deleterious effects on the aquarium inhabitants including reduced growth, reduced disease resistance, and metabolic stress. When the compounds contain nitrogen, they are mineralized by bacteria present in the tank, into ammonia. The ammonia is utilized by plants, leading to excessive growth, or oxidized by nitrifying bacteria to the final product, nitrate, which tends to accumulate in the aquarium. Unfortunately, many DOCs are not mineralized and tend to build-up in the aquarium. That is why water changes are usually advocated. Many people think that water changes are designed to lower the nitrate concentration. While this may occur to some extent, the real reason is to lower the DOC content of the water. Since nitrate and DOC concentrations are sometimes directly related, and nitrate is easy to measure, it is often used as a yard stick to determine when to make a water change on a standard saltwater fish aquarium. In reef aquariums DOC may accumulate while nitrate does not, because of denitrification and utilization of ammonia by algae, corals and clams. If we could remove much of the DOC before it accumulates or becomes converted into ammonia, we could reduce the need for water changes, lower the load on our filter, lower nitrate levels, and improve the growth and health of our organisms. These are the primary reasons behind the use of chemical filtration.

At this point we would just like to clarify that we are not saying that if you use chemical filtration, water changes will no longer be necessary. That is, of course, not true. First of all, no method of chemical filtration is 100% efficient and many substances are difficult to remove by chemical filtration. Secondly, water changes provide other benefits besides removing and diluting DOC, such as providing pH control, a balance of trace elements, and replenishment of other elements such as calcium that are lost to chemical filtration or biological processes. Even in the best maintained aquariums, the affects of a water change on the inhabitants can be quite stunning (see chapter 7 for water change recommendations). What chemical filtration does is help maintain a much lower concentration of DOC in the tank. This becomes extremely important when maintaining invertebrates such as stony corals. Corals, algae, and other marinelife leach many types of DOC into the water. The density of these organisms in a closed system reef aquarium results in heavy accumulation of DOC unless the filtration is there to remove it.

Activated Carbon

Many of us are familiar with the use of charcoal in the corner-box filters in freshwater aquariums. These usually consisted of small, shiny, irregularly-shaped pieces of bone or wood charcoal. This type of carbon is not really suitable for use in marine aquariums, and has been replaced by "activated" carbon. In this form the carbon has been subjected to extremely high pressures and temperatures to drive out all impurities and gases leaving behind extremely porous and pure grains of carbon. Particle size, type of gas used, activation temperature and, in some instances, inorganic salts of zinc, copper, phosphate, silicate and sulfate added before activation, provide carbon with specific adsorption characteristics (Moe, 1989). Therefore activated carbon can be tailored to the specific type of impurities that one wishes to remove. By creating such extremely porous structures within the carbon grains we have, in effect, created a gigantic sponge that can absorb many impurities from the passing water.

Activated carbon will remove a wide variety of organic molecules by simply trapping them in the carbon pores (absorption) or by chemically bonding them (adsorption). Adsorption relies on the fact that many organic molecules are polar in nature. This means that the two ends of a molecule differ in their affinity for water. One side is repelled by water and is termed hydrophobic ("water hating") while the other end is attracted to water and is called hydrophilic ("water loving"). When a polar molecule comes close to a polar surface such as GAC, it becomes attached, effectively removing the molecule from solution. Moe (1989) and Hovanec (1993) give detailed discussions of the properties of GAC and the factors that determine its efficiency, so we will not go into great depth on these topics.

A common point of disagreement about GAC use concerns its placement in the system. Naturally, there are several options available. Most aquarists have a sump design that includes a chamber for activated carbon, with all or some of the water flowing through the chamber. Alternately, one may hook-up a canister filter to the sump, filled with GAC. Another option is to build an in-line contact chamber. This consists of a section of PVC pipe with hose fittings at both ends. The pipe is filled with GAC and is located on the return line so that all the water returning to the tank passes through it. Moe (1989), Spotte (1979), and Thiel (1988,1989) describe the construction and placement of such units.

Although forcing all the water to pass through a chamber filled with GAC is the prescribed method, and is supposed to be the most effective means of using GAC for polishing the water, it may not be the best way. When water is forced through a chamber containing bags of GAC, the bags tend to become clogged by particulate matter on the surface because the fine mesh necessary to keep carbon particles inside acts as a mechanical filter. This effectively isolates much of the carbon, puts back-pressure on the line, and slows the water flow. Much of the water then ends up channeling between the bags. This problem can be solved by avoiding the use of bags, or avoiding the use of too fine a mesh size. Still, the bags make carbon easy to handle, and placement of bags of GAC in such a way that water flows around them, but is not directly forced through them, may be all that is necessary to maintain the water free of yellowing compounds. That is the primary reason for using activated carbon anyway. Using the carbon most efficiently is not the goal in reef aquariums.

Yellow water is a symptom of accumulated dissolved organic compounds. These can be removed from the water by chemical filtration such as activated carbon.

One may simply place bags of GAC in the sump. Through contact with the water flowing over the carbon, and through diffusion, organics will be taken up, keeping the water colorless with less rapid depletion of trace elements than by the prescribed method (J. Sprung, pers. obs.). Caution: be sure that the bags cannot slide around in the sump and become drawn up against the suction intake of the pump(s). Plastic baskets, eggcrate light diffuser material, or plastic panels can be used to create a carbon chamber. In natural system aquaria (see topic: natural systems, this chapter) using no external filters, one or more bags of carbon can be placed behind the rock-work in a location where they may easily be retrieved, near the water movement generated by the rising air from the bubbler(s). This passive flow technique will keep the water from becoming yellow for several months. Incidentally, bags of activated carbon buried in the gravel next to the lift tubes in undergravel-filtered fish tanks will likewise keep the water from becoming yellow for several months.

One of the most common questions concerning the use of GAC is how much to use and how often it should be replaced. These questions are very difficult to quantify simply because no two systems are identical. Differences in bioload and the type of organisms being kept greatly influence the type and amount of DOC required. For example, aquariums filled with macroalgae will produce a greater variety of DOCs than systems with very little algal growth. Thiel (1988) recommends using 36 ounces of GAC per 50 gallons, while Wilkens and Birkholz (1986) recommend 500 grams per 100 litres, which is roughly equivalent. Although this recommendation does seem excessive, one could use it as an upper figure and work downwards. The real indicator is the appearance of the inhabitants in the system and the colour of the water. As aquarists we should strive to be more in-tune with what is really going on in the tank by observing the animals contained therein. Too many aquarists today are seeking technological wizardry to maintain their aquariums. People are constantly talking about ozone, redox potential and carbon dioxide systems when they cannot correctly identify their tank inhabitants, or don't fully understand what pH is. The occupants of our aquariums are far more sensitive to water chemistry than any meter, and one should spend more time watching them instead of the flashing lights and numbers on the instruments.

It is difficult to recommend a specific time period after which the GAC should be replaced because of the differences in carbons, and the population composition in different aquariums. However, various authors state that GAC remains active for 5-7 months before needing replacement (Moe, 1989; Wilkens and Birkholz, 1986). Generally the presence of yellowing substances in the water can be used as a guide to determine if your GAC needs replacing, since these compounds are easily removed by GAC, and start to accumulate when the GAC begins to lose its activity. Moe (1989) describes the following method. Obtain a strip of white plastic and colour one half a faint yellow with a non-water soluble marker. Place the strip in the water at one end of the tank and observe from the opposite end. When you can no longer distinguish the yellow half from the white half, your water contains yellowing substances and you should replace your GAC.

Since GAC is a very porous substrate, nitrifying and denitrifying bacteria will quickly colonize it. If you use large amounts of GAC, replacing all of it every six months or so could affect the denitrification potential of the aquarium when little substrate is used. It might be wiser to replace 30% with new carbon and rinse the remaining 70% with seawater (Wilkens and Birkholz, 1986). Put the new carbon in a separate bag and place it in-front of the old carbon. This will preserve a large amount of the bacteria that have colonized the GAC. Wilkens and Birkholz (1986) recommend that if GAC is to be added to an established aquarium, it should be done gradually, say 20 grams per 100 litres added monthly, until 500 grams per 100 liters is attained.

There are numerous brands of GAC being marketed today. Unfortunately, not all GAC is created equal, and the levels of efficiency and quality vary greatly. The grains of GAC should be small, dull black in colour, and as dustless as possible, though the presence of dust is more of an inconvenience than an indication of quality or performance.

Certain brands of GAC will actually add phosphate to the water, which is something we are trying to avoid. The reason why some GAC leach phosphate is that they have been treated with phosphoric acid in order to increase porosity. The phosphoric acid etches holes in it, making it a really efficient super activated carbon, but it makes it a really bad one for a reef tank. These carbons are intended for the purpose of air purification, not water purification. Manufacturers of such acid washed carbon will tell you that a few rinses in freshwater will eliminate most of the phosphate, and that the amount leached is not significant. This advice contradicts our experience. We could not rinse such carbon sufficiently to affect its ability to leach phosphate, and the amount leached is significant (based on the test procedure described below). Some GAC that has not been acid washed may still contain phosphate or high levels of ash, both of which can contribute to undesirable algae growth.

The best way to determine if a carbon releases phosphate is to use a phosphate test kit. Add the reagent to purified freshwater and then add a few grains of your GAC to the test vial. If you see blue trails coming off the pieces, you know it releases phosphate.

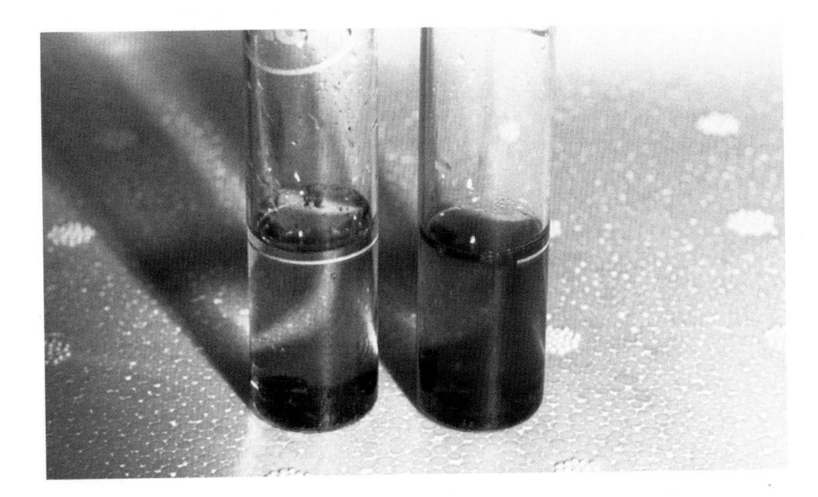

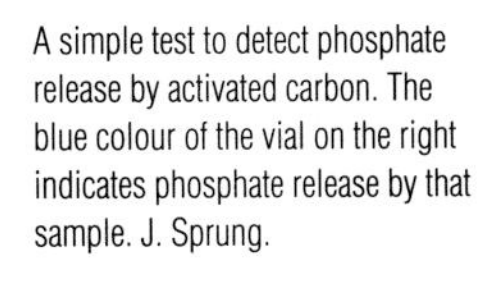

A simple test to detect phosphate release by activated carbon. The blue colour of the vial on the right indicates phosphate release by that sample. J. Sprung.

Another potential problem with activated carbon is that as it ages, some of the substances it has adsorbed and absorbed can be released back into the water (Spotte, 1979). However, if you change your GAC on a regular basis, you should be able to avoid this. A final caveat concerning GAC is that along with the other forms of chemical filtration mentioned here, it uncontrollably removes substances from the water, including some useful ones like trace elements. Therefore, regular water changes or trace element additions take on added importance when chemical filtration is present. Some aquarists use carbon only a few days per month for this reason, but others use it continuously with no ill effects. Apparently the most important trace element depleted by carbon and other means of chemical filtration is iodine. Refer to the section on trace element additions in chapter 8 for a description of dosages and the symptoms of iodide deficiency.

Not all aquarists use activated carbon. In fact some aquarists are vehemently opposed to its use, citing trace element depletion and the occurrence of mysterious ailments in corals or fishes as reasons to avoid using it. Most vehement opponents of activated carbon use have limited experience with it, and base their opinions on rumor and fear of the unknown more than first hand observation. We have pointed out the problem of trace element depletion. That is easily solved by supplemental replenishment or water changes. Further, we have pointed out that not all activated carbon is created equal (see Hovanec, 1993), and that some carbons leach phosphate. Therefore choosing the right carbon is important. Mysterious ailments in fishes, such as the stimulation of head and lateral line erosion sometimes linked with activated carbon use (T. Frakes, pers. comm.), may be the result of substances leached by particular types of activated carbon, and not a feature of all types (Sprung, 1993). That is a subject worthy of research.

Aquarists follow a spectrum of activated carbon use, some never using it, some using it periodically, only a few days per month, and some using it continuously. In our experience, continuous use is fine, and it is what we recommend, bearing in mind the cautions and parameters mentioned earlier. Ultimately you will decide what works best for your aquarium based on the health of the inhabitants, and the aquarium's aesthetic appearance.

Molecular Adsorption Filters

Although this form of chemical filtration is a relatively new addition to marine aquariums, there has been a rapid proliferation of products for the removal of various organics, nitrates, and phosphates. This type of filtration usually consists of styrene or acrylic polymers that selectively adsorb polar organics and nitrogen containing compounds onto their surface (Moe, 1989). Molecular adsorption media should be situated so that water is forced through the medium, not around it, with adequate prefiltration to prevent clogging of the media. It is not known whether long term use of such filters will lead to trace element depletion. In general, these products are best for temporary use in fixing problem situations. They only treat the symptoms of the problem. It is best to identify the factors that are causing the problems and deal with them directly (see Chapters 8 and 9).

Protein Skimming

If you've ever visited a beach on a windy day, you may recall seeing foam washing up on shore. This foam is produced by the

External protein skimmers can also be installed in such a way that they receive water directly from the pump that sends water back to the aquarium from the sump, with a gate valve on the feed to regulate the volume the skimmer receives. Be aware that when the power goes off, some water will drain back to the sump from the skimmer. Be sure that the sump is of sufficient size to handle this back flow. The water exiting the skimmer may be directed to the aquarium or back to the sump, depending on the location of the skimmer and its height.

Alternately, a dedicated pump can be used that draws water from the overflow chamber, sending it to the skimmer and back to the aquarium. In order for this design to work, a second hole or a siphon can be installed in the overflow chamber to feed the dedicated pump. The main drain of the overflow is separate, and simply drains water to the sump. A standpipe on the main drain maintains a static water level in the overflow chamber. With this design it is possible to achieve two advantages. One is that the dedicated pump can be of greater capacity than the pump that returns water from the sump, allowing the tank volume to be processed more quickly through the skimmer. The second advantage is that the surface skimming action will be enhanced by the increased volume of water flowing over the overflow. This will also raise the water level in the aquarium slightly. Don't forget to have a strainer on the suction for this pump to prevent mincing fish or invertebrates that wander over the overflow.

The Skimmer in Operation

Once you have the skimmer connected in-line, if it is an air-driven model, connect the air supply to the air blocks and turn on the air pump first before letting water into the skimmer. This prevents the wooden air block from being presoaked; it's much harder to start a wet wooden block under 120 cm (4 ft.) of water than a dry block. You will have to watch your skimmer closely at first to make sure that the flow is just right. Too great a flow will result in a full foam tube and collection cup, a wet floor and an empty aquarium! To guard against a mess on the floor, drill a hole near the top edge of the collection cup, and install a drainage tube of at least 1/2 in. diameter that leads to a large waste container. Do not allow the skimmed off material to flow back into the aquarium; it is toxic when so concentrated.

Figure 5.3a
External Air-driven Skimmer Installation
After Nilsen 1993

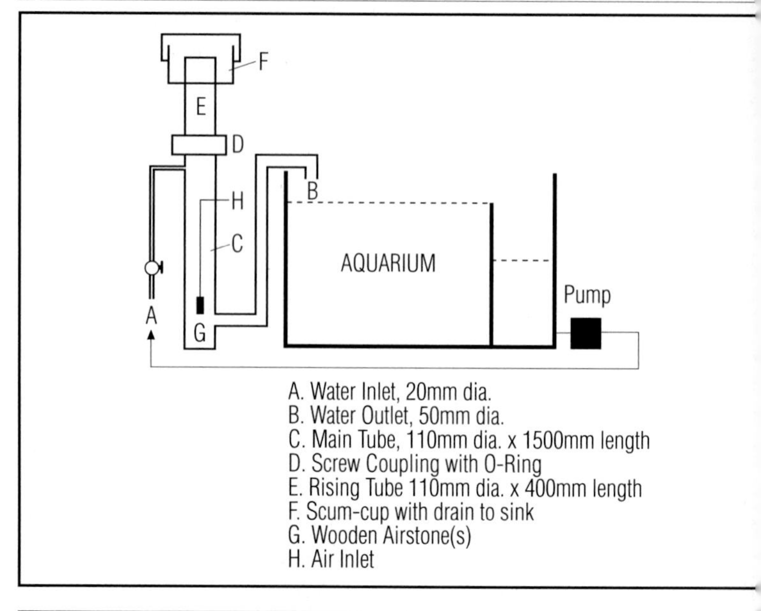

Figure 5.3b
External Venturi Skimmer Installation

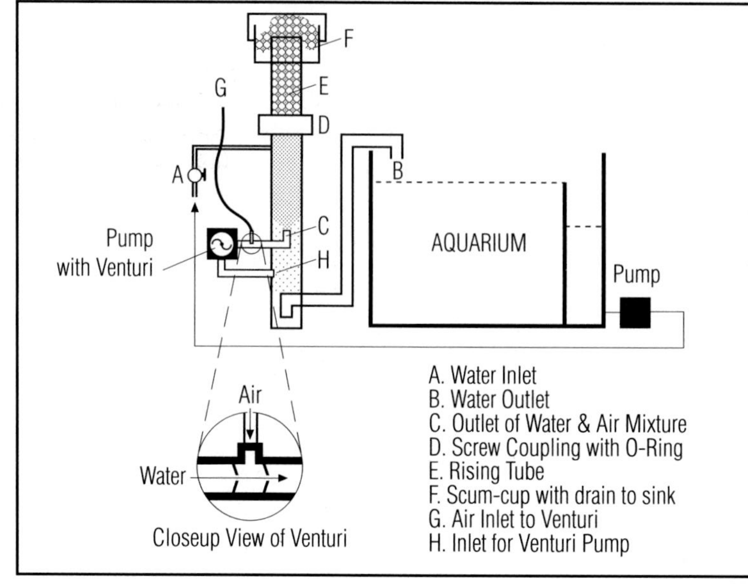

You should notice a dark brown fluid and sludge accumulating in the collection cup after only a few days. Initially you might have to empty the collection cup 2 times a day, but the output will soon slow down appreciably. The skimmer is still working, there just isn't as much left to remove, that's all. You'll notice that the skimmer will begin to work furiously after feedings, water changes, addition of trace elements, or the addition of fresh live rock (boy will it ever foam after you put in live rock!). A reduction in output from the skimmer is not always caused by the lack of substances to remove from the water. In time the skimmer becomes dirty, and this impedes its performance.

In order for your skimmer to keep working at peak efficiency it must be kept clean. Every few days or about once a week, depending on how quickly scum accumulates, the foam collection tube should be scrubbed clean. Wooden air blocks should be examined and replaced if necessary, about once per month. Every 3 to 4 months the main body tube may require some scrubbing. A bottle brush works fine for the main and foam collection tubes while a coffee percolator brush works great on smaller tubes. The inside of the skimmer must be kept clear of algae and any build-up of organic material on the sides. For transparent skimmers, algae growth can be impeded by covering the main body with some opaque material. For additional information on protein skimmer maintenance, see appendix B.

There are a few things to keep an eye on when using a protein skimmer. First of all, the continuous removal of small amounts of seawater by the skimmer, along with replenishment of evaporated water with freshwater, can lead to a gradual lowering of salinity. Therefore, the periodic addition of seawater to the make-up reservoir may be necessary to maintain the desired level of salinity. Secondly, efficient skimmers can remove some trace elements as we mentioned already. The regular addition of trace elements may be especially necessary when protein skimming is used (see chapter 8). Finally, the addition of certain buffers, vitamins and molecular adsorption filter pads can cause the skimmer to foam excessively. Rinse prefilter material and molecular absorption pads in pure freshwater before using them, and add buffers or vitamins very slowly, in small amounts.

At the introduction of the topic of protein skimming, we described the natural process of wind and waves generating foam that washes up on shorelines. The foam that collects around coral reefs, best observed at low tide on exposed reef flats, plays a role in nutrient exchange of carbon, nitrogen, and phosphorous. Some of the foam may be broken down by intense UV wavelengths at the water surface that are capable of breaking chemical bonds (much as ozone does), but most of the foam is washed by tides and wind to inner areas of the reef and associated seagrass and mangrove ecosystems. Proteins, carbohydrates, trace elements, bacteria, and other "stuff" attached to bubbles may be a significant source of food to filter feeding or particulate feeding organisms in these areas, and they may use the nutrient and element concentrating feature of bubbles to advantage. Gorgonians, for example, do capture and ingest tiny bubbles from the water (J. Sprung, pers. obs.). They may derive significant nutrition from the substances attached to the bubbles, and they can easily expel the excess gas collected. It is not known whether other filter feeders such as clams or tunicates utilize tiny bubbles this way. By demonstrating its obvious occurrence in the natural setting, we wish to emphasize that protein skimming is a natural process, and that it is easily duplicated in the care of aquarium systems.

We are satisfied that protein skimming is the simplest and most efficient means of water purification for the maintenance of live corals and for the creation of model ecosystems. So much benefit to the aquarium is provided by a device that simply mixes the water with fine bubbles of air.

Ozone

Ozone is a naturally occurring gas in the upper atmosphere, where its UV absorbing properties have been given wide exposure in news reports, regarding its recent depletion caused by chlorofluorocarbons (CFC's). Ozone is a powerful oxidant since it consists of three atoms of oxygen (O_3) and readily releases the extra oxygen atom to become the more stable form of oxygen gas, O_2. It is this property that we utilize in the aquarium to oxidize organic compounds. Unfortunately, other potentially harmful by-products such as hypochlorite and hypobromite can also be produced, and these can damage delicate invertebrates and fish gills (Moe, 1989).

Ozone is generally used in conjunction with a protein skimmer or a pressurized air reactor. Ozone is mixed with air and is introduced into a contact chamber. The ozone-air mixture combines with the aquarium water passing through the chamber and organics are oxidized. Water leaving the skimmer or reactor is then passed through a container of GAC before being returned to the aquarium, to remove any residual ozone and most harmful by-products produced. Used in conjunction with a redox controller, precise control over a system's redox potential can be obtained (Burleson, 1989) (see redox in chapter 8 for additional information about this parameter).

In our experience, hobbyists tend to use excessive amounts of ozone in their skimmers. Ozone use was originally suggested for skimmers in order to increase their efficiency. This was achieved by using small amounts of ozone, approximately 2-5 mg/L/hr (Wilkens, 1973). When much larger amounts of ozone are introduced into a skimmer, many of the organics usually extracted become oxidized into other forms that are more resistant to removal. Common symptoms of excessive ozone use are a dramatic drop in skimmer production, a much lighter coloured effluent and, in some cases, an increase in nitrate. If you wish to use ozone in a skimmer, you should use only small amounts (<10 mg/L/hr.).

A more efficient alternative is to use an ozone reactor where small amounts of ozone are injected into a canister under pressure. The decreased turbulence and pressurization result in a very efficient use of ozone, requiring very small amounts to achieve the desired results. Using an ozone reactor results in a more efficient administering of ozone, and frees the skimmer to perform the function it was designed for; the removal of waste from the aquarium. Not many home aquarists use ozone reactors any more, though these units really can benefit commercial holding facilities that receive frequent, large shipments of fish and other marinelife. Burleson (1989), Moe (1989) and Thiel (1988, 1989) go into more detail about ozone reactor construction and implementation, and redox potential.

Although many periodicals and books suggest that ozone use in Europe is quite common (see Moe, 1989), our personal contacts there indicate that ozone is reserved for emergency use only, primarily for its germicidal effect, and that skimming is normally employed without ozone. The main criticisms appear to be that ozone is not necessary to have a successful tank, and its use might cause problems in the long run. The various by-products produced (especially hypobromite) are potentially dangerous to the inhabitants if allowed to accumulate, and the oxidation of trace elements (e.g. iron and other chelated elements) by ozone renders them useless since they become insoluble and fall out of solution (Hebbinghaus,

1989; Stüber, 1989; Wilkens and Birkholz, 1986). Yet we have both seen many beautiful aquariums here in North America that use ozone in conjunction with redox controllers on a continuous basis. One thing that we have noticed though, is that reef systems that use ozone tend to run at higher nitrate levels than reef systems that do not. This may be a reflection of the increase in nitrate production caused by the oxidation of nitrite into nitrate by ozone.

Whether used singly or in conjunction with other forms of biological, chemical and mechanical filtration, it is safe to say that chemical filtration techniques are an important component of reef aquarium filtration systems. We prefer the most simple applications and, as we shall describe, successful reef aquariums can also be established by "natural" means, utilizing no chemical filters.

Methods for Establishing Reef Aquaria

Reef aquariums were introduced to North American hobbyists as the product of superior filtration provided by a gadget called the trickle filter. However, as we have mentioned a few times already, excellent success is attained without a trickle filter, and this is how we prefer to run our systems. This is by no means a new concept and has been practiced in Europe for more than ten years. In fact, it is a modified version of the natural system promoted by Lee Chin Eng over thirty years ago. Today many North American hobbyists have put away their trickle filters and run their aquariums with protein skimmers only.

Live Rock

Many of you are probably wondering where biological filtration will take place if there is no trickle filter. The answer is in the live rock. The use of live rock as a biological filter medium within the aquarium allows for the establishment of natural biological processes such as occur on the reef. That marine aquaria could be simple and very natural was recognized by Lee Chin Eng, the "father" of the natural system aquarium (Eng, 1961; 1976). The establishment of biological filtration and a diverse food web of bacteria, plants, microorganisms and invertebrates occurs whenever significant quantities of live rock are used, whether separate forms of filtration are used or not. However, we believe that a trickle filter can detract from the efficiency of these processes, as we shall soon explain. The incorporation of protein skimmers provides for the removal of various noxious organic compounds released into the water by the aquarium inhabitants, and removal of some items that enter the tank from the air. These substances otherwise tend to accumulate in the closed system aquarium.

invertebrates, then it is essential that the rocks be well-seeded in a separate container or aquarium before they are placed in the display aquarium. Finally, operate your protein skimmer from day one, but caution: it will be pulling out excessive amounts of material during the seeding process. Make sure you empty the collection cup regularly and clean the skimmer tube frequently to maintain peak efficiency.

As soon as ammonia and nitrite reach acceptable levels (less than 1 ppm), you can add herbivores to the tank. In our opinion the snail *Astraea (Lithopoma) tectum* is ideal, and we recommend approximately 1 snail per 4 L (1 gal). Introduced as soon as possible, these snails effectively limit the development of microalgae. *Turbo* sp. snails are also very effective algae grazers. They grow larger, so fewer of them are needed, but their size and habit of "bowling" over invertebrates can be a problem in smaller aquariums.

Sand can be placed in the aquarium right from the beginning, but it is better to wait a month or so first before adding the sand or other bottom media. After the rocks have seeded, the initial heavy release of detritus should be siphoned away. At this point one may add sand to the aquarium, either "live sand" or coral sand. Silica sand may also be used, but coral sand is more natural in appearance and plays a role in the precipitation of phosphate and in the buffering of the pH, especially when deep layers are used. See natural systems, this chapter, and chapters 7, 8 and 9 for details about the management of this important element of the reef aquarium.

Starting an aquarium without a trickle filter but with skimmers and well-seeded live rock is the easiest technique. It is possible to create a balanced aquarium in a day with this method, and if the fish are quarantined in advance and properly selected, the margin of safety is quite good. While it is possible to create a balanced environment rather quickly, it is always best to be patient and proceed slowly. In the goal of achieving a wonderfully beautiful and educational natural ecosystem, it would be a shame to risk killing organisms needlessly because of a lack of patience.

As a final note, we wish to make it clear that properly planned "natural system" or "Berlin" aquaria are NOT more fragile or difficult to balance and maintain than their trickle filtered counterparts. In fact they offer better long term success with stony corals, in our experience.

Alternative Filtration Systems

Natural Systems

In the late 1950's and early 60's Lee Chin Eng, an aquarist who lived in Indonesia, reported great success with a method of aquarium keeping that he called "nature's system" (Eng, 1961). Mr. Eng used unfiltered seawater, and circulation in his aquariums was achieved with air bubblers within the tank, either airstones or open ended air hoses for larger bubbles. The aquarium decor consisted of live rocks with attached plants and invertebrates, and live corals. His technique, now called the "natural system" or "Lee Chin Eng method", did not receive much acceptance at the time since it was believed to depend on too delicate a balance to succeed. It was believed that since there was "no filter" in these aquariums, they could not support a very high population of fish.

Furthermore, many aquarists who attempted to duplicate this system at home met with failure and a smelly mess. They published such results in aquarium literature, which effectively discredited Eng's methods and reputation. The reason for their failures is plain to see now for aquarists familiar with modern "living reef aquaria": the fresh live rock fouled, and these aquarists did not know enough, or were not patient enough to allow the rocks to become seeded for several weeks before adding fish or invertebrate specimens. We now know that biological filtration in these systems occurs within the live rocks, where nitrification and denitrification proceed side by side, and that these systems are not fragile or unstable. We know this because Eng's method does not differ much from our "modern" techniques.

The Berlin method of aquarium keeping employs natural system philosophy, combined with the additional benefits of protein skimming, and calcium and trace element additions. Eng used natural sunlight supplemented with fluorescent tubes, while Berlin methodology uses metal halide lights with supplemental blue fluorescent light. Eng lived by the sea and could change water to replenish depleted trace elements and calcium, though he reported success without regular water changes (Eng, 1961). In addition, vigorous aeration by means of airstones within the aquarium, as Eng practiced, does achieve some protein skimming. This can be witnessed along the aquarium walls just above the water line.

Natural system aquaria are both simple and inexpensive to create and maintain. To avoid the failures initially encountered with this

Figure 5.4
Natural Systems

Berlin System

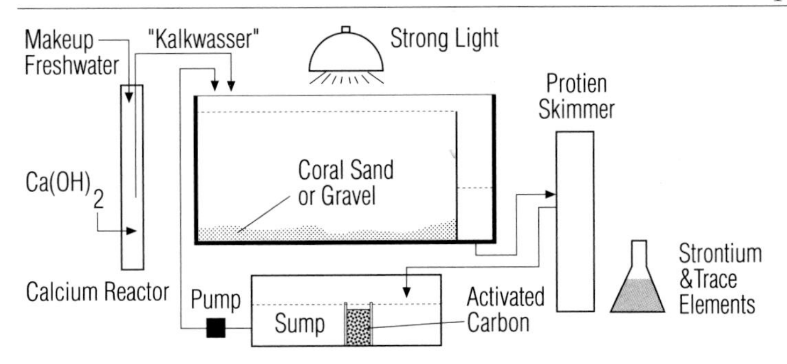

Dr. Adey's System

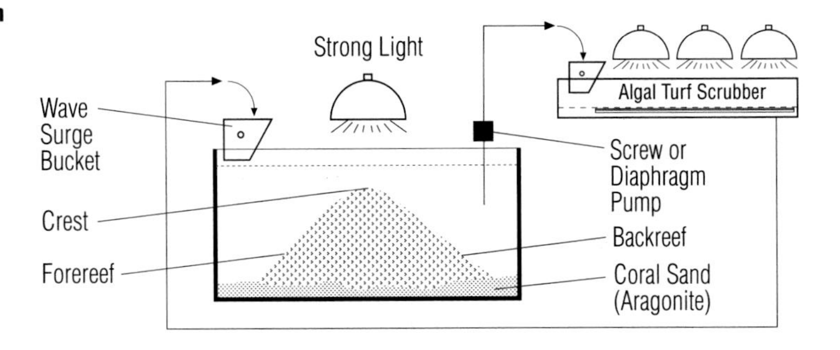

Dr. Jaubert's System

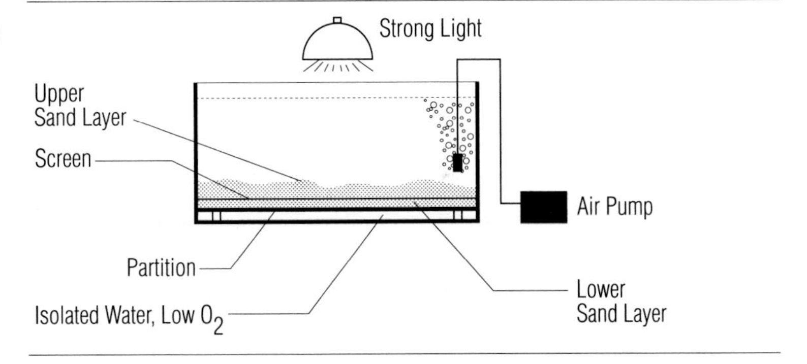

Lee Chin Eng's System

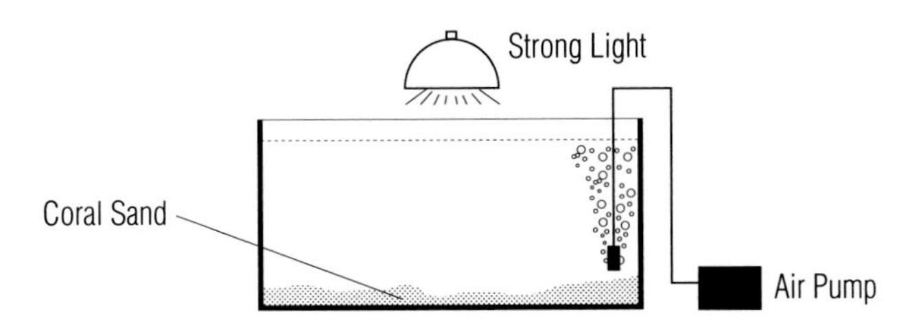

method, one must start by using cured or seeded live rock. See the sections "Methods for Establishing Reef Aquaria" and "Establishing an Aquarium Without a Trickle Filter", in this chapter.

As with other systems for maintaining reef ecosystems, one must not overlook the importance of water motion. One airstone in the aquarium may keep it alive by providing sufficient gas exchange for survival of the organisms, but to make the captive ecosystem thrive, the water must really move. It is quite a surprise for aquarists familiar only with centrifugal pumps for moving water, to see how efficiently rising air bubbles create substantial currents throughout the aquarium. In addition, the size of the bubbles dramatically affects the manner in which the water motion is generated, and the velocity of the resulting currents. Smaller bubbles set up laminar current flow similar to that generated by pumps. Larger bubbles generate turbulence, and really big bubbles create small waves! In a description of one of Eng's aquariums in the lobby of a hotel in Indonesia, "large flat bubbles" were seen percolating up from beneath large rocks (Emmens, 1975). Eng clearly recognized that larger bubbles made the entire volume of water oscillate; an efficient means of achieving water motion throughout the tank. His critics probably didn't understand the importance of this simple application.

In an air conditioned or naturally cool room with good sun exposure, artificial illumination is not necessary, and the aquarium can be placed near a window. Most aquaria will require supplemental illumination, however. Smaller aquaria must be observed carefully since they can rapidly change temperature, depending on exposure to the sun. A heater may be necessary as well to prevent wide temperature fluctuation from day to night in a cold climate.

A lid over the aquarium is essential to slow evaporation and prevent the settlement of dust on the surface. These features can also be managed by means of surface skimming and level switches with automatic water make-up, but those techniques we reserve for Berlin methodology and, though they are not complicated, we wish to emphasize the simplicity of the natural system and keep its definition intact.

Nevertheless, some additions to the purest form of this method do afford better results, and merit recognition here. For example, the placement of a bag of activated carbon beneath an easily moved live rock adjacent to the circulation from an airstone will maintain the water colorless. Although this is not the most efficient means of employing activated carbon, it achieves the principal desired effect of colorless water without rapidly depleting trace elements. The carbon may be left unattended until the water becomes noticeably colored, usually after about six months.

Mr. Eng died in the early 1980's shortly after he moved from Indonesia to Australia. His input surely would have made an impact during the rapid development of the reef aquarium hobby in the USA and Australia, which occurred not long after his death. We don't know if he was aware of the hobby in Europe that developed while he was still living. He might have been flattered to learn that the most successful techniques developed in Europe in the late 1970's and early 1980's were based on his methods.

A version of Eng's system has been developed and studied by Dr. Jean M. Jaubert at the University of Nice, France. Jaubert has set up several exhibits, ranging in size from 1 to 40 m^3. No external biological trickle filters, activated carbon, or foam fractionators are used. Circulation is from large airstones and water pumps. No water changes are performed for some systems, though some at the Aquarium in Monaco receive 5% per month water changes. The system employs a unique form of biological filtration (French patent number 03 28474, U.S. patent number 4,995,980) within the aquarium, or in separate connected reactor tanks (Jaubert, 1991). A perforated false bottom partition is used, similar to an undergravel filter plate, under which a body of confined water exists. A thick layer of coral sand and gravel lies on top of this partition, retained by screens to prevent fish or other marinelife from disturbing its uniform thickness. Another layer of sand covers the screen (Jaubert, 1991). The aquarium (or reactor) is heavily illuminated to encourage the growth of photosynthetic corals, anemones, and algae. This maintains supersaturation of oxygen during the day above the sand, and low oxygen content in the confined water below the plate. Aerobic nitrifying bacteria and algae colonize the upper layer of sand, heterotrophic denitrifying bacteria exist in the lower layer, and an intermediate region between them has a mix of aerobic and anaerobic organisms (Jaubert, 1989, 1991). The confined water space maintains permanent stratification of aerobic and anaerobic layers in the sand. Movement of oxygen, ammonium, nitrite, nitrate, nitrogen, and carbon dioxide through the substrate occurs by diffusion (Jaubert, 1989). Reactor tanks that can be used as accessory biological filters for any aquarium may have vertical partitioned chambers, using screening to retain the sand and gravel, or may simply use a fibrous matting material instead of sand (Jaubert, 1991). Water enters and exits such a chamber, but is not forced through the media, encouraging the diffusion process and preventing the media from becoming clogged (Jaubert, 1991). The confined water behind the partition in these reactor tanks has an average oxygen concentration of 1 mg/L. If it should rise above 1.5 mg/L, denitrification is incomplete producing excess nitrite. If it should fall below 0.5 mg/L, the production of toxic hydrogen sulfide occurs. The thickness of the sand layer or fibrous material used determines the oxygen level, because the reactors are also illuminated and aerated, to maintain high oxygen content in the circulating water above the sand or fibrous material (Jaubert, 1991).

Burrowing animals in the sand break down large organic debris into DOC, and dissolved nitrogenous compounds. The nitrogenous waste is mineralized by aerobic bacteria, producing nitrite and nitrate that diffuses downward into the anaerobic layers. There the heterotrophic denitrifying bacteria convert nitrite and nitrate to nitrogen gas, while using organic matter from the water as a food source (Jaubert, 1989). Acid secretions by bacteria and other organisms in the substrate are neutralized by the calcareous sand, causing it to dissolve, and the resulting input of calcium ions in the water seems to be adequate to maintain high, stable calcium levels despite excellent growth of the hard corals in the system. Some sand must be replenished periodically, of course (Jaubert, 1991). The use of DOC for food by heterotrophic anaerobic bacteria, and some "unknown process" removes yellowing substances from the water. (Jaubert, 1989, 1991).

It is interesting that Jaubert (1989) refers to the substrate in the tank as "living sand". This is the same term used by Riseley (1971), in a book describing the methodology of Lee Chin Eng. Riseley stressed the importance of lighting, live rock, and "live sand" taken directly from the reef. Presumably this sand contains organisms like the ones Jaubert uses in his system. Walter Adey, whose system we shall describe shortly, also is a proponent of live sand. Some of the microorganisms and worms that live in the sand should also be present in live rocks, which are often collected in sandy areas. These organisms will colonize the bottom sand layer. We should point out as well that thick live rocks with large hollow spaces inside will work to denitrify the water by the same diffusion principle as Jaubert's system, and thick sand layers on the bottom without a partitioned off confined water compartment will also denitrify the water. We have not experimented to determine whether the confined water space offers special advantages.

Water chemistry parameters for a 2 m^3 system set up by Jaubert were recorded for a 4 year period with no water changes. This test aquarium contained numerous growing hard corals, soft corals, anemones, and several large fish. Lighting was from natural sunlight for 4 to 5 hours per day, and from 5 Osram HQI(R) TS 250 W/D lamps (Jaubert and Gattuso, 1989). Nitrate levels, after an initial value of 0.350 mg/L, dropped steadily to a level of 0.013 mg/L after four years. Ammonium and nitrite levels remained close to 0.001 mg/L. The pH of the water varied daily, with a typical morning low of 7.8 and an afternoon high of 8.25. Calcium ion concentration peaked at 480 to 520 mg/L, but never fell below 460 mg/L (Jaubert, 1989, 1991). Jaubert (1992) discusses variation in the pH and calcium ion concentration in his systems, and proposes that low pH at night further allows some calcium carbonate from the coral sand substrate to dissolve.

Dr. Jaubert's experimental system photograped in 1989. T. A. Frakes.

The success of this system is evident, and the growth of the hard corals has been well documented (Jaubert and Gattuso, 1989). A 40 m^3 aquarium has been set up with coral fragments cemented to the walls, and growth is so good that the corals are being farmed (Jaubert et al., 1992).

Another system that borrows from Eng's natural system philosophy employs algae for water purification. Using separate aquariums or troughs as algae filters has been known and described in the aquarium literature for quite some time (Brandenburg, 1968, Wilkens, 1975, deGraaf, 1968). For an excellent review of the development of this technique and the history of maintaining corals in aquariums see Carlson (1987).

Algal Turf Filtration and Microcosm Management

One filtration system that has received attention in the scientific community is the algal turf scrubber, part of the microcosm system developed by Dr. Walter Adey of the Smithsonian Institution's Natural History Museum in Washington D.C. Dr. Adey began his development of a coral reef aquarium system in 1974 (Miller, 1980). These systems have been installed at numerous public and private aquariums in the United States, Canada, and Australia. Algal scrubbers are basically shallow troughs with a plastic mesh screen illuminated by intense lighting. Water pumped to the troughs enters them by means of a dump bucket, generating a surge that helps the algae exchange gases and take up metabolites, while preventing over-illumination or over-shading. Various turf-forming algae are grown on these screens, and they remove ammonia, nitrate, phosphate, and heavy metals from the water (Adey and Loveland, 1991). The screens are periodically "harvested" by removing them and scraping off the excess growth with a plastic wedge. The harvested algae is thus removed from the system. It may be discarded, analyzed for nutrient content, or returned to the aquarium to stimulate higher productivity when nutrient levels are very low. The harvested screens are reinstalled with the still living cropped algae adhering to them.

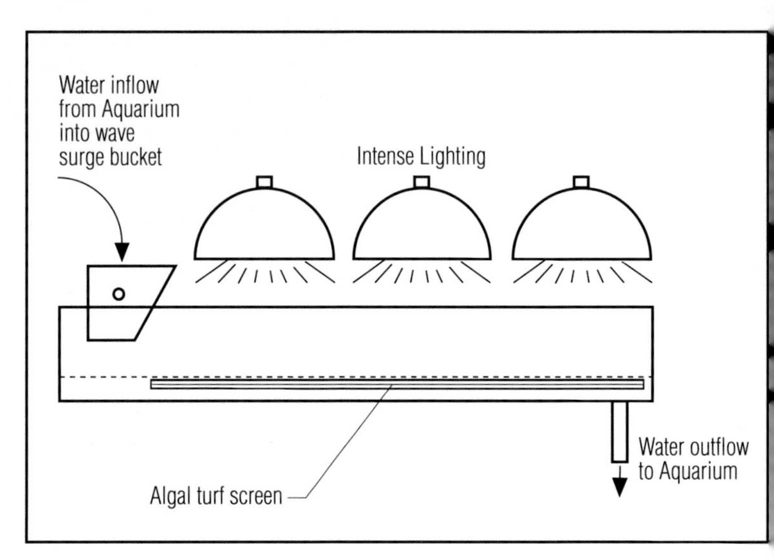

Figure 5.5
Algal Turf Scrubber
Modified after Adey & Loveland, 1991

Rear view of the Smithsonian reef microcosm. Photo taken in August 1988, using available light. Note the colour of the light above the aquarium and the colour of the light in the aquarium.
J.C. Delbeek.

Recently, small dosage of ozone has been used to remove the organic tint from the water at the Great Barrier Reef Aquarium in Townsville. Despite natural sunlight, the exhibit always had a greenish cast. Now, the water is blue. We expect this should improve the results initially, up to the point that trace elements are depleted by natural processes and oxidized by the ozone. If trace elements are not replenished, the corals and other creatures will not thrive. This brings up another important issue.

Some of the public exhibits using algal turf scrubbers also employ other forms of water purification, including protein skimming or, as at the Great Barrier Reef Aquarium, large sand filters to maintain water clarity (Adey and Loveland, 1991). Adey (pers. comm.) believes that supplemental filtration of this kind compromises the resulting ecosystem, primarily because of the impact on plankton, and that poor results with corals in systems employing other means of filtration cannot be blamed on algal turf scrubbers. While we don't feel there is a connection between plankton availability and coral health, we do agree that mixing systems and varying techniques makes it impossible to determine to what extent husbandry, system design, or biology is responsible for the results seen. Furthermore, of the 12 exhibits we have seen that used turf scrubbers, only 8 were reef ecosystems, and each one was maintained differently. This is too small a sample number to really make any definitive conclusion, though we must point out that the results have been very consistent.

Certain major, minor, and trace elements are critical for certain invertebrates, and success with these species in our systems depends on periodic replenishment. Teh (1974) reports that boron, bromine, strontium, phosphorous, manganese, molybdenum, lithium, flourine, rubidium, iodine, aluminum, zinc, vanadium, cobalt, iron, and copper are essential for most invertebrates. With algal filtration, the algae used to purify the water also remove trace elements, and as they are harvested from the system, the elements are removed with them. They must be replenished. Trace element depletion is a feature of many forms of filtration, not just algal filtration. Protein skimming and activated carbon also remove trace elements, and so do the invertebrates, plants, and microorganisms. In a closed system aquarium it does not take long for trace elements to become depleted. The loss of trace elements is not entirely an undesirable feature. In fact, it is a desirable feature of algal filtration because it can be employed to advantage in the removal of toxic heavy metals that could enter the system with make-up water, or from tank construction materials (Adey and Loveland, 1991). The disadvantage of trace element removal with any form of filtration is realized only when the philosophy of the aquarium keeper is to avoid replenishment via supplemental additions or water change.

The exact method of trace element replenishment in the different systems we have observed is not consistent. Some references suggest no supplements added and no water change performed (see Adey, 1993). Others describe micro-nutrient input via food, top-off water, and the return of some harvested turf algae to the system when nutrient levels near the lower values found in the natural ecosystem being modeled, or when no food inputs are made to the system (Adey and Loveland, 1991; Adey 1993; W. Adey, pers. comm.). Adey and Loveland (1991) recommend 1 to 2% water changes per month to replace micronutrients. The splash of saltwater from numerous wave dump-buckets in large exhibits must result in some loss of salt from the system, and this should be calculated in the total water exchange figure. The 3000 gallon Smithsonian Reef exhibit has water exchanges of 5 gallons per day (=5% per month), using water collected from the Gulf Stream, about 50 miles off the Maryland/Virginia coast (T. Goertemiller, pers. comm.). The unfiltered water is stored in Nalgene containers.

At least some systems utilizing turf scrubbing also incorporate additions of prepared trace element solutions as a part of maintenance. At the Smithsonian, keepers of the reef exhibit began adding a trace element supplement around January 1993 (T. Goertemiller, pers. comm.). The Pittsburgh Aquazoo does not use a supplement, but is considering this option. The Townsville Aquarium also does not yet use a supplement; very unfortunate for the corals considering the recent addition of ozone to this aquarium.

In his own home reef aquarium, Dr. Adey does not use a trace element supplement, but replenishes trace elements with the top-off water. He prefers natural water collected from streams, but suggests tap water or well-water can be used if they are not contaminated (W. Adey, pers. comm.). When natural water is not obtainable, and the tap or well-water is contaminated, then a water purification system must be used for the top-off water. Sometimes the top-off water is first treated in a freshwater system using algal turf scrubbers, to remove excess plant nutrients or heavy metals, while in some urban localities, other methods of water purification may be used to treat polluted tap water (W. Adey, pers. comm.). Dr. Adey uses Instant Ocean™ salt mixed with filtered water to replace water lost due to splashing (T. Goertemiller, pers. comm.). In the Smithsonian coral reef system, make-up water (Washington D.C. tap water) is filtered through two large ion-exchange cartridges and one activated carbon cartridge from Millipore™, then through a Milli-Q™ water filter, and finally, it pours through a column of dried Halimeda plates (calcium carbonate skeleton of Halimeda algae). This last step changes the pH of the filtered water from 6.0 coming out of the Milli-Q™, to 8.2 or higher before it is stored in a 120 gallon reservoir. This technique supplies additional calcium carbonate, and perhaps some strontium. The make-up water is added to the aquarium on demand by a pump and level sensing device.

To grow stony corals in captivity, the culture water should maintain a calcium level of about 400 mg/L, a carbonate hardness of about 8 dKH or greater, and a strontium level of about 8 mg/L.Without additions of calcium and strontium, these levels usually fall in closed systems, and this deficit is detrimental to stony corals and coralline algae. There is documented evidence that the calcium and strontium values in some of Dr. Adey's systems are NOT depleted (Meyer, 1991). Replenishment is possible with hard make-up water, and also via the gradual dissolving of the aragonite (coral or oolitic) sand substrate. In his home aquarium, Dr. Adey adds several hundred milligrams of aragonite sand every few weeks.The calcium and strontium apparently are supplied by acid secretions dissolving the sand in the deep substrate, and with the other minerals in the top-off water. Jaubert, (1991), manages the calcium level by the same process. This process needs to be examined and described in a manner that can make it simple for all aquarists to reproduce. We are most interested in the results achieved because we are proponents of making aquariums simpler, not more complex.

In concentrating on the problems for stony corals with this system we are targeting it's weakest point. Actually, microcosm and mesocosm systems employing algal turf scrubbers exclusively with no other forms of filtration work beautifully for nutrient rich ecosystems such as salt marshes and seagrass meadows, and can be employed with success for the maintenance of highly productive tropical reef ecosystems, but since we have observed that stony corals grow better in systems employing protein skimming and trace element addition, we feel it is important to point out the shortcomings of algal turf scrubbing as the sole filtration for a model coral reef ecosystem. When other forms of filtration are employed to remove the organic leachates from the water, and trace element supplements are added, corals do thrive in systems using algal turf scrubbers (J. Sprung and J.C. Delbeek, pers. obs.) The philosophy of creating complex food webs and whole ecosystem models is good, and we wish to promote it. However, in our opinion the use of algae as a filter does not create a reef ecosystem, it hinders it.

It must be pointed out that fish do extremely well in these systems, often with no feeding at all. If combined with protein skimming and GAC or ozone, algal scrubbers may yet prove to be very useful for large fish displays. We would like to see more fish exhibits either using algal turf scrubbers or encouraging more natural growth in the aquarium, instead of the current trend in public aquaria of maintaining nearly sterile rock-work. The benefits are many. The fish stay healthy, they hardly need to be fed, and since nitrate doesn't accumulate, less water needs to be changed. All of these benefits afford a significant cost savings on maintenance. Furthermore, less labor is involved in the cleaning of the aquarium, though the cleaning of algal turf screens is admittedly labor intensive on large systems. Live rock can provide most of the same benefits without the maintenance time. A system with living rock and a deep sand substrate does not require algal turf scrubbers to maintain low nitrates and phosphates, and when protein skimming and GAC are employed, accumulation of organic substances is avoided.

Refugia

The use of a refugium has important benefits to the aquarium. Adey (1983) uses the term "refugia" to describe chambers separate from the main display but connected with it where populations of organisms can develop in the absence of predation. The term also applies to places within the tank where these organisms may develop prolifically without predation, such as within a stack of rocks or in the sand substrate. This concept is typically employed within other natural systems, but Adey's use external reservoirs is an ingenious way to accomplish two tasks. Such refugia can be used to settle out detritus, thereby eliminating the need for a mechanical filter. One can also use these refugia to cultivate various micro-crustaceans that can serve as food for the system. Grass shrimps, mysid, peppermint shrimp, and others may also be maintained in these refugia where the fish won't eat them, and their offspring will drift back into the main display where they are eaten by fish and invertebrates. One of the nicest elements of the reef display at the Smithsonian Institution's Natural History Museum is it's seagrass (*Thalassia testudinum*) tank attached to the reef tank. This large refugium harbors great numbers of mysid shrimp that feed on detritus produced in the system. Some mysids can also be seen in the reef tank.

Aquarists can duplicate such a system quite easily by constructing a second, smaller aquarium plumbed with the main aquarium, and adding live or dead rock, with or without illumination. This becomes a constant source of food in the form of larvae for the main system. The use of sand in such a refugium can lower nitrate too. A "v" shaped bottom allows the refugium to be a settling trap.

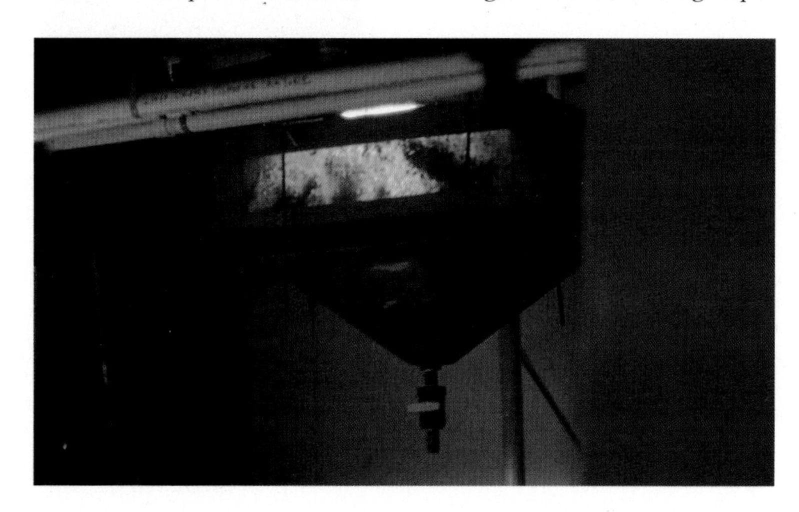

A refugium connected with the reef microcosm at the Smithsonian Institution. Photo taken in August 1988. Note the sloped bottom and drain. This design allows the refugium to be a settling trap to remove detritus from the system. J.C. Delbeek.

Algal turf scrubbers are themselves refugia. The screens develop nice diversity of algae species in nutrient poor systems seeded with live rock from coral reefs. Tiny herbivorous amphipods and copepods also develop on algal turf screens and reduce their efficiency slightly by grazing the algae. Scraping the screen controls the population of amphipods on it, and is a disturbance that promotes diversity of algae species (Adey, 1983; Adey and Loveland, 1991). During the growth phase, the algae in properly maintained scrubbers are not heavily grazed, allowing them the opportunity to grow and release reproductive spores that flow back into the aquarium, simulating phytoplankton input (Adey and Loveland, 1991). If screens from an established reef microcosm are installed on a new system with bare limestone, the algae spores rapidly colonize the bare rock. Therefore a potential use of algal turf scrubbers is in the mariculture of live "plant rocks".

This photo shows the diversity of algae growing on the screen of an algal turf scrubber. Photo shown with permission of Space Biospheres Ventures. J. Sprung.

Some Other Useful Applications of Turf Scrubbers.

We have already mentioned the value of algal turf scrubbers as filters for estuarine and mangrove ecosystems, and their value as refugia and seed for mariculture of plant rocks. There are additional benefits to their use that we want to mention as well. The incorporation of a flip-flop day/night cycle, which has the scrubbers illuminated at night while the display tank is dark, achieves some nice stabilizing benefits (Adey, 1983). The oxygen produced by the photosynthesizing algae boosts the level of dissolved oxygen in the system at a time when it would otherwise drop. Photosynthesis also removes CO_2 produced by the organisms in the dark tank, thereby stabilizing the pH. Finally, turf scrubbers do remove heavy metals and other toxins from the water (Adey and Loveland, 1991).

These applications are useful and effective, but turf scrubbing is not the only means of achieving these ends. In the "Berlin method", the addition of calcareous water by a water level sensing system stabilizes the pH at night. Jaubert's system actually encourages the pH drop at night to help dissolve calcium in the substrate, thereby maintaining the dissolved calcium level and alkalinity. Protein skimming helps maintain the level of dissolved oxygen by removing oxygen-scavenging compounds, and removes heavy metals and other toxins. Activated carbon also removes heavy metals and other toxins.

Of course algal filtration can be employed within the aquarium. With strong illumination and enough herbivores such as tangs and the small herbivorous snails *(Turbo* and *Astraea* spp.), the algae growing on the rocks and glass can be maintained at the highly productive state where it is almost not perceptible because it is being consumed as quickly as it grows. If the system employs strong water motion, then the herbivores' lightweight fecal pellets can easily be trapped in a mechanical or settling filter, and removed from the system. With protein skimming and activated carbon the leachate from such algae is easily managed.

Outside of the aquarium realm turf scrubbers have advantages too. The ability to concentrate heavy metals and other toxins from the water is a key feature in what we believe is a bright future for algal turf scrubbing in water purification and sewage treatment applications. Scientists, including Dr. Adey, are studying the amazing abilities of plants and microorganisms to purify both air and water. We believe the future standard in "filtration" devices for water purification and treatment of sewage or industrial waste will include a form of turf scrubbing and duplication of wetland or marsh ecosystems with rich populations of microorganisms that consume toxic compounds and convert them into harmless ones. Such wastewater treatment technology combines low cost, low maintenance, and results better than any artificial filtration. The use of huge protein skimmers would enhance this simple system of water treatment. The use of algae, higher plants, and protein skimmers (foam separation columns) is already in practice in wastewater treatment, but the state of the art has yet to reach its full potential.

We wish to dispel two myths about nitrate that Dr. Adey's philosophy proposes. The first is that even a low concentration of nitrate is toxic to corals (Adey, 1983). As a general statement this claim is certainly false. It is possible that certain species may be claim is certainly false. It is possible that certain species may be affected by nitrate more severely than others, but we have observed many delicate species of Indo-Pacific and Caribbean corals growing extremely well at nitrate concentrations as high as 40 mg/L as nitrate ion. We have not witnessed a direct relationship between nitrate level and coral health, but acknowledge that low nitrate levels in the aquarium are better for the stability of alkalinity and the whole ecosystem. We are not promoting the intentional maintenance of high nitrate. We only wish to point out that it can be demonstrated that nitrate itself does not harm corals. The other myth is that one can only achieve super low nitrate levels by using algal turf scrubbing (Adey, 1993). When protein skimming is combined with sufficient photosynthetic invertebrates, live rock, and anaerobic zones (sand bottom, live rock anaerobic cores), nitrate does not accumulate. The nitrogen can be removed by the photosynthetic organisms and protein skimming before it is converted to nitrate, and any nitrate produced in the system is easily denitrified in the anaerobic zones. It is true, however, that the natural ability to reduce the nitrate can be exceeded by excess food inputs. Turf scrubbing could be employed to solve such a problem, but so could additional protein skimming, or denitrification. The denitrification that takes place within the rocks and deep sand beds is a most efficient means of removing nitrate from the system, and it works fine even in the absence of protein skimming, as we showed in the example of Jaubert's system. Denitrification alone in this system works to keep the nitrate level low, consistently below 0.015 mg/L (Jaubert, 1991).

The use of dump buckets to return the water to the main aquarium, and on algal turf scrubbers is also problematical. Some dump bucket designs require frequent maintenance to ensure that they continue to operate properly. The salt from splash gets into the pivoting parts, and wears the surfaces until they don't turn as easily. Furthermore, the fine salt spray generated by these devices promotes corrosion problems in the immediate vicinity of the aquarium. The surge provided by the dump bucket, however, is very beneficial, and a significant improvement over mere circulation. Some mesocosm exhibits employ a different surge device, using air displacement in a chamber attached to the aquarium (see topic: water motion, next section).

A scene from the Great Barrier Reef Aquarium in Townsville, Australia. A.J. Nilsen.

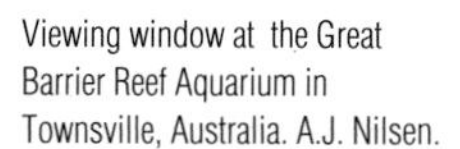

Viewing window at the Great Barrier Reef Aquarium in Townsville, Australia. A.J. Nilsen.

A portion of the enormous array of algal turf scrubbers located on the roof of the aquarium. A.J. Nilsen.

Mid 3/4 of Walter Adey's 130 gallon reef. Invertebrates, R to L: *Palythoa caribaeorum*, *Eunicea* sp., *Diploria clivosa* with *Porites asteroides* behind, *Diploria strigosa*, *Montastrea cavernosa*, *Sinularia* sp., *Tridacna maxima*, *Sarcophyton* sp. Algae: *Halimeda discoidea*, *Caulerpa racemosa*, *C. sertularioides*, *Amphiroa fragillissima*. S. Gill.

Close up of *Diploria clivosa* (brain coral) and *Eunicea* sp. (gorgonian) with *amphiprion ocellaris* in Walter Adey's 130 gallon reef microcosm. S. Gill.

Cave entrance in the same tank with plate-forming crustose coralline *Mesophyllum mesomorphum* on wall. Dark red patches are *Peyssonnelia sp.* and *Schizothrix. Ricordea florida* (corallimorpharian) is upper right; *Pseudochromis* sp. (fish) below. S. Gill.

Reef microcosm at the Ontario Science Centre, Toronto, Canada. Note the construction of the reef, the colour of the lighting, and the colour of the water. J.C. Delbeek.

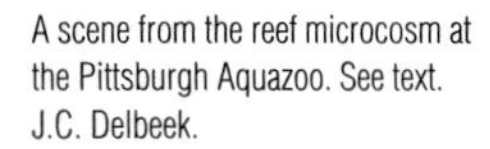

A scene from the reef microcosm at the Pittsburgh Aquazoo. See text. J.C. Delbeek.

Accessory Pumps and Other Devices

Different types of accessory pumps may be used for additional circulation or variation of the flow within the aquarium. These include external dedicated circulatory pumps and powerheads. The powerheads may be hooked up with wave-maker switching devices, "pulse timer" switching devices, or with simple timers.

External dedicated circulatory pumps have the advantage of not being an obstruction within the aquarium, and not contributing as much heat to the water as submersible powerheads. When using an external pump, the location of the intake is important. We have seen many aquariums equipped with recirculating pumps with the intake plumbed through the wall of the aquarium (either a side or the bottom), and outfitted with a strainer to prevent fish from being slurped into the motor. This technique works fine for fish-only aquariums, but is a disaster in the reef tank. Don't make this mistake! Wandering clownfish anemones, loose mushroom anemones, algae, sea cucumbers and other creatures end up against the strainer and often, through the pump. They become reef puree. The design of choice with this arrangement is simply to put the intake in the overflow. This arrangement can also be used for feeding the protein skimmer from the overflow.

Powerheads

Powerheads are usually an obstruction to the decor in small aquariums less than 200 L (50 gal.), but are most useful for added circulation in really big tanks. Manufacturers have been building smaller powerheads that are less noticeable, and these can be incorporated in smaller exhibits, concealed within the rock-work. One very significant disadvantage to powerheads is the risk of them sucking up small fish, wandering anemones, nudibranchs or sea cucumbers, or algae fragments. A mechanical filter cartridge attachment on the bottom of the powerhead prevents this possibility, but adds a potentially unsightly item that could detract from the decor. Mechanical filter cartridges also need servicing. Locating a powerhead strategically among rocks can effectively conceal its presence and prevent stray creatures or algae from getting sucked into the intake grid or port, but the powerheads themselves need to be serviced occasionally, and it is not desirable to have to pull apart a section of the reef to service them. We have seen aquarists who have simply left powerheads (unplugged) in place in the reef after they ceased working!

One significant advantage to the use of powerheads is the ability to change flow direction or create pulse waves with timers or switching devices. With simple household utility timers one can turn powerheads on or off for set intervals, simulating the change in water flow and direction associated with tidal change.

Powerheads come in two types: submersible synchronous motors and non-submersible non-synchronous motors. Both types work well, but the switching devices for them are different. The submersible types, most popular in North America, can only be used with so-called wave makers that turn the pump on and off intermittently. Some submersible pumps do not tolerate these continuous hard starts and stops, and need to be restarted occasionally. They may stop working altogether. Recently, wave-makers for these powerheads have been introduced that feature a softer start mechanism that is kinder to powerhead operation.

Non-synchronous motors can be used with a different kind of wave maker, the so-called "power" or interval timers that pulse the motor speed. By accelerating and slowing the impeller, interval timers and non-synchronous powerheads produce very natural wave-like currents, either with a pause or without one.

Motorized Ball Valves

Some aquarists equip their tanks with motorized ball valves on the pump return lines that provide currents in the tank. These can be employed with timers to achieve a change in the direction and intensity of the flow as one valve opens and another closes. The effect mimics the change in water motion associated with tides.

Solenoids

The use of solenoids for controlling the water flow has been explored by some crafty hobbyists, and has potential use for the generation of pulsed wave-like currents or, like motorized ball valves, tidal currents.

The generation of real surge within the aquarium is a challenge with numerous rewards, including improved health of the organisms, rapid growth, and the virtual elimination of dead spots and detritus accumulation. Pulsed water flow from powerheads achieves the same effects, but less dramatically than the techniques we will briefly describe that are not without their own disadvantages.

Wave Buckets

Surge motion in the aquarium can be generated by means of a simple device called a wave or "dump" bucket. See diagram. Water being returned to the aquarium is fed into the bucket. When the weight of the water it contains creates an imbalance, the bucket tips over. The bucket pivot points can be made from plastic bearings or just pvc pipes of different diameter. Salt entering the bearings or rubbing parts will hinder the performance and may erode the plastic with time and the continuous turning of the pivot. A well-designed pivot avoids this problem. A poor design will necessitate constant re-starting of the dumping action. Lubricants should be avoided since they can easily enter the aquarium from this location. Additionally, it is difficult to build a dump bucket that does not generate at least some splash, which is undesirable mainly because splash produces salt creep and gradual loss of salt. Despite the disadvantages, the effect in the aquarium is wonderful, and really makes the animals and plants move to and fro in the same manner as waves in the sea.

Figure 5.6
Wave Bucket

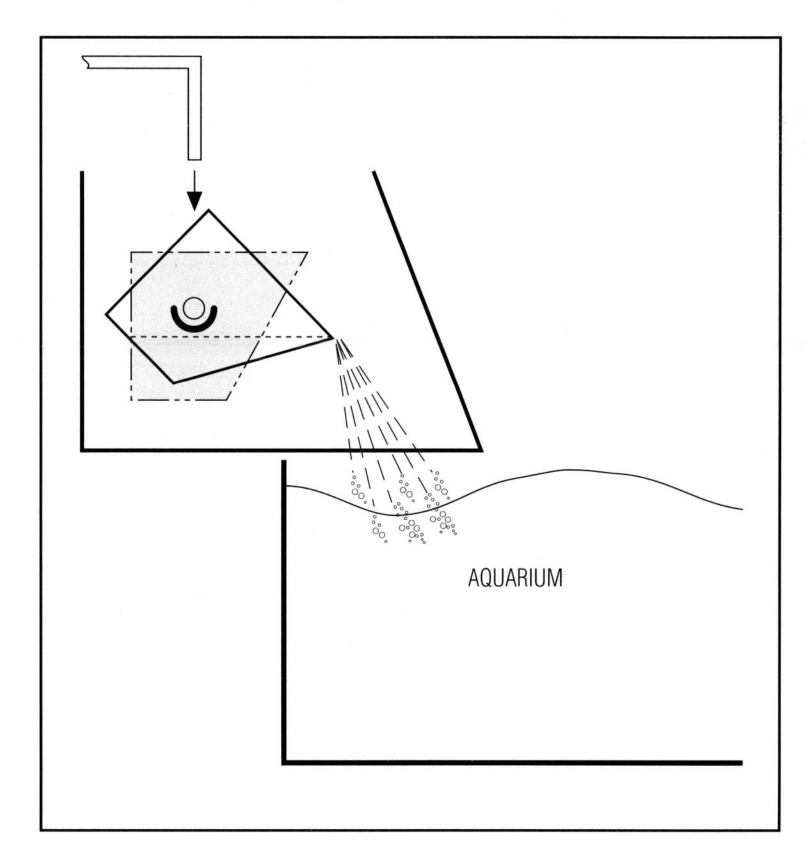

Other Wave Making Devices

Dr. Bruce Carlson, the director of the Waikiki Aquarium, uses another type of surge device that mimics the long duration surge found in channels on reef flats. See figure 5.7. He uses a stationary bucket above the aquarium into which water from the aquarium is fed. During the time this water is feeding into the bucket the water level in the aquarium falls, simulating the water drawing back before a long wave. When the water in the bucket reaches a critical height, the large automatic siphon built into it rapidly empties the water back into the tank, generating lots of turbulence and powerful currents.

Figure 5.7
Surge Device

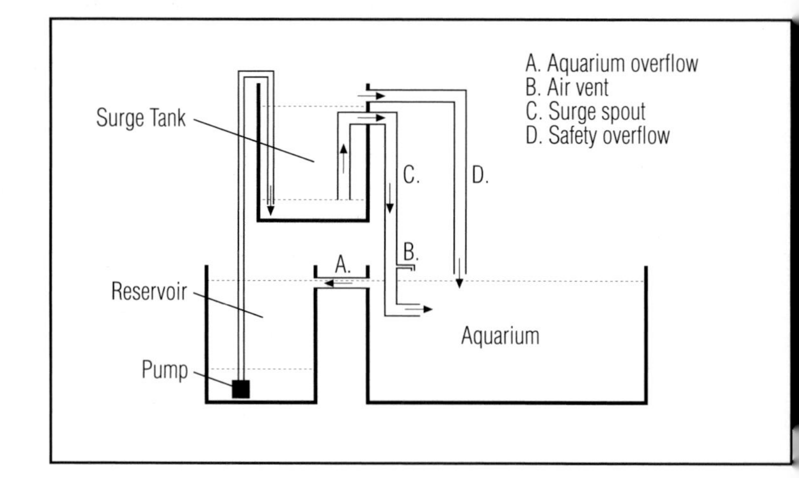

Pistons and Paddles

We have seen three other surge devices employed in aquariums in our travels to different public aquaria. The National Aquarium in Baltimore uses a paddle system controlled by an hydraulic arm to generate a beautiful back and forth surge in their living coral reef exhibit. The paddle is located behind the artificial rock-work. We have seen similar paddle systems moved by motor or air compressor at the Virginia Beach Science Museum and the New York Aquarium.

The most magnificent surge device at a public aquarium is used on the giant kelp tank at the Monterey Bay Aquarium in Monterey, California. Thirty foot growths of living kelp sway back and forth behind the tall windows in front of you, with natural sunlight streaming in as well. The device used to create the surge in this tall display is a kind of piston that is periodically lifted and dropped.

of artificial light duplicates the light attenuation with increasing depth on a coral reef, but on a much smaller scale. For example, depending on the intensity of the light source, 1 cm increase in distance from the bulb might equal 1 m depth increase in the sea.

In order to maintain the useful colour temperature and intensity of fluorescent tubes, it is best to replace them at least once a year, but perhaps as often as every six months for high intensity lamps. You should judge this requirement by the appearance and health of the invertebrates. Metal halide bulbs should also be replaced about once per year, since their colour temperature and intensity change with use as well.

Colour Rendition Index (CRI)

Another description used in association with artificial light sources is colour rendition index. The colour rendition index is a standard measurement used to describe how closely a lamp or other artificial light source renders the natural colour of an object, compared to the colour rendered by sunlight. Sunlight has a CRI value of 100. CRI is related to spectrum, of course, but the value is not an indication of the lamp's exact spectrum, only what the human eye perceives. While lamps with a CRI of more than 90 are most desirable because they provide a natural appearance, blue tubes do not have a high CRI. This is not an indication that they have no use. On the contrary, the light on coral reefs, particularly at depth, is quite different from sunlight above the water.

Types of Lighting Systems

There are many options available now for illuminating an aquarium, but they all fit conveniently into two categories: fluorescent and metal halide. Within these two categories there are a few different formats, and plenty of bulbs to choose from.

Forms of lighting that are not recommended for reef tanks include some mercury vapour and sodium vapour lights, as well as HQL and HQI-NDL lighting which have colour temperatures (4300 K) and spectrums that aren't ideal. These lights can be used successfully for reef aquariums, provided UV emissions are blocked and temperature is managed. With 4300 K light, the colours of the animals and appearance of the tank is not as good as with daylight spectrum, but most corals will still grow as long as they receive bright light. Quartz halogen lights that are inexpensive and readily available from hardware or department stores are also unsuitable because of the spectrum and the tremendous amount of heat they produce.

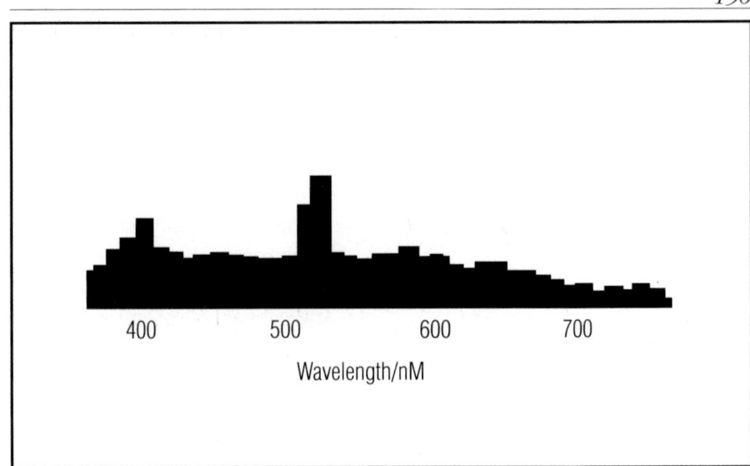

6.3c
Relative Spectral Distribution for HQI.../D daylight metal halide bulb
After Osram.

Fluorescent Lights

Fluorescent tubes have been used for many years to illuminate aquaria, and the recent interest in reef systems has stimulated the development of many new types. While it is possible to maintain a successful low profile (height) reef aquarium with standard output tubes, it is better to incorporate some of the high output or very high output bulb formats. These require special, expensive ballasts, but the benefit of the extra intensity is worth it.

Fluorescent lights are quite handy to use since they are easily installed and maintained by the average hobbyist, and they are less expensive. There are so many different types of bulbs available that an aquarist can easily develop a combination that will duplicate the exact spectrum desired. Since light intensity attenuates with respect to distance from the bulb, and since fluorescent bulbs are typically mounted closer to the water than metal halide bulbs are, the upper regions of a fluorescent lighted aquarium can be quite bright. For shallow tanks, less than 18 inches (45 cm), four to six standard fluorescents with reflectors provides enough lighting for most corals.

Using high output (H.O.) or very high output (V.H.O.) fluorescent lamps provides more intensity than standard output bulbs. These bulbs require special high output or very high output ballasts, respectively. When using H.O. and V.H.O. bulbs you must be aware that some of these can also produce potentially harmful amounts of UV, and may need to be shielded appropriately. One of the problems with fluorescent light sources is that both their intensity and spectrum change with age. This is especially true of

H.O. and V.H.O. fluorescents. The decrease in intensity could cause reduced growth rates of both corals and algae, while the spectrum usually shifts towards the red end which can lead to breakouts of undesirable algae.

The use of a good reflector on fluorescent bulb may increase the light that it casts into the aquarium by as much as 50%. A low profile aquarium using standard output tubes with reflectors could achieve results close to that of high output tubes without reflectors, for instance. The V.H.O. tubes are still brighter, and with a proper reflector, the intensity of light cast into the aquarium compares with metal halide light sources. Polished aluminum reflectors are available from aquarium industry manufacturers for fluorescent bulbs, and metal halide fixtures typically are manufactured with some kind of a reflector. Some fluorescent bulbs, in addition, are manufactured with an internal reflector as part of the bulb. If your canopy is custom made, you can make a good reflective surface by painting the interior with a flat bright white paint, or installing some of the polished aluminum reflectors available from your pet dealer.

The accepted recommendation for fluorescent lights has been to use a combination of Daylight and Actinic type bulbs, primarily to ensure a spectrum heavy in the end blue and low in the red (Burleson, 1987). However, there are now a great number of fluorescent lights available from aquarium industry sources, and the choice of bulbs may seem bewildering. Not to worry, they all work. We suggest that aquarists choose bulbs based on the satisfactory (or magnificent) appearance of aquariums using particular types of bulbs. Moe (1989) gives a complete listing of the more common fluorescent lamps, including colour temperature, colour rendition index, and intensity.

The manufacturers of lighting fixtures for aquariums now make canopies with four or six fluorescent bulbs. These commercial units are easy to install. Many aquarists build their own custom canopies, and we feel that this is fine, but certainly more complicated for someone who isn't familiar with proper wiring techniques. Building your own fluorescent lighting system typically involves the use of waterproof end-caps and bulb-holding clips. These two items make it possible to fit the bulbs right next to each other, so that six or more lamps can be mounted over a 12 inch wide aquarium. End-caps do not last forever, especially with V.H.O. tubes that quickly break down plastic with heat and ultraviolet wavelengths. Changing the bulbs when end-caps are used can be a real chore, and they become brittle after a couple of years. Sometimes plastic clips become brittle and break, but there are strong, long-lasting types available.

Because fluorescent bulbs are located so close to the water, splash results in the deposition of salt on the bulbs. While this doesn't harm the bulbs, it reduces the light entering the tank dramatically. A removable sheet of acrylic to protect the bulbs from splash will reduce some of the light intensity (by about 15%), but makes maintenance easier. It is a simple matter to wipe the sheet down about once per week with a damp cloth. If you must wipe the bulbs down, turn them off first and let them cool down.

To cool the air in the canopy, fans are typically installed that push air through the space. Fans should always be installed in such a way that they are pushing the air into the canopy rather than pulling it out, because the relatively high humidity of the air inside the canopy can corrode the fan. One or more vent ports should be installed in the canopy to allow the air to flow out. The upper edge of the port(s) should be as close as possible to the top of the canopy to help vent out the hottest air.

Selecting the proper ballasts for the lighting system can be a bit confusing. H.O. fluorescent lamps require an 800 milliamp (mA) ballast to fire them up to the proper colour temperature. V.H.O. fluorescent lamps require a 1500 mA ballast. Aquarists often confuse the bulb wattage with the ballast wattage. For instance, a 48 inch 60 watt lamp is an H.O. format, requiring the 800 milliamp ballast. A 24 inch 40 watt bulb is also an H.O. format, requiring the same ballast as the former example, not a 40 watt ballast. A 48 inch 40 watt lamp is a standard output tube, that uses a 400 mA ballast. A 48 inch 110 watt lamp is a V.H.O. format requiring a 1500 mA ballast. In addition to the milliamp rating, one must also refer to the minimum bulb length description printed on the ballast.

Electronic ballasts are now available that run cooler and are much smaller than the comparable H.O. and V.H.O. ballasts. Some electronic ballasts enable the fluorescent bulbs to be dimmed, simulating the change in intensity and spectrum with the change in altitude of the sun. It remains to be seen how well these ballasts perform in the long run, and for now they are more expensive than conventional ballasts. Since the traditional variety of H.O. and V.H.O. ballasts are like large, heavy, hot bricks, the new generation of small, cool and lightweight electronic ballasts are a welcome development that we hope will prove reliable.

Fluorescent fixtures can be cheaper to install initially than metal halide systems, but the replacement of multiple bulbs and, ultimately, end-caps, may make them more expensive in the long run than metal halide.

Metal Halide

Metal halide lights are available in several formats and wattages. In the globe format, the light element is surrounded by a glass envelope, which effectively reduces the ultraviolet emissions. Newer "metal-arc" lamps that are available in 6500 K colour temperature also have UV absorbing glass as a part of the bulb

Relative Spectral Irradiance
100
90
80
70
60
50
40
30
20
10
0
350 400 450 500 550 600 650 700 750
Wavelength/nm

6.3d
Relative Spectral Distribution Diagram for Coralife® 5500K 175W metal halide bulb
Courtesy of Coralife/Energy Savers Unlimited, Inc.

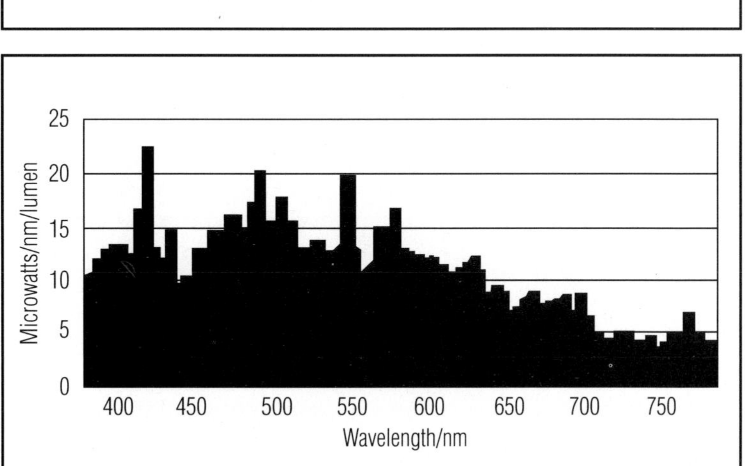

6.3e
Spectral Output Diagram for ULTRALUX® 6500K metal halide bulb
Courtesy of Ultramarine Enterprises ®

envelope. In the Osram HQI (mercury quartz iodide) Powerstar™, a glass lens must be incorporated in the fixture to protect the bulb from water splash, and to prevent the passage of too much ultraviolet light. These bulbs come in double or single end formats.

Globe type HQI metal halides, especially those with daylight colour temperature, offer intensity and spectrum similar to V.H.O. daylight fluorescents, and depending on the wattage, may be significantly brighter. In a tall aquarium that is short on length, metal halide is really necessary to provide adequate intensity. Metal halide is required, in our opinion, for tanks taller than 76 cm (30 in.), but can be used on small, shallow aquariums as well, provided the aquarist manages the temperature of the aquarium.

Aquarium industry metal halide lights range from 75 watts to 400 watts. Most aquarists are using either 175 or 150 watt formats, which are ideal for average aquariums of 100 gallons or less, not taller than 30 inches. For aquariums taller than 30 inches, 250 or 400 watt lamps should be used. Some aquarists in Europe are using 1000 watt fixtures with great success, but we cannot comment on their electric bill.

An important factor is the degree of heating that the light will cause in the tank. Metal halide lights should be about 30 cm (1 ft.) above the water surface to prevent overheating of the water. They can be located closer to the water when a chiller is used to maintain the water temperature. As we mentioned earlier, infra-red light emitted from the bulbs does have the potential ability to burn or irritate coral tissue close to the water surface when there is not sufficient distance between the bulb and the specimen.

Some European aquarists recommend that one 250 W HQI/D lamp, at a height of 90 cm (35 in.) above the bottom of the tank, is sufficient for a tank length of 130 cm (35 in.) (D. Stüber, pers. comm.). This is a minimum value, and depends on the use of a reflector to cast the light across this span. A good rule of thumb is one metal halide bulb per two feet of tank length. A six foot long tank with a width of two feet or less would therefore use three metal halides; one in the middle and two on each side about a foot from the adjacent end.

Some authors (Thiel, 1988; Adey, 1989) maintain that metal halide lamps alone provide enough blue spectral intensity. Our experience also shows that supplemental blue lighting is not

Chapter Seven

Aquascaping

These living ecosystems we create, these aquariums, are really living works of art; each living plant or animal a combination of colour, pattern and texture. In creating a natural environment we make a sculpture that grows, and this creative process is perhaps the most satisfying aspect of the hobby. In the creation of your living masterpiece you have choices to make, and it is wise to plan from the start what kind of reef you wish to create, what it will look like and what you plan to grow.

Start this planning by making a rough sketch of how you would like the layout of your aquarium to appear, a blue-print of sorts. Too often hobbyists limit their tank's potential by building a pile of rocks along the back wall, and sorting the corals and other specimens like so many items placed on shelves. As you will see, it is possible to build far more dramatic aquascapes using the materials now available to hobbyists. At the end of this section we provide a few rock aquascaping ideas, and at the end of our book we offer a view of some very beautiful and successful aquariums.

Live Rock

Live rock is composed of old coral skeletons and shells that have become encrusted with coralline algae and a variety of other plants and invertebrates. Worms, large and small crustaceans, clams, and sponges inhabit the holes in its porous structure. It also carries nitrifying and denitrifying bacteria, as well as heterotrophic bacteria and other microorganisms.

Types

Florida Keys

Nice aquascape! A section of John Burleson's reef aquarium
J. Sprung.

In North America much of the live rock collected for reef aquariums comes from the Florida Keys. It is generally the least expensive rock available. The availability of this live rock is being regulated by the state and federal government, and the plan is for a complete shutdown of collections or "landings" effective within about two years. The collectors are fighting this action, and it is possible they will succeed within limits (see Introduction).

Rocks from the Florida Keys are collected mostly in shallow rubble zones behind the main reef. Storm surge creates huge piles of sand

and coral rubble, and these are constantly churned up with each storm, maintaining a high diversity of plants on the rocks, which receive intense illumination when they are not buried with sand. Beware that the calm conditions in an aquarium may encourage a tough green brush algae, *Cladophora* sp. to dominate. Most of the material is branchy Elkhorn or Staghorn (*Acropora*) coral skeletons, and the shapes are nice. The pieces are fairly dense however, and have mostly smooth surfaces. Rocks collected from deeper rubble zones are generally lighter in weight, from *Porites, Millepora,* and *Agaricia,* and are encrusted with more species of coralline algae. Less disturbance there affords better surface texture and porosity, a desirable attribute.

Gulf of Mexico

Another source of live rock from Florida is from hardbottom areas in the Gulf of Mexico. Gulf of Mexico rock is also available from Texas and Alabama, but most comes from Florida presently. When the new legislation takes effect, the only rock that will be allowed to come from this area will be aquacultured on leased seabottom. Natural Gulf of Mexico rock is mostly composed of hardened oolitic sand mixed with silica sand, and sometimes a significant portion of the rock is made of the skeletons of bryozoans and coralline algae. The pieces are usually chunky or flat, not branchy. Their surface is extremely irregular and coated by the most brilliant pink, red, and purple coralline algae, and hardy, encrusting sponges, bryozoans, and tunicates in red, orange and yellow. Some of the rock may have the lovely "red grape" algae, *Botryocladia* spp., or *Halymenia floresia,* a slippery red leafy alga. The rock is very heavy, but the life it contains is spectacular. The coralline algae on pieces from deep water can be fluorescent pink.

Marshall Islands

Marshall Islands rock is lightweight, open and branchy in structure, and coated by a diversity of coralline algae and macroalgae. It is composed mostly of the skeletons of *Porites, Pocillopora,* and *Acropora.* Based on the attached macroalgae, it appears to be collected in shallow rubble zones similar to where Florida Keys rock is collected. Rocks from these areas have many species of algae on them, including undesirable ones such as *Derbesia, Valonia,* and *Cladophora.* With herbivores, control of nutrients, and care of the calcium and alkalinity levels, coralline algae can be encouraged to dominate, and problem algae growth can be curbed. The shapes of these rocks makes it easy to build a beautiful open structure. They fit together like pieces of a puzzle.

Marshall Islands live rock. Note the Mexican *Turbo* sp. snails for size reference. J. Sprung.

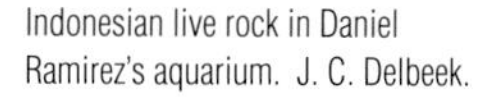

Indonesian live rock in Daniel Ramirez's aquarium. J. C. Delbeek.

Indonesian

Indonesian rock is choice material on the basis of colour, weight and shapes. The coralline algae are more predominantly purple, often very dark shades, and the surface may have nice irregular texture. Small colonies of soft coral, zoanthids, or even stony corals may also appear after several months. The iridescent blue alga, *Ochtodes* is a common plant with this rock, as are a few lovely *Halymenia* spp. algae and an encrusting brown *Lobophora* sp. A form of *Cladophoropsis,* tough green tufts, should not be allowed to grow unchecked where it occurs on this rock. Sometimes Indonesian rock has hitchhikers, and these range from extremely good to bad. On the good side, a small herbivorous snail, *Stomatella varia,* which resembles a tiny abalone is a great algae eater. It is small, and reproduces within the aquarium (see chapter 9). What more could you want for a perfect herbivore? On the bad side, Indonesian rock sometimes brings with it the commensal flatworms described in chapter 10. The cost is high, but the weight is so low that it usually compares with less expensive but denser rock.

Tank Raised

Cultured live rock is presently available from a few sources, and the quality and availability is expected to improve in the future. The tank raised material now available is primarily "plant rock", with one or more species of conspicuous algae. In the future all kinds of rock should be available, with coralline algae and even attached invertebrates. Rock farms are being set up, as bottom leases in the natural environment, as land-based open system culture, and land-based closed system culture. The future for the hobby looks bright. Whole exhibits of entirely tank raised reef life will be commonplace within a few years. Already exhibits containing mostly propagated specimens exist (e.g. Bochum Tierpark, Germany; New York Aquarium for Conservation, U.S.A.).

Nearly all of the corals in this aquarium at the Bochum Tierpark are from cuttings donated by hobbyists. Julian expresses his admiration of the expanse of tank-raised *Sinularia asterolobata*. S. A. Fosså.

Construction of the Reef

Amount of Live Rock Required

How much live rock should be used is a commonly asked question. The old rule of thumb is 1 kg (2 lbs.) of live rock per 4 L (1 gal.). This recommendation is not adequate for the simple reason that live rock varies greatly in density. If you use very dense coral rock, 2 lbs./gal may only give you five rocks in a 50 gal tank, whereas if you use very porous rock, the same 2 lb./gal figure might yield 20 pieces of rock, enough to fill the entire aquarium, with some pieces sticking out. A more realistic recommendation for an aesthetically pleasing display would be to use enough rock to fill about 1/4 to 1/3 of the visual space inside the aquarium. This amount is well in excess of what would be required to biologically filter the aquarium, so even less can be used. Any porous rock provides a substrate for nitrifying bacteria, so with respect to the ability to serve as a biological filter, the quality of the rock is not important, and the rock need not be "live rock". However, quality live rock does afford biological stability with respect to the populations of coralline algae, sponges, crustaceans, worms, and microorganisms it contains, and this has a side benefit of providing an aesthetically pleasing structure and a balanced ecosystem.

Arranging the Live Rock

Not only is the quality and type of rock important but also the way that the rock is arranged in the aquarium can have a profound influence on the long term success and maintenance of a reef tank. The old rule of thumb often results in people trying to cram in as much rock as possible to reach the "ideal" weight of 2 lbs of rock per gallon. The danger in this approach is that water flow around the rocks becomes greatly restricted and detritus eventually accumulates between the rocks, which could result in serious problems with algae later. Don't forget, many of the specimens you will add come on their own rock, so space allowances must be made for these too. We have seen many reef aquariums with a haphazard conglomeration of rock piled into a brick-like wall, with very little regard given to water circulation, or detritus build-up and removal. When arranging live rock it is much better to construct a loose arrangement, with many overhangs and bridges. Couple this with small contact points between the rocks and the substrate. Do not pile the rock up against the back of the aquarium, leave enough space behind the rock for water circulation and for detritus removal. Some aquarists suspend their live rock above the bottom of the tank with bases of lava rock, PVC pipes, sheets of acrylic, or light diffuser material (eggcrate). This allows detritus to accumulate below the rock for easy removal by siphoning. The same effect, however, can be obtained by judicious placement of live rock or through other techniques (see additional descriptions of Eggcrate and other materials in this section).

Of course arranging the rock in such a manner is not easy to do when the live rocks offered for sale are smallish, rounded pieces.

The ideal shape for loose arrangement of rock is elongated flattened pieces or clusters of loose finger-like branches. Rocks of these types can be easily arranged to form platforms and bridges. By arranging the rock in this manner, organism placement is easier. Water circulates freely around the rock on all sides, and detritus is carried away from the rock, collecting in the prefilter, on the bottom of the tank, or in a settling filter. It can be easily managed via siphoning and the stirring and feeding habits of gobies, sea cucumbers, brittle starfish, worms, etc.

Other Options— Alternatives to Live Rock

Limestone and Cement

It is possible to build the reef structure dry, with pieces of limestone fused together with pure portland cement. Coral skeletons can also be incorporated in the structure. Once the

This reef aquarium at Jago Aquaristic in Berlin was set up with a limestone and portland cement structure. No live rock was used, but the invertebrates have grown onto the rocks, and coralline algae have spread over the remaining bare surfaces. J. Sprung.

cement dries and hardens, the structure must be soaked for a couple of weeks in a heavy brine solution to cure it. Once the structure has cured and the brine solution is drained and flushed out, the tank can be set up. A few small pieces of live rock with coralline algae will seed the system, and with herbivores, control of nutrients, addition of trace elements, and care of the calcium and alkalinity levels, the coralline algae can be encouraged to dominate, and problem algae growth can be eliminated. Using monofilament line, one can attach stony corals, soft corals and zoanthid anemones to the reef structure, thereby covering much of the exposed, bare surface.

Coral Skeletons

One can build a really nice reef structure using dead coral skeletons instead of live rock. Many Dutch hobbyists build their reef tanks this way (Delbeek, 1992). As we said before, once the tank is set up, a few pieces of live rock with coralline algae will seed the system, and with herbivores, control of nutrients, and care of the calcium and alkalinity levels, the coralline algae can be encouraged to dominate, and problem algae growth can be eliminated. Using monofilament line, one can attach stony corals, soft corals, and zoanthid anemones onto the reef structure, thereby covering much of the exposed, bare surface.

This 2100 L (525 gal.) aquarium of Mr. Tini Broeders, Oosterhout, Netherlands, contains coral skeletons as a reef base. L.N. Dekker.

Polyurethane Foam

Some large public and private reef aquariums in Europe have very little limestone rock at all in them. What looks like rock in these tanks is actually a polyurethane foam substance which has been shaped to look like rock. The surface is apparently very attractive to coralline algae which rapidly colonize it, making it virtually indistinguishable from the "real McCoy." The Aquaria-Vattenmuseet in Stockholm, Sweden, Hagenbeck's Tierpark in Hamburg, Germany and the Lobbecke Museum in Dusseldorf have large reef displays with such material used as a substrate. We do not know much about the material or whether it can be acquired easily, but it may be an important aspect of large public aquarium reef displays in the future. We caution aquarists about the use of such materials because of the possibility of toxic compounds leaching into the water. The success at the aquariums mentioned above, however, indicates that acceptable inert versions of these materials do exist.

Ceramic

Some ceramics are also a suitable material for building reef structures, though we caution that heavy rocks should not be supported by ceramic pots or forms. Some ceramics may leach harmful substances into the water, but most are inert when fired. Ceramic rock forms have good potential for reef aquarium building, since they would be porous like limestone, and the shapes can be quite marvelous.

Materials That Facilitate Reef Construction

Cable Ties

Solid plastic cable ties, commonly used by electricians and mechanics to secure bundles of wire, also have great use for securing rocks together when building the reef structure. These ties seem to have a million uses! Cable ties come in a variety of lengths and thicknesses, and in either opaque white, grey or black. Caution: be sure that the ties are solid plastic. Some ties are made with a metal core.

Alf Nilsen shows how it's done! He uses plastic cable ties to secure the live rock structure in a large reef aquarium in Oslo, Norway. Courtesy of A.J. Nilsen.

Close-up view of a gorgonian, *Pseudopterogorgia* sp., that has been attached to a live rock by means of a plastic cable tie. Coralline algae and the gorgonian's tissue have already grown onto the plastic, obscuring its presence. J. Sprung.

After planning the layout of the reef structure and selecting the rocks, a drill and masonry bit can be used to bore holes into the rocks to slip the cable ties through. Be careful when drilling holes, since limestone rock is brittle and easily crumbles from the vibration of the drill. Do not drill holes too close to the edges of rocks. The leverage of the attached rock pulling on the hole near the edge of another rock can cause the cable to break through the hole, sending the rock(s) tumbling. Cinch up the cable ties tightly, and cut off the end with a sharp scissors. In one aquarium in Norway, a wooden beam across the top was used to support large rocks, forming an otherwise impossible overhang/cave. The rocks were tied to each other and to the beam with long, heavy- duty plastic cable ties. Cable ties may also be used to attach the uppermost rocks of a wall to plumbing along the back or sides of the tank, thereby affording greater support.

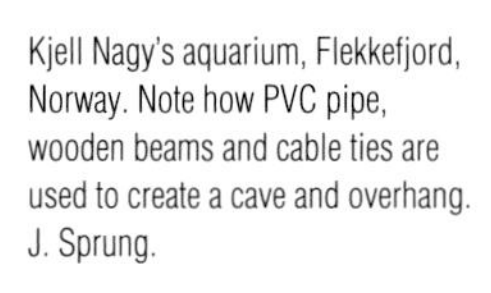

Kjell Nagy's aquarium, Flekkefjord, Norway. Note how PVC pipe, wooden beams and cable ties are used to create a cave and overhang. J. Sprung.

Monofilament Line (a.k.a. "fishing line", nylon wire)

Monofilament is one of the most versatile construction aids for building the aquascape. Its advantages include, that it is inert, inexpensive, strong, and invisible. It is easy to work with if you have good manual dexterity, can tie knots, and use a scissors. In addition, its thin diameter allows you to use it in place of cable ties where the ties would be too broad. It is the ideal aid to attaching zoanthids, soft corals, or cuttings from corals to live rock, and when the specimen has made a strong attachment, the line may simply be cut with a scissors and removed, like pulling out stitches after the wound has closed. The main disadvantage to monofilament line is that it can loosen up if the attached pieces are frequently disturbed, or if the knots aren't very tight.

Stainless Steel Wire

We have seen some hobbyists and public aquaria in Germany using a fine stainless steel wire instead of monofilament line. The wire is easy to cut, like monofilament, but does not need to be tied. It can be looped around and twisted to secure a specimen in place. We caution aquarists that stainless steel comes in different grades. Not all stainless steel can withstand the corrosive effect of saltwater. As with monofilament, the wire may be removed after the specimen has attached to the rock.

Plastic Toothpicks

When company comes over for dinner, your guests may smile knowingly as they eat hors d'oeurvres with their fingers and admire your reef. Your spouse will wonder why sooner or later everything in the house is used for the reef tank. Just as the plastic toothpicks are useful for spearing those little hot dogs, they work really well for soft corals since one end pierces and holds the coral while the other end can be inserted into the rockwork. Small, lightweight stony corals can also be positioned with toothpicks, as can small live rocks. It may be helpful to combine the use of toothpicks with cement, gum or epoxies, described later in this section. One of the best uses for plastic toothpicks, and this is quite a secret we are revealing to you, is for the placement of sponges that were collected without a base. As with soft corals, the toothpick(s) can be inserted into the sponge, and the opposite end(s) pushed into the rock.

Eggcrate

In order to minimize the rocks' contact with the bottom and prevent the formation of dirt traps, some people use eggcrate as an elevated, open base. "Eggcrate" is manufactured as a diffuser for overhead fluorescent lights. It is a grid-like plastic material with squares approximately 1 cm (0.5 inch) wide. The typical color is white, but it is also available in black. Metallic eggcrate, manufactured to lend a high-tech look to overhead fluorescents, is probably unsuitable for use inside aquariums. Eggcrate is rigid but easy to cut with a hack saw or a table saw, and pieces can be glued together with PVC cement or, preferably, secured with solid plastic cable ties. It is a versatile material for aquarists, for use in the fabrication of undergravel filters, mechanical or chemical filter chambers, as a strainer to block the flow of curious fish over the overflow and to provide elevation to the live rock in reef aquariums, keeping it off the bottom so that detritus can easily be swept away. For really big exhibits, a tougher, larger material called Chemgrate fiberglass grating, is also useful. It is used primarily to make lightweight but strong walkways, but has many applications to construction of the decor in really big aquariums. You can obtain a catalog from Chemgrate at 1-(800) 345-5636, or (206) 483-9797.

For dealers, eggcrate "staircases" are ideal for live coral tanks, as they allow the corals to be sorted by species, prevent them from falling, and afford good circulation around the specimens, not to mention easy tank maintenance. Eggcrate is fine for dealer's holding tanks, but it is not the most flexible construction device for building a natural-looking display tank. Eggcrate limits the decorative potential of the aquarium. It is difficult to build a reef that doesn't look like a wall on eggcrate. There are better ways to achieve the same positive aspects of eggcrate, in our opinion without the limitations. (see next item).

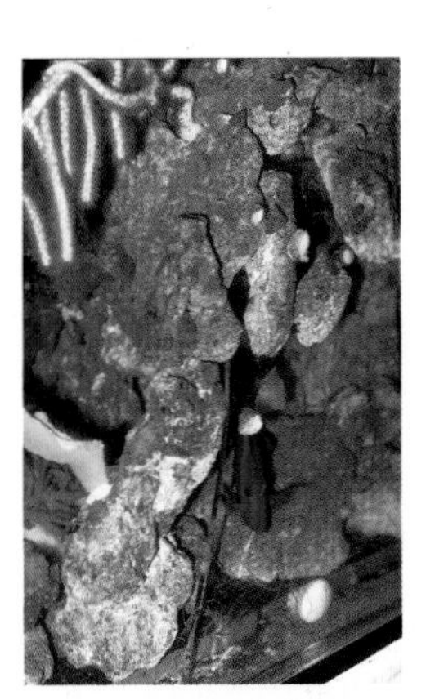

In this aquarium set up by Frank Massacesi, a curved acrylic sheet was used to build a cave. Rocks positioned over and in front of the sheet obscure it from view. The photo was taken from the side of the aquarium. J. Sprung.

Plastic Screws, Acrylic Sheets or Rods, PVC Pipe

Rick Graff of the Bucks County Aquarium Society built his reef in a unique way that makes it look like the rocks are floating above the bottom by some means of levitation. His trick is to use clear acrylic cubes as feet for the large bottom pieces. The feet are attached to the rocks with acrylic rods. Holes are drilled into the cubes, the lengths of acrylic rod are inserted, and then these assembled feet are inserted into the rocks. Unlike eggcrate, there is no limitation to the shape of the reef you can create with this technique.

A variation of this method could be used with sand on the bottom. Instead of using rods and cubes to make feet, the higher elevation necessary to "float" rocks above sand requires the use of cut lengths of acrylic tubing, at least 3/4" diameter. With this method it is possible to have either portions or the entire reef appearing to be suspended over the sand, with terrific caves underneath.

You can use lengths of clear, grey, or white PVC similarly to make feet, attaching them to the rocks with cable ties cinched up through holes drilled into the rock with a masonry bit. Wider pipe diameters provide greater stability. Even thick pieces of acrylic can be used, or pieces fitted together to form an "x" (a suggestion offered to us by Martin Moe). If you can find thick plastic screws, these too can be used to elevate rocks, the head serving as the foot in this case. Plastic screws or rods can also be used to position coral heads or to link rocks together. Be sure that whatever you use to support rocks above the bottom of the tank is structurally suited to this purpose, and will not break down, crumble or shift and give way.

Dental Cement

Some aquarists use a two-part dental cement for securing corals in position. The cement does not stick to the rocks when wet, so it is not used to adhere anything. Rather it is used to make a mechanical locking fit between pieces, allowing them to be held together sufficiently for growth of the specimen to make a solid attachment bond. For example, a branch of coral pushed into a hole in the live rock without any locking device will shift in position and may fall out of the hole. If the putty-like dental cement is pushed into the hole first, and surrounds the branch, it will hold it in place until the coral tissue grows over it and new skeleton is laid down onto the live rock.

Chewing Gum

Instead of dental cement, it is possible to use chewing gum as a locking device. Chew it first to remove the flavoring, and then soak it in ice water to give it firm consistency, (Don't tell anyone what you're doing).

Epoxy Cements

There are some non-toxic (when cured) epoxy cements that are useful both for locking devices and for actually forming a bond between pieces. The following recommendations for Z-spar and Sea repair compounds are provided by Dr. Bruce Carlson, who has used both products extensively (and very successfully we might add) at the Waikiki Aquarium, Hawaii. Both products and Devcon Underwater Repair Putty, described afterward, can be obtained from marine supply and yachting supply stores. We caution that for the sake of safety, one should wear old rubber gloves or disposable latex gloves when mixing or handling epoxy cements, or just mix and apply the epoxy with an old knife or thin spatula. It is very important to protect your skin from all epoxies since they contain many sensitizers. It is best to wear two layers of gloves because as you are working with the epoxy the gloves may become fouled. Simply remove the outer glove and you are still protected and can continue to work. An individual may work with epoxies for years with no trouble and then one day develop severe rashes and medical problems, and no longer be able to be in the presence of epoxy.

Z-Spar Splash Zone Compound (A-788)

This compound is a two-part epoxy putty. After mixing equal parts of A and B according to the directions, some heat will be generated, but not enough to really be a problem underwater. Virtually all epoxies are very sticky above water, but do not adhere well when wet. Z-Spar loses enough stickiness to make it easy to handle if your (gloved) fingers are wet, but one of the unique characteristics of this compound is that it still stays relatively sticky and adheres well to rocks even underwater. With this advantage comes one minor disadvantage: it takes a long time to set up (harden). You will have to prop up the pieces of rock or coral, or hold them together by some mechanical means for at least a few hours. You can use some 20 minute underwater epoxy (which doesn't stick too well) to just hold the rocks together long enough until the Z-Spar sets.

Z-Spar is best for attaching rocks to each other, or to rocks with living coral on them. First brush clean the surfaces that are to be bonded. Next cover both surfaces with Z-Spar and press them together. After the surfaces are mated, spread the excess, extruded Z-Spar over all the adjacent surfaces to get a good set, and feather it to give a natural, rock-like appearance. See caution about use in closed systems below.

Sea Repair High Density Epoxy Putty Stick (20 minute epoxy)

This 20 minute underwater epoxy has parts A and B conveniently in stick form. Simply cut off the amount you need and mix them together in your (gloved) fingers or with a metal spatula. You can push a piece of the mixed Sea Repair putty into a hole in the rock and then also cover the base of the branch (it's ok even if it covers a little of the living tissue). Next insert the base of the branch of coral into the hole, and to secure the fit, push the excess epoxy firmly around the coral and the rock, filling any gaps.

Devcon Underwater Repair Putty

Joseph Yaiullo from the New York Aquarium uses this epoxy cement to secure corals and rocks in place. Joe first heard about it from the director of the Monaco Aquarium, Nadia Ouinais, who works with Dr. J. Jaubert. The following procedure works best: Mix equal parts by volume of the pre-thickened parts A and B and let the mixture cure until it has the consistency of thickened clay or silly putty. It can then be pressed onto a scrubbed clean surface of the coral skeleton, and the coral with the attached putty can be pressed onto a scrubbed surface of reef rock or almost any other substrate, including the walls of the aquarium! It is best to work with small batches since the epoxy has a limited working time and you are letting it cure a bit prior to using it. Speeding up the curing time can be accomplished by preheating the epoxy mixture and allowing it to cool before contact with corals. The coral with epoxy attached can be held in place temporarily by propping it with toothpicks or any wooden or plastic sticks. It will set in 10-20 minutes depending on the working temperature. Corallines and coral tissue grow rapidily over the epoxy, and in several weeks the "perfectly balanced" coral will have grown in place. Epoxying the pieces in place avoids repeated damaging falls, and promotes increased growth rates as the corals are stabilized. It also allows decoration of vertical walls and very natural placement of corals with ease.

Caution in Closed Systems

Both Z-spar and 20 minute epoxy produce a milky substance in the water when first spread onto the rocks. In the 250 gallon open system aquariums at the Waikiki Aquarium this substance has not bothered fish or invertebrates at all (B. Carlson, pers. comm.). However, it is possible that fish and invertebrates in smaller, closed system aquariums could be affected by the milky substance. We recommend that you set up a separate aquarium, bucket, or other container with seawater, for adhering rocks and corals together. The pieces can be allowed to set up in the separate container, with an airstone providing circulation and adequate gas exchange. If you need to glue pieces within the display aquarium, be careful not to use too much. Again, Dr. Carlson's experience has shown these substances are benign, (in fact he even observed Lemonpeel angels, *Centropyge flavissimus,* picking repeatedly at Z-spar while it was still soft!), but in small closed system aquariums one should use them with caution.

Quick-Setting Portland Cements, Thorite

The quick-set type of pure portland cement is an ideal material for permanently cementing corals in place because it is very adhesive even to damp surfaces, sets in about one hour, is non-toxic, and strong (Bronikowski, 1982; 1993). Bronikowski exposes the horny skeleton of gorgonians at the base, and dries it with paper towels. Then he positions the gorgonian on a rock base outside the aquarium, applying a thick mix of the quick-setting cement on the exposed part to hold it in place. The cement can be worked to give it a texture like the rock, and it can be applied up to the living tissue. During the one hour setting period, the coral is laid horizontally, and is kept damp by laying plastic sheets over it and periodically wetting it (but not the cement) with seawater administered with a pipette. Although Bronikowski's descriptions are for gorgonian placement, quick-setting cements can be used in the same fashion for stony corals, particularly for branching pieces that can have one end pushed into a hole in a rock, where the cement can be applied to make a secure bond. The product Bronikowski used, Thorite, may no longer be manufactured, but other quick-setting portland cements exist.

Cyanoacrylate

Cyanoacrylate glues can be used for attaching small coral fragments to substrates, including glass, and for reattaching light-weight broken branches. Unless the two pieces fit together perfectly, the bond that forms is brittle, and easily broken. The value of its use is to hold a piece in place until it has the chance to grow and form a more permanent attachment. Cyanoacrylate works best if the items being attached can be partially dried, as we described for quick-setting cements. It does work underwater, but not as well. We don't know whether the leached compounds from un-cured Cyanoacrylate pose a threat to marinelife, but we suspect they do. Cyanoacrylate cures quickly, and is non-toxic once cured. We recommend that the pieces be joined outside of the aquarium. The glued items can be placed in the aquarium after the glue has mostly cured, in about half an hour.

Natural Ways to Glue Rocks Together

In addition to the use of cements and mechanical locking devices, structural bonding and attachments can be made by natural means. These are the biological ways that the substrate becomes fused together on coral reefs. In our aquaria we can combine mechanical locking devices and biological ones for the best means of achieving the desired aquascape. Mechanical locking with monofilament, cable ties, or cements affords the necessary short term stability for subsequent growth of the following organisms to form a natural bond.

Sponges

While some sponges do not survive long in aquariums, there are many sponge species that not only grow, but reproduce in our aquariums. Some of these are quite useful for attaching other organisms because growth is so rapid that firm attachment to the substrate is achieved within 24 hours! By far the best species for this purpose is the "Chicken Liver" sponge, *Chondrilla nucula*. It is a common sponge on live rocks, and has a thick consistency like rubber and an appearance that confirms the reason for its name. It is also a common sponge in shallow grass flats and bays, where it is loosely attached to the grass and algae, making collection simple. On live rock it adheres tightly, so collection involves some injury to the sponge. Such tearing of the sponge tissue may not harm it, but does increase the possibility of an infection that could kill it. Therefore it is best to use whole, intact specimens from seagrass areas. The blue sponge from Indonesia is also excellent for adhering pieces, though it is unfortunately available only infrequently. It is possible for some sponges to encroach upon living coral tissue and cause it to recede, but this seldom occurs in our aquariums. It is also possible for some corals to sting sponges and kill the affected portions, but usually the sponge and coral tissue can touch without harm to either. Another caution is in order with using sponges for attachment of specimens: fouling sponges can be toxic, especially to fish. Use only small sponges, approximately the size of your big toe or smaller to minimize the risk.

To "glue" a specimen in place with a sponge, simply position it and lay the sponge next to it. Sometimes it is helpful to hold either the sponge or the specimen in place with monofilament line until the attachment is complete. Within a few days the line can be cut and removed.

Zoanthid Anemones

The encrusting anemones of the genus *Zoanthus* and *Palythoa* rapidly attach to the substrate and, placed between two rocks, can securely fasten them. *Palythoa* species are especially good for this purpose since they grow rapidly, adhere strongly, and their tissue is especially tough.

Leather Corals and other Soft Corals

Various soft corals, including *Sarcophyton, Sinularia, Clavularia, Briareum, Erythropodium,* and the photosynthetic gorgonians, form attachments to the rocks and effectively glue them together with living tissue. The gorgonians, in addition, secrete a tough, horny material that provides a strong lasting bond even if the coral tissue dies.

Stony Corals

Stony corals will grow and attach to the rocks, forming solid connections between adjacent pieces. Initially brittle, the bonds become stronger with time and the deposition of layers of calcium carbonate. Attachment of pieces this way is a slow process, but it is common in older reef aquariums which have had time to grow.

Coralline Algae

As with the stony corals, fusing of the rocks by coralline algae is a slow process. Coralline algae are considered the natural cement of the reefs, though truly their work is much assisted by the faster growing sponges that fuse the rocks and sediments together, in turn becoming covered by a veneer of coralline algae. Many coralline algae will form fast growing circular crusts on surfaces within the aquarium, but a few types, such as *Mesophyllum,* will actually form attachments between rocks.

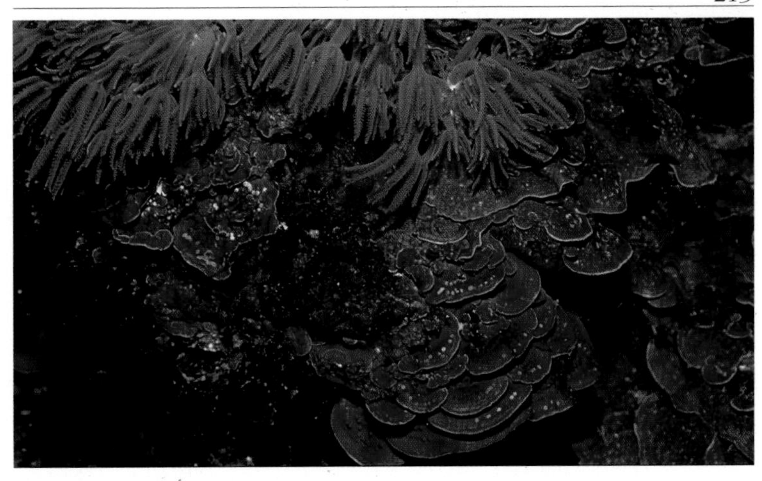

Plating growth of coralline algae *(Mesophyllum)* in a 10 year old Dutch reef aquarium. J.C. Delbeek.

Clams

A minor sort of attachment occurs where the byssus filaments of clams are adhered to several rocks. This effectively holds the rocks in position.

The Use of Coral Sand and Other Bottom Media

Sometimes people think that adding coral sand to their reef tank is taboo. This notion is probably partly owing to old recommendations in aquarium literature that said, in effect, that a sand or gravel bed without an undergravel filter to pull water through it would become foul and release toxic hydrogen sulfide into the water. In more recent recommendations in North American reef aquarium literature, sand or gravel on the bottom has been discouraged because it is perceived as a detritus trap that can promote the proliferation of undesirable algae. These partly factual notions have resulted in an unfortunate proliferation of tanks with nothing at all on the bottom! We hope we can change that trend with our recommendations in this book.

The "bare-bottom" idea affords a good measure of the detritus production by the rocks, usually an awful lot, and it is easy to siphon this away or prevent its accumulation with bottom water jets. Such an arrangement certainly prevents the accumulation of detritus, but it does not make for a natural aquarium ecosystem.

Sand has some important functions in the reef aquarium. It is decorative, of course, affording a realistic appearance and a complimentary, lighter colored horizontal contrast to the upright structure of the reef. Beyond that, sand serves some wonderful biological functions that can be used to advantage in closed system aquariums, provided simple rules are followed.

The rules include using Sleeper Gobies *(Valenciennea* spp.), sea cucumbers and possibly small pistol shrimps (for large aquariums only since they can bother fish in small aquariums), to maintain the substrate. Goatfish are also good at stirring up the bottom, as are a number of the small Partner or Watchman gobies (*Amblyeleotris* spp., *Amblygobius* spp., and *Cryptocentrus* spp.), *Pholidichthys leucotaenia*, and *Istigobius* spp. gobies (see Delbeek and Michael, 1993). Certain polychaete worms and serpent stars also help keep the sand clean, and they are generally introduced to the aquarium with live rock. All of these organisms constantly sift the sand in search of food or in the construction of burrows, thus turning it over and preventing algae from coating the surface. This action also prevents detritus accumulation and the formation of dead spots.

Valenciennea puellaris. J.C. Delbeek.

A mated pair of *Valenciennea strigatus.* J.C. Delbeek.

Valenciennea wardi. J.C. Delbeek.

Istigobius ornatus. J.C. Delbeek.

Cryptocentrus cinctus. J.C. Delbeek.

Pholidichthys leucotaenia is a strange, eel-like fish that digs burrows in the sand, excavating and aerating the zones beneath rocks. J. Sprung and J.C. Delbeek.

A goldenheaded sleeper goby sifts sand through its gills. J. Sprung.

Note: The gobies can only sift fine sand, as gravel may injure their gills. The sand should not be powder fine, but coarse, about the size of sugar crystals or a little larger. The most desirable and natural looking material is real coral sand, which is collected from beaches in the vicinity of coral reefs. A few coral fragments or small stones thrown in provide a natural appearance as well as useful building material for fortification of burrows. Pet dealers should be able to obtain coral sand for you from a number of different suppliers, though it may become scarce in the future as a result of restrictions on the importation of coral.

Certain *Holothuria* spp. sea cucumbers which resemble terds (yes, we said that) are also desirable for keeping the sand clean. As they move through the sand, they ingest it and digest any algae, detritus

A small sea cucumber, *Holothuria* sp. feeds on detritus by mopping up sand. J. Sprung.

and bacteria attached, which provide them with nutrition. Their digestive acids even liberate a tiny amount of calcium to the water. Sea cucumbers are really best at mopping up detritus from between grains of gravel. Beware! avoid large sea cucumbers (greater than 15 cm; 6 in.) and never include a "Medusa Worm" *(Synapta* sp. sea cucumber) in your aquarium. *Synapta* sp. are toxic to fish when injured, and large sea cucumbers also can poison an aquarium if they become injured or suctioned by a powerhead. Small sea cucumbers present no danger. The colorful Sea Apples *(Pseudocolochirus* sp.) and other species that are just filter feeders, are best maintained by themselves as an exhibit, as they too can be toxic to fish when injured, and their eggs are so poisonous that they are deadly candy for the fish that eat them.

The benefits derived from sand are its effect on water quality and the refuge it presents to microorganisms, worms and crustaceans (see chapter 5). When the sand is thick enough, 2.5 cm (1.0 inch) or more, anoxic zones develop adjacent to aerobic ones, and biological processes occur as in live rock but more efficiently. The bacteria that colonize the sand denitrify the water. The bottom substrate is a most efficient, built-in denitrifying filter! Though some people prefer to maintain thin layers of sand, there isn't a limit to its acceptable, safe depth. The thicker it is the greater potential there is for denitrification. One must remember, though, that a living sand bed consumes a lot of oxygen. As long as there is strong circulation within the aquarium, the aerobic sand layer on top is maintained, and the aquarium will remain healthy. The microorganisms, worms, and crustaceans living in the sand also digest detritus settling there, and their reproduction in this refuge generates food for the filter feeders and fish. Additionally, the digestion of detritus liberates carbon as food for the facultative anaerobic bacteria in the sand, fueling the denitrification process.

Some aquarists like the look of gravel on the bottom of the reef tank, but gravel is a little more difficult to maintain than sand. The same is true of coral fragments or small live rocks on the bottom, which are effectively big gravel. The depth of the gravel bed is not critical, though thin layers are easily serviced by manually stirring them. When the bed is thicker than about 2 cm (0.8 in.), certain creatures play a key role in its successful maintenance. Fortunately there are some natural cleaners that can be employed to keep the gravel bed healthy. The small sea cucumbers mentioned earlier are useful for this purpose, but Sleeper gobies cannot help here. Another useful creature is a small, harmless type of bristleworm (about 5 cm (2 in.) or less and as thick as spaghetti) that multiplies prolifically in the gravel. It resembles the undesirable *Hermodice carunculata* that grows large and eats coral. The small worms often find their way into the reef tank with live rock and, though they are really benign, aquarists often worry themselves sick over the sight of them because of the generic equation that bristleworm = bad. Unlike their larger cousins, these small worms eat detritus and uneaten food, and thus help to keep the spaces between gravel grains clear. They reproduce rapidly in gravel beds without any assistance from the aquarist. Their population explosion can appear unsightly to some aquarists, and it can be checked by including in the aquarium a pair of banded coral shrimp, *Stenopus hispidus*. Certain fish such as *Pseudochromis springeri* and *P. aldabraensis,* and some of the small wrasses will also eat these bristleworms and prevent the population from becoming too large. Commercially available worm traps may also be used to control the population (see Chapter 10).

Several other types of polychaete worms are introduced with live rock or with "live sand", if it is available. These multiply in the aquarium both sexually and asexually, and the result is a very live sand bed full of worms that actively feed on detritus and prevent the sand from becoming a dirt trap. Terebellid or "spaghetti worms" live in or beneath rocks, or in gravel-lined tubes they construct in the substrate. Their tentacles stretch several centimeters to over one metre in some varieties, such as *Amphitrite* sp. Aquarists often believe the tentacles themselves are long thin worms. The tentacles "crawl" along the rocks and bottom, and trap particulate matter that passes down a thin groove running their length. Periodically the worm will defecate a large pile of detritus that it has collected and ingested. Similar to the terebellid worms are spionid worms. Like terebellids they live in rocks, beneath them, and in tubes in the sand, and they have long tentacles that collect particulate detritus. The distinguishing characteristic of spionids is that they have only two tentacles, whereas terebellids have many.

Sandy areas in coral lagoons, around seagrass beds, and intertidal mudflats often have another type of substrate dwelling worm that can be beneficial in large reef aquariums with a thick sand bottom. The "lugworm", *Arenicola* spp. makes what looks like a small volcano in the sand, with the occasional plume coming out the top and all. It lives deep in the sand, and has two "openings" to its burrow. The volcano mound opening is the side from which the worm defecates ingested sand, and the pile created rises like a cone with a hole in the center. Water normally enters through this hole, and the worm sends a current toward the exit of its burrow, causing the sand at the surface to cave in like a sinkhole. The sinking sand from the surface is rich in detritus, bacteria, and algae on which the worm feeds. When the worm defecates, the current is reversed. These worms literally turn the sand over.

Brittle starfish, a.k.a. "serpent stars" are also useful scavengers that eat feces or any missed food particles, and seek out and consume any organism that has died. Several should be included, especially in aquariums with gravel or coral fragments on the bottom. Serpent stars come in many colours and varieties. Having different types in the aquarium can be visually dramatic. When food is added, their writhing arms of many colours and patterns come out from the rocks. Some brittle stars remain beneath the rocks or gravel, where they continuously move the substrate in search of edible detritus; sand grains moving along their legs as if on a conveyer belt.

Serpent starfish are excellent scavengers. They feed on fish feces and uneaten food. J. Sprung.

Tiny hermit crabs described in chapter 9 under the topic of herbivores, are good scavengers and very effective at preventing algae from coating sand or gravel.

Finally, the use of alternating currents, surge waves, or strong circulation will greatly assist the management of a thriving sand or gravel bed. The water circulation maintains the healthy population of microorganisms and invertebrates living there, while keeping the oxygen levels in the aquarium above saturation. We present here some decorating ideas for the creation of an aesthetically pleasing display.

7.1

Aquascaping plans, viewed from above

a. A reef built in the center of a long rectangular aquarium affords viewing from at least three sides, the fourth side being obstructed by the overflow chamber, if present. Note direction of water flow creates circular motion and sweeps surface water over the overflow wall.

b. In this plan the reef is shaped in a spur-and-groove fashion, providing many surfaces for decoration. A corner overflow is installed with the water return directed to sweep surface water toward it. A powerhead (or two) controlled by automatic pulse timers sends surge-like water motion over the reef front.

c. This reef aquarium is poorly planned and aquascaped. The rocks are stacked like a wall against the back of the aquarium- easy to build but BORING! The water flow is directed away from the surface skimming overflow, resulting in accumulation of a surface slick at the opposite end.

d. This reef aquascape was planned to model on a small scale the shape of the whole reef. A wave bucket sends a surge of water over the reef. See the description of Dr. Adey's system in chapter 5.

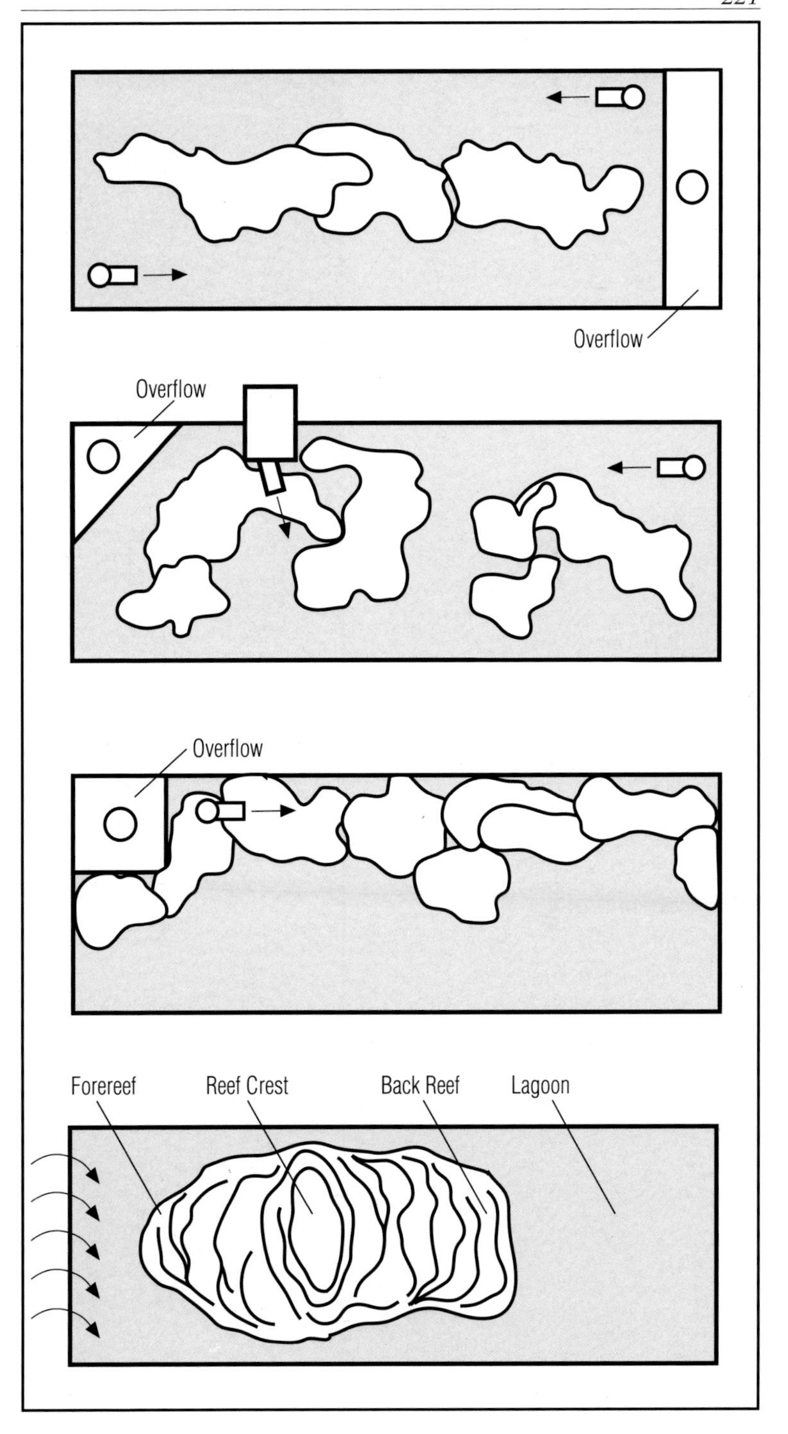

A reef aquarium aquascape similar to the plan in 7.1 b. Notice the open construction of the live rock structure.
J. Sprung.

7.2
More Aquascaping plans, viewed from above

a. This square aquarium has a reef structure built in the center. Two powerheads provide circular movement of water. The reef can be viewed from all four sides. An overflow could be installed in the center. We have left out the overflow to demonstrate the possibility of a reef aquarium without one. An internal protein skimmer could be used, and a level switch could be installed in the aquarium for top off of evaporated water.

b. This wide reef aquarium has the reef structure built along the sides with a coral bommie in the center. With this design the effect of a coral grotto is created.

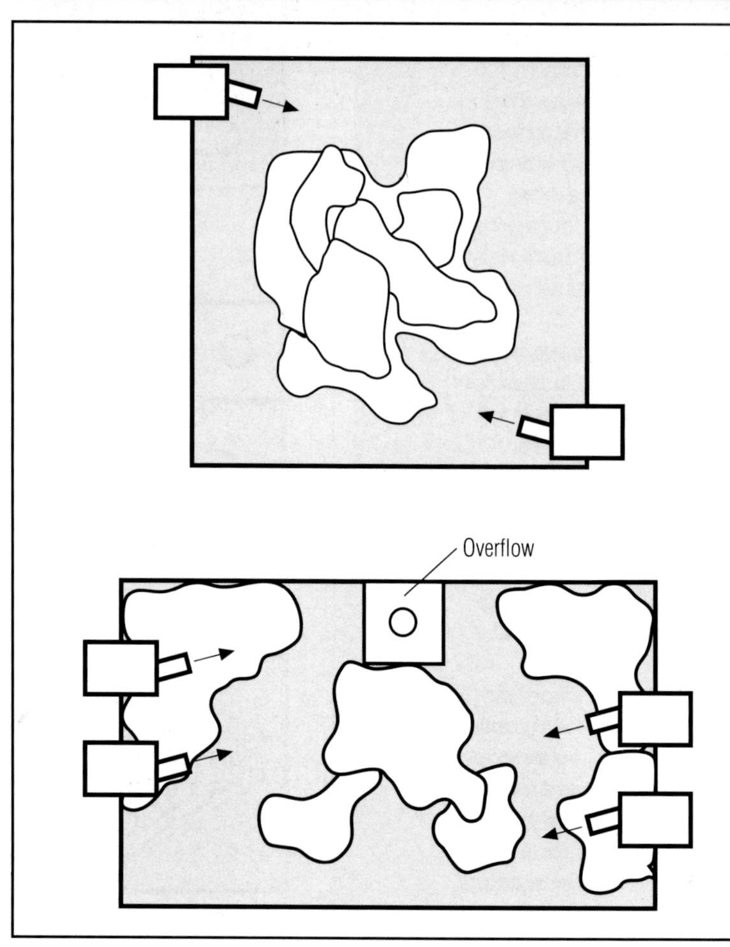

Construction of a Pacific reef exhibit at the Waikiki Aquarium, Honolulu, Hawaii. Concrete cinder blocks were used beneath the live rock structure. The cinder blocks first had to be cured in freshwater with weak acid for several weeks to remove excess alkaline substances they initially leached into the water. Note use of plastic cable ties. B. Carlson.

The reef structure is complete, and some corals are added. B. Carlson.

After several months the exhibit is a growing, healthy slice of reef. B. Carlson.

Chapter Eight

Water Quality Parameters

Temperature

Temperature is the most critical physical parameter for captive reef systems. If all other parameters are ideal, but the temperature is not, the reef will not thrive (see chapter 2). The temperature in reef aquariums should be between 21-27 °C (70-80 °F) for the best results, and as stable as possible. In our experience, a temperature of 23-24 °C (74-76 °F) is ideal. If the temperature varies plus or minus one or two degrees Fahrenheit during the course of the day, this is not a problem. The most common problem is high temperature; above 27 °C (80 °F). Wide fluctuations are also harmful, especially to the fishes' health, since temperature fluctuations are commonly associated with the incidence of *Cryptocaryon irritans* (saltwater "ich").

You should ensure that the top of the tank has adequate air ventilation, either by having a hanging light fixture over the aquarium, or by having a well-ventilated (or fan-cooled) light hood. This will aid gas exchange and evaporative cooling, and will retard heat build-up in the hood. You could also design your sump so that it is open and provides a large surface area for evaporative cooling. Evaporative cooling can be further enhanced by placing a fan over the water surface.

The most expensive but effective solution to temperature control is to purchase an aquarium chiller designed for saltwater use. Many styles and capacity ranges are available now, and both the price and quality of the units have improved in recent years. Somewhat less expensive is to purchase an air conditioner unit for the aquarium room. This has an added comfort benefit for the aquarist of course, which may help you convince your spouse of the extra expense! Still, it may be cheaper in the long run to use a chiller, considering the cost of electricity to run an air conditioner constantly, set cold enough to keep the tank cool. Chillers do add considerable heat to the room, causing the room's air conditioner to work harder in a silly, wasteful battle. Therefore it is best to locate the chiller in the garage, basement, attic, or other location where the heat will not be noticed.

Photo: Allan Storace.

pH

The pH of aquarium water is a measure of the concentration of hydrogen (H^+) ions in the solution. The concentration of hydroxide ions (OH^-) is inversely proportional to the concentration of hydrogen ions. When the hydrogen ions are most abundant, the solution is acidic. If hydroxide ions are in greater abundance in a solution, the solution is alkaline or basic. Values of pH range from 0 to 14, or at least that is what we are always told. Actually, it is possible to have a negative pH, (hydrogen ion activity greater than molar), or a pH greater than 14 (hydroxide ion activity greater than molar). It has only become customary to use the scale between 0 and 14. The scale is logarithmic. Therefore, each gradation represents a factor of ten. For example, a pH of 8 is ten times more basic than a pH of 7 (i.e. a solution with a pH of 8 has 1/10 the H^+ ions of a solution with a pH of 7). A pH of 9 is 100 times more basic than 7. If the pH is less than 7 the solution is acidic, if it is greater than 7 it is basic, and if it equals 7 it is neutral. See table 8.1. The pH can be measured with reagents and pH sensitive dye, or more accurately with an electrode and pH meter. A pH meter measures the voltage produced by hydrogen ions diffusing through the proton-permeable glass of an electrode immersed in the solution. A pH meter is a most valuable aid for the reef aquarist, but it must be calibrated often to maintain accuracy.

Table 8.1

pH	[H^+] concentration in mol/L	[OH^-] concentration in mol/L
0	10^{0}	10^{-14}
1	10^{-1}	10^{-13}
2	10^{-2}	10^{-12}
3	10^{-3}	10^{-11}
4	10^{-4}	10^{-10}
5	10^{-5}	10^{-9}
6	10^{-6}	10^{-8}
7	10^{-7}	10^{-7}
8	10^{-8}	10^{-6}
9	10^{-9}	10^{-5}
10	10^{-10}	10^{-4}
11	10^{-11}	10^{-3}
12	10^{-12}	10^{-2}
13	10^{-13}	10^{-1}
14	10^{-14}	10^{0}

Seawater is a basic solution with a pH ranging from 8.0 to 8.25 typically, but in the vicinity of high rates of photosynthesis and respiration, as on coral reefs, the pH may fall below 8.0 at night and rise above 8.4 during the day. In reef aquariums, the ideal pH does not fall below 8.2, nor climb above 8.5 (A. Nilsen, pers. comm.). A daily variance within these limits best promotes calcification. The range of acceptable tolerance is between about 7.6 and 9.0, but at the extremes calcification is impeded, and some organisms may not be healthy. The ideal pH for calcification is about 8.4. Low pH inhibits precipitation of calcium carbonate by the organisms, and high pH tends to lower the calcium ion concentration. Some tanks will show wide fluctuations of pH from day to night, while others will remain stable. High carbonate hardness (and alkalinity) promote pH stability.

Carbon dioxide dissolved in the water from the atmosphere and respiration by animals and plants, is responsible for the daily variance in pH. As CO_2 dissolves in seawater, it immediately forms carbonic acid (H_2CO_3), so that little CO_2 is actually present in the water. The carbonic acid dissociates into carbonate (CO_3^{-2}) and bicarbonate (HCO_3^-) ions. The seawater buffering system works because these compounds shift back and forth in equilibrium reactions when changes in the concentration of hydrogen ions occur, tending to maintain a pH of 8.2. Removal of CO_2 by photosynthesis results in less formation of carbonic acid, and therefore raises the pH. Retention of excess CO_2 in the water because of poor circulation, or administration of CO_2 by means of a dosing system, causes the pH to decrease. As the CO_2 equilibrates with the atmosphere, the pH naturally returns to about 8.2 if the carbonate/bicarbonate buffer system is not depleted.

Organic acids and phosphate that tend to accumulate in closed systems, and the formation of nitric acid from biological filtration (nitrification), deplete the buffer capacity or alkalinity of the water (see next section). The acids deplete bicarbonates and calcium carbonate from the water, while phosphate lowers alkalinity by precipitating out of solution in compounds with calcium and magnesium. This is the cause of the long term decline in pH characteristic of many closed system aquariums. Denitrification helps reverse the loss of alkalinity, as the bacteria involved liberate carbonates and bicarbonates from the breakdown of organic matter (see Jaubert's system, Chapter 5).

Alkalinity

This term confuses both novice and advanced hobbyists. Stated simply, the alkalinity of a solution refers to its capacity to buffer against drops in pH. Higher alkalinity affords greater ability to prevent rapid pH swings. Once the alkalinity is exhausted, the pH can fall rapidly. Alkalinity is provided in the aquarium by various negatively charged ionic compounds (anions) such as carbonates, bicarbonates, borates and hydroxides. The rather confusing term "carbonate hardness" has also been used to describe alkalinity, but this refers only to the carbonate and bicarbonate portions of alkalinity and does not take into consideration the other compounds involved. Therefore alkalinity is generally slightly higher than carbonate hardness (Spotte, 1979).

To further add to the confusion, there are several scales used to measure alkalinity. Two different units of measurement are commonly used by aquarists, milliequivalents per litre (meq/L), and degrees of carbonate (German = karbonat) hardness (dKH). There are numerous test kits available and they are all simple to use. Some test kits for alkalinity use the metric unit (meq/L), while others use the German unit, (dKH). To convert meq/L to dKH, multiply by 2.8. Natural seawater has an alkalinity of 2.1 to 2.5 (6-7 dKH). Alkalinity values in the aquarium should be maintained between 2.5 and 3.5 meq/L (7-10 dKH). Note: most alkalinity or KH test kits use exactly the same titration.

Alkalinity can be maintained via addition of kalkwasser (saturated calcium hydroxide solution) as make-up water, and through the use of commercial buffers. These buffers are also used to maintain pH. There are numerous powdered buffers available that will maintain both alkalinity and pH, and the majority work very well by raising the alkalinity without causing rapid increase in pH. Avoid buffers that cause rapid increase in pH, and follow manufacturer's instructions.

Kalkwasser maintains alkalinity by supplying both hydroxide ions and calcium ions. The hydroxide ions neutralize acids and therefore prevent loss of alkalinity. The calcium ions can combine with dissolved CO_2 to replenish dissolved calcium carbonate. However, kalkwasser can deplete alkalinity when administered improperly. Too much kalkwasser added at once to the aquarium will raise the pH high enough to cause calcium carbonate to precipitate spontaneously from the water. For this reason it is best to add kalkwasser via a dosing system. Old kalkwasser typically has precipitated calcium carbonate settled on the bottom of the container. If this white powder is dosed into the tank with kalkwasser, the high pH around the calcium carbonate particles floating in the water will cause further precipitation of calcium carbonate onto the particles that act as nuclei for crystallization. Use fresh kalkwasser only, or periodically remove the accumulated precipitate from the kalkwasser reservoir. (see additional information under calcium additions, this chapter).

Specific Gravity

Specific gravity is used to measure the relative salinity of seawater compared to distilled water. Distilled water has a specific gravity of 1.000 while seawater ranges from 1.022 to 1.030. The acceptable range for specific gravity is between 1.015 and 1.032. Specific gravity values depend on the temperature of the water, and so aquarists' measurements of specific gravity are usually only approximate. Estimates of salinity based on specific gravity maybe much lower than the actual value (see Moe, 1989).

We have seen healthy corals and other marinelife in nature, and in captivity, at specific gravity values within the range of 1.010 and 1.035. Full strength seawater is about halfway between the extremes of this range, about 1.025, and this is a common value on many open ocean reefs around the world that are not influenced by run-off from rains, and not greatly influenced by evaporation, due to the high rate of exchange of water from the open sea.

Most reef aquarists maintain their aquariums at a slightly lower specific gravity, approximately 1.022. This is fine. The most important aspect of specific gravity to remember for the reef aquarium is to maintain stability at whatever value you choose. Constant fluctuation of specific gravity is stressful to invertebrates, plants, and microorganisms including bacteria. Stability is the key to making them thrive.

Sudden change in the specific gravity is occasionally used to control disease outbreaks in the fish, and is particularly effective against *Cryptocaryon irritans*, saltwater "ich". Causing the specific gravity to drop from about 1.022 to 1.017 over the course of a few hours by the addition of freshwater will not harm the reef aquarium, and can effectively control the disease. After a few weeks with low specific gravity, when the fish have recovered from the disease, the specific gravity can be gradually increased back to its normal value and maintained there. Something else worth noting with respect to specific gravity is its affect on the concentration of calcium ions and alkalinity. Full strength seawater has a calcium level of about 400 to 450 mg/L, and an alkalinity of about 7 to 8 dKH. Specific gravity less than 1.025 is diluted from full strength seawater. When full strength seawater calcium and alkalinity values are achieved at diluted specific gravity, they are really of higher concentration compared to the other ions. This is fine. It is acceptable to maintain the calcium ion concentration a little shy of full strength values when the specific gravity is less than full strength. Alkalinity, however, should always be maintained at or above 7 dKH, to prevent harmful, large pH drops at night. This is especially important to mind when the specific gravity is lowered for the purpose of disease treatment, as we just described. The simplest way to manage the alkalinity when the specific gravity is dropped is to add some buffer.

The specific gravity of the aquarium water increases when water evaporates. Pure freshwater is used to replenish what evaporates. This replenishment can be done manually by adding the lost water daily either with a scoop or via drip feed, or automatically by using a system with level switches, a dosing pump, and a freshwater reservoir. Level switches can be simple mechanical float valve assemblies that open a valve attached to a freshwater reservoir. Under no circumstance should the valve be attached to the main water supply to the house, like a toilet float valve is installed. In that arrangement it is a safe bet the tank will one day be "flushed" with freshwater. Other types of level sensing switches include mercury switches, floating collar switches, and optical infrared sensors. The level sensing switches are located where the water level changes with evaporation, either in the sump, or directly in the aquarium if there is no external sump. Two switches may be used as a safety precaution. One switch turns the dosing pump on when the level in the sump or tank falls, and it shuts the pump off when the correct level (or higher) is reached. The other switch is located above it, and turns the dosing pump off if the level in the sump is too high, in the event of failure of the lower switch. Level switches are also useful when attached to the main circulating pump, for preventing it from running dry if the sump level is too low, and they can be installed with an alarm to notify the aquarist of high or low water. Variable dose metering pumps can also be used for the automatic addition of top-off water, with or without level sensing switches.

Ammonium and Nitrite

Normally, ammonium and nitrite levels in an established aquarium will be zero. If you can measure detectable ammonia or nitrite, your aquarium is new and has not yet cycled completely or you have something decaying in the aquarium. You should carefully inspect your aquarium for any decaying animals, food or live rock, and remove it immediately. Decaying live rock is easily recognized by the smell and the white film that usually accompanies it. Ammonium and nitrite can also be produced in the substrates (sand/gravel/rock) from incomplete denitrification, but this does not produce significant measurable values in the water column in well-established aquariums.

Nitrate

Nitrate is discussed in detail in chapters 2 and 9. The nitrate level in reef aquariums should be less than 1ppm as nitrate-nitrogen ideally, but need not be maintained so low. In fact, elevated nitrate levels as high as 10 ppm nitrate-nitrogen (approximately = 40 ppm nitrate ion) may encourage more rapid growth of both soft and stony corals (D. Stüber, pers. comm.). The principal disadvantage to higher nitrate levels is the affect on alkalinity and pH described in the section on control of nitrate. Nitrate-stimulated growth by the corals causes greater usage (removal) of calcium from the water, while the accumulation of nitrate depletes the alkalinity in a closed system through the accompanying formation of nitric acid. We bring this up to emphasize that if high nitrate levels are maintained, it is especially important to observe the calcium and alkalinity levels in the aquarium. Use of calcareous water for water make-up effectively counters the depletion of calcium and alkalinity.

Phosphate

Phosphate can cause problems in marine aquaria, especially reef aquaria if allowed to build to levels above 0.1 ppm as orthophosphate. Elevated phosphate levels will fuel unwanted algae growth and interfere with the calcification processes of corals and coralline algae.

Phosphates are present in many forms in the aquarium, and not all of them can be easily measured. The majority of phosphate test kits used by aquarists measure only inorganic phosphate (orthophosphate) and ignore organic phosphates. This often leads to the observation that although there is no measurable "phosphate" in the water, an aquarium still has microalgae growth. These organic sources of phosphate can easily be converted into inorganic forms by the algae and can thus be utilized. The real trick to maintaining low phosphate levels is to minimize the inputs and maximize the removal.

Top-off water can be a significant source of phosphate, as municipal and rural water supplies often contain phosphate in a variety of forms, both organic and inorganic. Use reverse osmosis or deionization to eliminate this potential source of phosphate. Some trace supplements and salt mixes may contain measurable levels of phosphate as well, though the quantity is small compare to inputs from feeding. Avoid adding excessive amounts of liqui food supplements. Also, always check your new activated carbo to make sure it does not release phosphate (see chapters 5 and 9 for additional info on phosphate control).

As a side note, excessive feeding of the fish can lead to accumulation of phosphate in the aquarium, but this should not encourage the aquarist to keep anorexic fish! Although some aquarists advocate minimal feedings of the fish in a reef tank, thi is not always a wise practice. Unless adequate food is available, some fish will slowly waste away. Careful, moderate, but freque feedings of high quality foods, should be carried out several tim a week, if not daily. Active fish such as *Pseudanthias* spp. requir small feedings several times a day to maintain their health. Tang and surgeon fish may not get enough vegetable matter to eat in a reef tank, particularly smaller aquariums, and these fish should b provided with a constant supply of vegetables such as fresh seaweeds, leaf lettuce, bok choy, zucchini or (uncooked) seaweeds such as nori, from Asian grocery stores. In larger aquariums with strong illumination and deep substrate, the grow of algae and the development of populations of crustaceans and worms provides enough food for many fish to thrive with little o no feeding. Each aquarium differs in its food input requirement and ability to generate live food.

Oxygen

The saturation level of oxygen in the water depends on the temperature, specific gravity, and concentration of oxygen in the atmosphere over the aquarium. Aquariums maintained at high altitude will have slightly lower oxygen levels. For most reef aquariums within the normal specific gravity and temperature rang given here the oxygen level at saturation is approximately 7 mg/L.

The oxygen level should be maintained close to saturation or a li higher. Supersaturation levels of oxygen are good to a point beyond which they are harmful. Harmful, high levels of oxygen c be reached within coral and anemone tissues as a result of photosynthesis (Dykens and Shick, 1984). They are not obtainab in an aquarium unless pure oxygen is administered under pressu in a contact chamber. Pressurized cylinders used for ozone conta known as oxygen reactors, should never be used with pure oxyg since toxic supersaturated oxygen levels are easily reached with this kind of device (Burleson, 1989). Photosynthesis during the day elevates the oxygen level in the water above saturation. At night, respiration tends to reduce it below saturation. Circulation is especially important to keep the oxygen level from falling far below saturation at night. Good circulation prevents stratification of the water, and exposes the volume to the surface where gas exchange occurs. Protein skimming and surface skimming are also means of maintaining the oxygen level at saturation. Surface skimming removes the surface film that prevents gas exchange across the water surface, and exposes the thin layer of skimmed water to air as it swirls down the drain to the reservoir below the aquarium. Protein skimming removes compounds that would break down and consume oxygen (BOD and COD).

Oxygen testing is not necessary in the general monitoring of the aquarium water quality, since the level tends to remain within the same daily limits. It is lower at night when the plants aren't photosynthesizing, and higher during the day when they are. This also occurs in the natural setting on coral reefs. In aquariums, the oxygen level changes appreciably only in the event of putrification from a dead organism, or the fouling of the bottom substrate.

Redox

Redox is an abbreviation for reduction/oxidation, referring to types of chemical reactions. Oxidation and reduction reactions occur by electron transfer and by atom transfer. Reducing compounds or agents are electron donors, and oxidizing compounds or agents are electron acceptors. The process of oxidation involves a loss of electrons to oxidizing compounds, and the process of reduction involves a gain of electrons from reducing compounds. The transfer of oxygen atoms may also occur in redox reactions. Oxidation occurs with the gain of an oxygen atom, and reduction occurs with the loss of an oxygen atom.

In aquariums, redox potential is useful to measure because it gives a relative measure of the water purity. The measurement is a potential based on the sum of redox reactions occurring in the water. The higher the redox value, the greater the potential for oxidation to occur, the lower the number, the greater the potential for reduction. Reduction occurs in deep substrates, where natural redox values are negative. Measurements of redox potential in the ocean vary from 350-400 millivolts (Moe, 1989) to as low as 160-190 millivolts (Wilkens and Birkholz, 1986). Caution is advised in any comparisons due to differences in measuring conditions, technique and equipment used. Recommended aquarium redox levels range from 350-450 millivolts but each aquarist must go by the appearance of his/her own aquarium. Differences in probe placement, frequency of cleaning the probe, bioload, fauna composition, etc. all affect redox readings. It is not so much the value that is important, but the appearance of your aquarium inhabitants and the trend in measured redox values. If you notice that the redox begins to decline rapidly, this is a sign that something is fouling in the aquarium and you should investigate the cause.

Aquarists should not forget that the appearance of the animals is the most important quality of the system. There is a common tendency for aquarists who have redox meters to become more concerned about the numbers on the digital display than the animals in the aquarium. It makes no sense to worry about redox numbers when the aquarium is fine. Don't ruin a good thing by striving for levels you hear are "best". Far more important is the pH and alkalinity level. A high redox level which might please the aquarist could occur with low pH and alkalinity, which does not please the corals.

Control of redox is achieved naturally through the techniques described in chapter 9. Artificial manipulation of redox is achieved through the use of ozone and a redox controller. See the topic ozone, this chapter for more detail.

Calcium

Calcium is the primary building block of the corals, clams, calcareous algae and many other organisms that we grow in our aquariums. Without adequate supplies of calcium these organisms will not flourish and most will eventually waste away and die. Calcium levels are measured in milligrams per litre (= parts per million, ppm). Calcium levels in natural seawater range from 380 to 480 mg/L depending on location. In our aquariums calcium levels should be maintained between 350 and 500 mg/L to ensure proper growth of the calcifying plants and animals. See calcium additions, later in this chapter, for information on maintaining calcium levels.

Water Changes

Water changes should be performed, in our opinion. The principal reason for changing water in the old days of marine aquarium keeping was to reduce the accumulation of nitrate and refractory organic substances, and to maintain adequate pH. Natural biological processes that occur in an ecosystem aquarium prevent the accumulation of nitrate and limit the drop in pH and accumulation of organic material. The use of protein skimming and activated carbon further removes organic substances from the water. Additions of calcium prevent the loss of this essential ion and help buffer the pH. With all of the "old" problems solved, it is not surprising that some aquarists have tried maintaining aquaria without water changes. Yes, it is possible to maintain a healthy aquarium for very extended periods (years) without water changes, but the value of this practice appears to be more a sense of dogmatic mental satisfaction than anything else, and to recommend it, especially to beginners who are less able to distinguish a situation which dictates the need for water change is irresponsible, in our opinion.

We recommend that the aquarist find a comfortable routine and follow it. For reef aquariums we believe the less you "get your hands in there" when things are doing well, the better. From this you can draw the conclusion that we don't recommend frequent water changes (i.e. daily or weekly ones as some aquarists perform). We recommend ballpark monthly (about every one, two or three months depending on your memory) that you change 10 to 25% of the water, all the while keeping up with the make-up water to replace evaporation, and the additions of calcium and trace supplements. When performing a water change, one can use the opportunity to siphon out detritus from between the rocks. See appendix B for additional information on general maintenance procedures.

Larger water changes can be performed in the event of an emergency, but for general maintenance, 10 to 25% changes are all that is necessary, and they are less disturbing to the aquarium. Don't forget too that some saltwater is removed by protein skimming, so that the tendency over time is for the aquarium to become less salty (due to freshwater evaporation top off). A little salt mix added to the make-up reservoir on occasion makes up for the loss.

Large water changes can leave some of the marinelife temporarily exposed to the air, simulating low tide on the reef flat. Though generally this is not harmful, some organisms may actually tear apart from their own sagging weight, or retain air bubbles that can cause tissue damage. Smaller water changes avoid this problem. If your sump (the reservoir below or behind the aquarium into which the surface skimmed water flows) has a volume 10 to 15% or more of the tank volume, then water changes can be performed by shutting off the pump(s) returning water to the display tank, and changing all of the sump water.

There is a way to perform a large water change without exposing the organisms to the air. Roger Bull explained to us the technique he uses when he makes a large water change. He prepares seawater to a temperature about 5 °F (2 °C) cooler than the tank. With the circulation pumps in the display aquarium turned off, the new water is pumped into the aquarium slowly at the bottom, where it stays because it is denser than the warmer, old water. The old water at the surface flows over the surface skimming overflow and into the sump. The sump water is simultaneously pumped out to a drain.

Aquarists often wonder whether ocean water is better to use than artificial seawater or vice versa, and many articles or comments in articles have expressed strong opinion about the subject. We are quite amused at comments regarding the dangers of using real seawater. We suggest that individuals who claim that seawater is dangerous to marinelife immediately go out and do a water change on the ocean! In our experience it makes little difference whether you use real seawater or artificial seawater. Many successful reef aquariums have been established and maintained with either solution.

Natural seawater can be collected from coastal regions, along the seashore, at inlets, or from bays near ocean inlets. Pollution can be a problem, but it is obvious when it is. If the marinelife where you plan to get your water looks ill (or dead), if there is a "red tide" or if there is an oil slick on the surface, don't collect the water. If the collection site is near the ocean, even in populated coastal areas, pollution is generally not a problem. We know of successful reef aquariums that utilize water from Sheepshead Bay, New York, for instance, where pollution certainly exists, but apparently not to such an extreme that corals or other delicate reef fish and invertebrates are affected by it.

The choice to use natural seawater is often an economic one. Natural seawater is cheaper to use than artificial seawater if you live by the sea. It can also be a matter of convenience. For some people, collecting 50 gallons of seawater and carrying it home is a real chore, and not worth doing. For others it is no problem at all. For most people, there is no choice. The ocean is too far away.
If you decide to use natural seawater, there are some precautions

about its use that you must know. Check the specific gravity of the seawater before you use it. The salt content of coastal waters varies as a result of freshwater springs, run-off, or evaporation. Full strength seawater is saltier than most aquariums, at a specific gravity of about 1.025. Some localities have even higher specific gravity. Another concern is that seawater contains live plankton. Proponents of its use declare that plankton is one of the benefits of natural water not offered by artificial seawater. Opponents, however, say that plankton may include parasites. In our experience, parasites are most often introduced with fish, not seawater. It is possible to introduce harmful bacteria with seawater, particularly if it is collected near a sewage or storm-water outfall, but disease introduction is rare. A real concern about plankton in natural seawater is that it dies rapidly if the water is not aerated. If you plan to use the seawater immediately, it must be aerated if the temperature is warm and your trip home will last more than an hour. If you live right by the sea in an area where the water is clean, you can collect the water and promptly pour it right in the tank without filtering or aerating it. Of course the new water's temperature and specific gravity should be close to the tank values to avoid shocking the inhabitants. You can filter the water as it is collected to remove the plankton before the water enters the bucket. This is accomplished by pouring the water through fine micron mechanical filter bags. It is not necessary to filter the water this way, but many aquarists prefer to do so, and it does prevent the water from becoming foul with dying plankton. If the water is neither filtered nor aerated, it will foul as the plankton dies. After about a day it will have a slight foul odor. It is unsuitable to use when it is fouling this way, as the bacteria which bloom to decompose the dead plankton, and ammonia and other toxins generated during the process, can harm fish and other marinelife. If the fouling water is stored in a closed container for two weeks or longer it becomes safe to use. During storage the bacterial bloom crashes and the water becomes clear, while brownish mulm settles on the bottom. See Moe (1982, 1992) for additional techniques for collecting and treating seawater.

In closed system aquariums, artificial seawater gives results on par with natural seawater, in our opinion. As the marine aquarium hobby has grown, the number of salt manufacturers also has grown. An important feature to look for when selecting a brand of salt is proper blending. The salt should have uniform crystal size throughout the mix. If there are chunks and bits of different size, then the salt is not homogeneous, and the resulting solution may vary. Aquarists argue about what brand of salt affords them the best results, and we cannot offer recommendations in this regard since the quality of salts has improved so much over the years that good results can be had with any brand. This doesn't mean that all brands of salt are the same. They aren't. The different brands are different, and you may find that you like the results with one brand better than the results with another. Use whatever salt that seems to make your aquarium's inhabitants thrive, and stick with it. All seawater (including natural) in closed systems does decline in quality with time, depending on the filtration, and load of marinelife. Periodic water changes and the addition of major, minor, and trace elements are necessary for long term maintenance of optimal water quality no matter what brand of salt is used.

A factor that does significantly affect the quality of the artificial seawater is the type of water you use to make it. Tap water and well-water varies substantially in composition. Therefore, the results with artificial water can be compromised when the source of freshwater is contaminated with heavy metals, nitrate, phosphate, organic compounds, and toxic elements. Sometimes the product is enhanced by trace elements or calcium in the freshwater. More often the impurity of the freshwater is not advantageous. To avoid the possibility of adding an unknown detrimental factor with the water, it is best to purify it. Reverse osmosis or deionization systems available through aquarium industry manufacturers, or through home water purification companies, make the water safe to use (and drink). Home water softening systems, used primarily to remove hardness from well or tap water, are generally unsuitable for the filtration of water to be used in marine aquariums due to the high level of sodium ions they may add to the water.

The water from a reverse osmosis unit is usually acidic, with a pH of about 6 or less. It is not necessary to buffer this water before mixing it with salt. The salt already contains buffer. Passing the acidic purified water over aragonite sand, however, may bring the pH up. The benefit of this practice is that small quantities of calcium and strontium carbonate may dissolve in the water.

Element Additions

Replenishment of the ionic forms of certain elements is required to maintain an environment ideal for the growth and reproduction of marinelife. Many organisms utilize specific elements for the formation of their tissues and skeletons, and for biological processes. In so doing they alter the composition of the water in a closed system. In addition to extraction from the water by microorganisms, animals, and plants, some major, minor, and trace elements are lost by means of the different forms of filtration employed. These include protein skimming, activated carbon and other chemical filter media, and algal turf scrubbing. A less recognized but still important depletion of elements occurs through adsorption on particulate matter floating in the water (Segedi, 1976). Since trace elements can be concentrated by bubbles or particulate matter, it is possible that the unexplained high concentration of trace elements in some invertebrates' tissues is achieved when they filter and ingest the bubbles and particles from the water. Still another means by which elements are lost is adsorption on the materials with which the water comes in contact, including even the walls of the aquarium! (Segedi, 1976)

Sometimes aquariums have significant element replenishment occurring without specific effort by the aquarist. For example, some elements can enter the system from the air, from food, and the use of un-filtered freshwater for make-up of evaporation. These sources can sufficiently replenish some elements and even create excesses of certain others (i.e. copper from copper pipes).

The replenishment of various major, minor, and trace elements lost from the system is a subject of much debate, and we do not intend to provide the final word. While some aquarists contend that their aquariums do not suffer for the lack of replenished trace elements, it is our opinion that these elements must be replenished somehow, and the best we can do is to provide a description of the many techniques and elements used.

Calcium Additions

So much confusion and discussion surrounds the notion of adding calcium, but the methods are pretty straightforward. Some aquarists are trying to think too hard about an easier way to add calcium, rather than accepting the simple steps outlined by numerous authors. It is our intention here to plainly suggest the proper means of maintaining a good calcium level, and to avoid confusing you with chemical formulas and theory. Imagine you are cooking, if you will, and just follow the recipe please.

First, a little background is in order. Away from the influence of land run-off, calcium in natural seawater is generally at a concentration of 400 mg/L. The level in the vicinity of coral reefs and in coastal waters tends to be slightly elevated, at around 424 to 480 mg/L (J. Sprung, pers. obs.). This simply means the aquarist should strive to maintain the calcium level in a reef aquarium at 400 mg/L at least, and ideally closer to 500 mg/L. In our experience successful reef aquaria can be maintained with deficient calcium levels (less than 400 mg/L), but only when the carbonate hardness (alkalinity) is normal or high. Troubles occur more as a result of low carbonate hardness than low calcium level. High alkalinity maintains a high pH, resulting in the ratio of calcium and carbonate ions in the water being closer to saturation, which makes calcification more likely. More on this in a moment.

The elevated calcium level around coral reefs seems a paradox; the place where calcium is being removed from the water most rapidly has a higher level of calcium in the water than the nearby open sea. The explanation for this phenomenon is both simple and complex. The general explanation for the elevated calcium level around coral reefs centers around the fact that the amassed skeletons of the calcifying organisms, as rock, sand and gravel, are a reserve of calcium that is constantly being eroded mechanically and chemically, elevating the calcium level in the immediate environment and making it easier for further calcium deposition to occur. The various mechanisms by which this occurs are many and complex, but what is known so far can be summed up with a few examples.

The motion of water and sand, and the grazing action of fishes and invertebrates grind the limestone of the reefs into fine particles. The tumbling of sand as the surge tosses it to and fro also generates very fine particles. Though basically insoluble as calcium carbonate, these tiny particles may still dissolve by means of carbonic acid formed from the respiration of microorganisms that are attracted to the particles.

Grazing fishes not only grind the limestone, but the sand passing through their gut is exposed to digestive acids that dissolve some of it. Numerous invertebrate and vertebrate grazers consume calcareous algae as well, and digestive acids could liberate calcium this way too. Furthermore, numerous invertebrates, particularly sea-cucumbers, ingest great volumes of sand to consume the microorganisms living on it, and so they too expose the sand to digestive acids.

The microorganisms and algae living in the sediments and rocks may emit special enzymes or acids that liberate phosphate adsorbed on the substrate, and in so doing, they may also liberate some calcium (see chapter 2). Deep within the sediments and rocks the accumulation of CO_2 from the breakdown of organic material, and respiration by crustaceans, worms, bacteria, and other microorganisms further dissolves the limestone. Boring organisms, in addition, employ both mechanical and chemical means to erode the limestone coral skeletons and shells that they inhabit.

Freshwater from rain, rivers, etc. passing over limestone picks up some calcium, particularly where peat or excess CO_2 deep below the ground acidifies it. This process can elevate calcium levels around coastal reefs or on reefs near large islands with accumulations of limestone. This process approximately mimics the method of water replenishment with calcareous water that we will describe shortly.

Finally, reef shallows exposed to intense illumination and ocean breezes evaporate water, and the resulting slight elevation of specific gravity compared with the surrounding deep sea reflects a slight increase in the concentration of all ions, including calcium. Comparisons of measurements of the calcium ion concentration in seawater must include a comparison of the salinity level.

In typical reef aquaria, when no calcium additions have been made, the calcium level usually falls to 250 to 350 mg/L between water changes, depending on the abundance of calcifying organisms and losses due to precipitation. If you have never added calcium to your reef aquarium, you can expect the level to be somewhere in this range. We mentioned under the topic of specific gravity that calcium ion concentration is lowered when the specific gravity is lowered. The standard value of 400 mg/L calcium ion is for full strength seawater, at a specific gravity of 1.025. Since most aquariums have specific gravity below full strength seawater, the calcium can be maintained at slightly less than 400 mg/L. At 400 mg/L or more, this water has more than 100% natural seawater concentration of calcium ion.

Before considering how to get the calcium level up, it is best to invest in a good calcium test kit. Actually, you may follow our "recipe" for adding calcareous water, and never use a test kit, but you will be sacrificing your confidence level and a real understanding of the affects of these additions to your particular tank. Some of the commercially available calcium test kits are sold as calcium hardness kits. One must be sure that a conversion factor has been included with the kit's instructions since these kits measure calcium ion, but express it in terms of calcium carbonate in mg/L. The standard conversion is to multiply the calcium carbonate value by 0.4. Also, some kits are not designed to measure high calcium levels (this is indicated in the instructions). Measurement of water with over 400 mg/L may require that the test sample be diluted. A dilution using pure freshwater should be performed, usually by half, which will require a multiplication of the final value by 2. Performing such dilutions additionally saves on the consumption of the titrant. Generally, calcium test kits are designed to give a range of calcium, but the accuracy is not precise, particularly for seawater. Seawater interferences can lead to false high readings. Diluting the sample helps, but the variable concentration of magnesium, strontium, and other elements in the water makes precise readings impossible. A calcium value in excess of 550 mg/L when the alkalinity is at 7dKH or higher, and the pH is above 8.2 is certainly a false reading, since this suggests highly supersaturated calcium carbonate, which should fall out of solution at the high pH. If the calcium level is really so high, the alkalinity must be very low. On the subject of test kits, it is also necessary to use a carbonate hardness (alkalinity) test kit to monitor that parameter as well, since calcium additions do affect the alkalinity, as we will shortly explain.

Several methods exist for raising the calcium level in the tank, but we will focus attention on the long term maintenance of the calcium level via additions of calcareous water. We will also discuss the use of calcium chloride, as well as water exchanges to elevate the calcium level. There are also numerous calcium supplements on the market with wondrous claims about their efficiency. It remains to be seen how well these products will perform over long term use.

Calcareous water, a.k.a. limewater, or "kalkwasser" was first suggested as a means of maintaining calcium levels in reef aquariums by Peter Wilkens (1973). It may be used as the sole source of make-up water for evaporation, or it may be added diluted with the make-up water, or merely on occasion. The best aquarists in Europe generally agree that using kalkwasser exclusively as the make-up water is the most practical and successful way to maintain the calcium level. Kalkwasser is most advantageous because it does not add anything to the water that accumulates, it precipitates phosphate from the water thereby enhancing corals' ability to deposit calcium, and it maintains alkalinity by combining with dissolved CO_2 or carbonic acid in the aquarium. Furthermore, OH^- ions added with kalkwasser neutralize organic acids, and thus help to maintain the alkalinity and pH. Wilkens (pers. comm.) has observed that kalkwasser addition enhances protein skimming, and he believes this results from the pH increase and reactions of calcium hydroxide with fatty acids and other organic compounds attracted to the air-water interface.

You can make the kalkwasser with water purified by reverse osmosis and/or deionization if your tap water is very polluted, as it often is in urban areas. The use of a purification system prevents the addition of plant nutrients with the make-up water, though some phosphate can be precipitated with calcium in the alkaline solution. Some aquarists do not need to use purified water, but many do.

Kalkwasser is a saturated solution of calcium hydroxide or calcium oxide in water. Either of these compounds may be used, and both barely dissolve at all. Only about 1.5 g (0.05 oz.) will dissolve in a liter of water, which means about 6 g (0.2 oz.) per 4 L (1 gal.), and even this requires vigorous stirring. Adding excess calcium hydroxide or oxide to the kalkwasser helps insure that a saturation state of calcium is maintained longer since the reaction of dissolved calcium hydroxide with carbon dioxide in the fresh water causes precipitation of calcium carbonate, lowering the amount of calcium ions dissolved in the water. The recommended procedure is to add excess calcium hydroxide or oxide, mix the solution vigorously and then allow it to settle for several hours until you see the clear, saturated "kalkwasser" solution above some undissolved white calcium hydroxide settled on the bottom (Wilkens, 1973). Siphon off the clear solution for use, and either discard the undissolved portion or add more water and calcium hydroxide or oxide to make another batch. The calcium will react with carbon dioxide in the air to form a calcium carbonate crusty skin on the surface of the mixing container, and some calcium carbonate will fall out of solution and settle on the bottom along with the undissolved excess calcium hydroxide. The older the solution, the higher the concentration of insoluble calcium carbonate on the bottom of the container.

This procedure works fine, but may be deficient for large aquariums that have low rates of evaporation. We wish to emphasize that it makes a big difference whether you use fresh, milky, saturated kalkwasser compared to clear, old, undersaturated solution. It is common experience that old kalkwasser does not maintain the calcium level. Old kalkwasser may not be a saturat solution because CO_2 from the air reacts with calcium ions to yie insoluble calcium carbonate. Furthermore, mixing the powder th settles on the bottom of the container in old kalkwasser may sometimes cause the alkalinity to drop in the aquarium. This powder is not undissolved calcium hydroxide, it is calcium carbonate which has precipitated from the water, and it will not r dissolve unless exposed to very low pH. It is hypothesized that when these fine particles of pure calcium carbonate are added to the aquarium, the alkaline micro-environment around them mak them a "seed" site for crystallization of calcium carbonate in the aquarium, and therefore causes a drop in alkalinity. Making only small quantities of kalkwasser so that the solution is used within a day or two, helps prevent this accumulation of a precipitate. The most trouble free method for adding kalkwasser is the system devised by Alf Jacob Nilsen that we will describe shortly. It provides highly saturated kalkwasser as the make-up water, and prevents accumulation of precipitate on the bottom. Aquarists wh have tried mixing the water in a calcium reactor by using a submersible powerhead have discovered that the pump soon becomes jammed with calcium deposits, or permanently damage by sharp flakes of calcium carbonate from the surface skin that break off and sink to the bottom. Using an air bubbler to mix the kalkwasser is an unsuitable practice because it continuously introduces CO_2, and causes most of the calcium to fall out of solution as calcium carbonate.

One might reasonably ask if it makes a difference whether one us calcium oxide or calcium hydroxide. Well, ultimately the results with either are the same, so no, it makes no difference. Calcium oxide is sometimes slightly cheaper than calcium hydroxide, and i also a little easier to handle because of its clumpy consistency, whereas calcium hydroxide is like talcum powder but worse, and the dust is very harmful to breathe. Still, calcium hydroxide dissolves more readily than the oxide, which simply means that calcium oxide requires a greater mixing effort. On the subject of handling either substance, a dust mask and gloves are good precautions. These substances are caustic, though getting some or your hands is quickly remedied by rinsing with running water. Keep them out of reach of the kids, of course. Calcium hydroxide, calcium oxide, and calcium chloride are available through chemic supply companies, though sometimes it is not so easy to order them to be sent to your home address. Several aquarium industry companies now also supply calcium hydroxide and calcium chloride to pet stores in dry and liquid form.

Aquarists may also wonder if it matters what grade of calcium hydroxide, oxide, or chloride is used. In our experience it is not worth the extra expense to use the purest grade. *Trace* impurities (i.e. strontium, magnesium, iron, silicate) are beneficial. Avoid grades with *high levels* of magnesium, phosphate, or heavy metals.

Another precaution is most important to understand: don't add too much to the tank at once! Kalkwasser has a pH of nearly 12, and even a little bit will raise the pH in the aquarium temporarily...a lot will raise the pH too high and could injure or kill the fish. If you have an automatic CO_2 dosing system installed, this is less of a concern since the pH controller will cause the administering of CO_2 to counter the rise in pH from the kalkwasser. If you don't have a CO_2 system, then it is best to add kalkwasser slowly, either by dripping it in or by means of an automatic water make-up system. If you choose to go with a drip method, dripping the kalkwasser in at night creates the least disruption of the pH, since there is more CO_2 in the water after the lights have gone off, when the algae and zooxanthellae stop photosynthesizing. If you are using an automatic water make-up system with a level sensing switch and dosing pump, you may simply add the kalkwasser to the freshwater reservoir. Nilsen (1990, 1991) describes a simple device that you can build for providing high-quality (fresh) kalkwasser as part of an

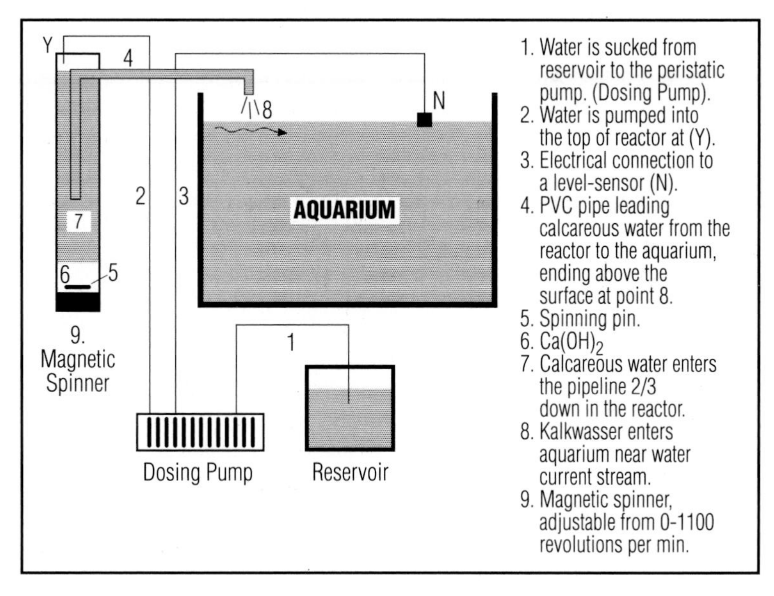

Diagram 8.1
Calcium Reactor.
After Nilsen 1992

automatic water make-up system (See figure 8.1). The design incorporates a magnetic spinner to keep suspending the excess calcium hydroxide in the reactor tube, thus maintaining the saturation state. The dosing pump is not in the reaction tube, but in a separate, pure freshwater reservoir, which eliminates the possibility of the dosing pump getting jammed by calcium deposits. We have also seen good calcium reactor dosing systems that use a level sensing switch connected with an air pump and diaphragm to mix and or/dose the kalkwasser. The air driven diaphragm displaces the water and causes it to flow out of the reactor until the level in the sump causes the switch to shut off the air pump.

Depending on the use of calcium by the organisms, the evaporation rate of the aquarium, and the quality of the kalkwasser used, the long term calcium level in the aquarium may vary from well below 400 to slightly over 500 mg/L, when kalkwasser is used as the make-up water. If you are starting to add kalkwasser to a large, established reef tank that is deficient in calcium and has a low rate of evaporation, it may seem to take "forever" to get the calcium level up, considering only a little over a gram per liter will be added with the make-up water while at the same time the invertebrates are extracting the calcium from the water. Some aquarists have been disappointed with the results because the calcium level did not rise, but continued to fall steadily. This situation occurs with large aquariums that have very little evaporation, and also with the use of undersaturated kalkwasser. To solve this problem the aquarist can use excess calcium hydroxide in the make-up water (e.g. 3.0 g/L) and keep the solution mixing to achieve a high saturation level of calcium. This is the purpose of Alf Nilsen's calcium reactor. Increasing the rate of evaporation when kalkwasser is used for make-up water will further help maintain the calcium level. It is also perfectly acceptable to add some calcium chloride into the container with the calcium hydroxide to make up for any deficiency. There is nothing wrong with being patient and allowing the calcium level to rise slowly, but there are other ways to get it where you want it quicker, though some caution is in order.

Water changes with a salt mix that produces water with a high calcium level are a practical means of raising the calcium level quickly. In fact, water changes alone can be used to maintain adequate calcium levels, but we feel this is less practical than calcium additions, and two important advantages of kalkwasser are its ability to precipitate phosphate from the water and maintain alkalinity.

Calcium chloride is another "quick-fix" solution to raising the calcium level, and some hobbyists use it alone to maintain the level. The chief advantages of calcium chloride are that it dissolves quickly in water, a lot of it will dissolve in the water, and the addition of calcium chloride solution will not raise the pH. There are two disadvantages to using calcium chloride. One is that such additions raise the chlorinity of the aquarium's water, meaning that in time the balance of sodium and chloride ions in the aquarium water is off. The severity of this imbalance depends on the frequency of water changes. The other disadvantage is that rapid elevation of the calcium ion level often causes the carbonate hardness and pH to fall, requiring the subsequent addition of a carbonate buffer. This leaves the aquarist baffled because the calcium level rises only a little, much less than it should have based on the weight of calcium chloride added, and the carbonate hardness falls. Kalkwasser additions also may cause this dKH drop phenomenon. If a large dose is added to the aquarium at once, and no CO_2 system is used, the resulting rise in pH will cause precipitation of both calcium carbonate and magnesium carbonate. If the additions are made slowly, as with an automatic water make-up system or drip system, this is not a problem. In the case of calcium chloride, if a lot is added at once, the influx of calcium ions combines with free carbonate ions in the water, resulting in both falling out of solution as calcium carbonate (it could look like it is snowing in your tank!). A similar situation can occur if you do a water change with a salt mix that has a high calcium content (R. Graff, pers. comm.). When the calcium level reaches 550 or higher at a pH of 8.2 or higher, supersaturation has been reached and white precipitation will occur. To sum up: go slow. While no immediate toxic effect results from adding calcium chloride solution all at once to the aquarium, if the same quantity is administered slowly over a period of days, the carbonate hardness will be more stable.

Once the calcium level is above 400 mg/L, it is best to begin using kalkwasser as the make-up water. In time this becomes routine: you'll be mixing kalkwasser every week, and occasionally checking the long term effect on the carbonate hardness and pH.

Some aquarists are fortunate to live in an area where the tap-water or well-water is extremely hard, rich in calcium carbonate, and not polluted. It is possible to maintain adequate or high levels of calcium in the aquarium by using this water. It is also possible that acids from respiration and metabolic waste of organisms in deep sand substrates in the aquarium can liberate some calcium carbonate. See the topic natural systems in chapter 5 for further details.

We briefly mentioned the use of a CO_2 dosing system. In marine aquariums, these systems are primarily used to control pH, as in the example we gave, when kalkwasser is used for top-off water. These systems were introduced to the aquarium hobby for the purpose of enhancing plant growth in freshwater aquariums. Administering CO_2 to the water when plants are photosynthesizing and it is scarce, achieves a more stable pH and more rapid plant growth. A pH controller is used to switch a magnetic valve, allowing the slow release of CO_2 from a cylinder via a pressure reducing regulator. The CO_2 is administered into a small reactor with either a coil pathway or inert packing media inside to assist the uptake of the gas into the water. The reactor is usually fed aquarium water by means of a small pump. These systems are not necessary for reef aquariums, but they do afford a margin of safety in preventing the pH from rising too high if a failure of the top-off dosing system occurred and too much kalkwasser was added to the aquarium. Of course if the CO_2 system malfunctions, and too much CO_2 is administered, the resulting pH drop would be fatal. Such mechanical failures are extremely unlikely, but possible. So we offer a word of caution: Although mechanical devices such as automatic water make-up and pH controllers can maintain a more stable environment and make aquarium keeping simpler, the less complex your system is mechanically, the less prone it will be to equipment malfunction. Also, if you decide to use such automated equipment, be sure that it is well made, by a reliable manufacturer. Cheap is dear! A final word of caution about CO_2 dosing: Just as in freshwater aquariums, the administering of CO_2 to saltwater aquariums can enhance plant growth. Stimulation of rapid algae growth is a possibility, but with control of other nutrients such as phosphate, nitrate, and silicate, and with sufficient herbivores, the growth of undesirable filamentous and slimy algae is limited (see chapter 9).

A CO_2 system can also be used to stimulate growth of the desirable coralline algae, by incorporating it in another device for adding calcium to the fresh top-off water, the so-called "lime reactor". In this system a reactor cylinder is also used, but instead of inert media for packing, the reactor has calcium carbonate chips or aragonite gravel. The CO_2 acidifies the freshwater trickling over the calcareous substrate in the reactor, causing some of the calcium carbonate to dissolve. The result is freshwater rich in calcium carbonate. A pH controller can be used to maintain the pH of the fresh water low enough to keep dissolving the calcium carbonate. The authors have not compared this method with the more widely practiced use of kalkwasser, but suspect that it works well.

If you are already maintaining a nutrient poor system, but have never added calcium, you will probably notice a change in your aquarium as a result of these additions: pink and purple encrusting coralline algae will spread and coat the bare surfaces, and your corals will continue to grow at an amazing rate. Please refer to the section on calcification in corals, chapter 3, for additional information about carbonate hardness and pH.

This rock photographed in a reef aquarium at Jago Aquaristic in Berlin has a thick coating of coralline algae and attached colonial anemones and soft corals. It looks like live rock and is live rock, but it was not taken from a reef. The stone is mined limestone from Germany, and the growth on it all occured in the aquarium. With the addition of kalkwasser, strontium, and trace elements, coralline algae can be encouraged to proliferate. J. Sprung.

Strontium

Strontium is a minor element occurring in sea water at a concentration of about 8 to 10 mg/L. It is chemically very similar to calcium, and many organisms incorporate it in their skeletons along with calcium. Its addition to reef aquariums aids the growth of corals, coralline algae, and other organisms that lay down a calcareous skeleton or shell, including tridacnid clams. The survival of *Acropora* species in closed system aquariums is very dependent upon strontium additions (J. Sprung, pers. obs.). Swart (1980) also found that increasing strontium levels 2 and 10X above natural levels increased the growth rate in *Acropora*. The role of strontium in calcification is unclear, however, strontium can be used in place of calcium, to build coral skeletons. Yet, the mechanisms by which strontium is incorporated into the skeleton appear to be different than those involved in calcium deposition (Ip and Krishnaveni, 1991).

Peter Wilkens was the first aquarist to discover and report the benefits of additions of strontium to the reef aquarium, and his recipe and dosage are the standard employed today by reef keeping aquarists around the world. Numerous aquarium industry manufacturers now supply strontium supplements, and we encourage hobbyists to purchase these from their local pet dealer, thus avoiding the hassle and potential health risk involved in preparing the solution.

Making your own stock solution:

Caution! Strontium is hazardous to your health. Handle only with gloves, wear a dust mask and avoid contact. Strontium is added to the aquarium in the form of an aqueous solution of strontium chloride. A 10% (approximately) stock solution is made by adding 25 grams of strontium chloride hexahydrate ($SrCl_2 \cdot 6H_20$) to 250 mL of purified fresh water. The stock solution is added at the rate of 2 mL per 100L of aquarium capacity per two weeks. Significantly more can be added, especially when there is heavy growth of coralline algae and many stony corals. When experimenting with the dosage, make increases slowly! Adding too much strontium to the aquarium can be detrimental. It is possible to over-dose it, and we have seen that the result is sick or dead corals!

It is acceptable to dose the tank initially as if it contained no strontium. Remember that it is a 10% solution of strontium chloride hexahydrate, not just strontium. Calculate how many mL of the solution must be added to achieve approximately 8 mg/L of strontium in the whole aquarium. For example, with the recommended dosage of 2mL per 100L, you are adding 65 mg of strontium per 100L each week. This means you are only replenishing 0.65 mg strontium per liter weekly. Therefore, to be sure you are starting with at least 8 mg/L in the aquarium, you need to add 25 mL of the 10% solution per 100 L as an initial dose. The dose should not be added all at once, but divided into four portions to be added over the course of four days. Thereafter, follow the regular dosage of 2mL per 100 liters per week, beginning one week after finishing the initial dosage.

Trace elements must be replenished to maintain healthy growth in a reef aquarium. J. Sprung.

Trace Element Additions

Iodine

Natural seawater has about 60 parts per billion of iodine, including all forms such as iodide, iodate, and organically bound iodine. This value is the standard for natural oceanic seawater, away from coastal influences. We suspect that iodine values in coastal water are somewhat higher, as a result of concentration of iodine by algae in their tissues, and subsequent release of organic iodine when the algae decompose. If this is true, then oceanic reefs may be more iodine limited than coastal reefs.

Iodine (as iodide) may be added in the aquarium in several different forms. The most commonly used form is potassium iodide. The stock is prepared by adding 25 grams of potassium iodide to 0.5 L of pure water. This is added to the aquarium at a rate of 0.5 ml per 100 L per two weeks (Wilkens, 1973). Other options for sources of iodine include Lugol's solution, organic sources of iodine, and tincture of iodine. Organic sources of iodine, as found in some supplements, afford continuous release of iodide without risk of overdose. Lugol's and tincture of iodine are solutions in which pure iodine has been dissolved in a solvent of potassium iodide solution. Tincture of iodine contains some alcohol in addition, but this is not a concern or disadvantage considering the small dosages. Lugol's solution, also known as "strong iodine" solution, was brought to our attention by John Burleson and Merrill Cohen. They both found that it stimulated growth in *Xenia* species, and helped prevent common "crashes" of this soft coral. Interestingly, they both observed that potassium iodide alone did not have the same effect. It is likely that the benefit of the mixed iodide/iodine solution over just iodide lies in its germicidal property (i.e. its ability to kill bacteria or protozoans that kill *Xenia)* rather than any nutritive value. However, iodine rapidly converts to iodide in seawater, and therefore it is possible that the benefit is due to greater dosage.

While the dosage for potassium iodide is known, as described at the beginning of this discussion, the limits for Lugol's and tincture of iodine are less clear. We have safely administered one drop from an eyedropper per 80 L (20 gal.) aquarium capacity weekly. Take one drop at a time diluted with some aquarium water in the eyedropper, and either administer it over the target organism, be it a *Xenia* sp., or *Dictyota* or *Sargassum* algae, or simply add it in the stream of current from your pump return. Lugol's solution is not a common item on the shelf at your local drugstore, but most pharmacies are able to order it for you, if you are patient enough to wait. Tincture of iodine is available at grocery stores and pharmacies everywhere. There are numerous aquarium industry companies now marketing iodine supplements, and these can be bought from your pet dealer. It is safest for the hobbyist to buy these and follow the directions.

Iodine (as iodide) appears to be essential for the long term maintenance of *Xenia* and certain algae. It is beneficial to other soft corals, especially *Anthelia* spp. and Clove polyp *(Clavularia* sp.). Iodine additions are also absolutely critical for long-term success with stony corals. Unfortunately, there are few scientific papers which describe the importance of iodine to coral health, so our comments will appear anecdotal to scientists. Iodine is an integral part of skeletal proteins in gorgonians (see Ciereszko and Karns, 1973), and it may be that it plays a similar role in other hermatypic organisms.

Furthermore, iodine may detoxify active oxygen produced by photosynthesis (R. Buddemeier, pers. comm.), This may partially explain the observed benefits to corals and anemones, particularly with respect to bleaching, as we shall describe. See the topic of toxic oxygen in chapter 6 and bleaching, chapter 10.

Based on observations in our own aquaria, confirmed by our communications with A. Nilsen and D. Stüber, and prior work by Peter Wilkens, it appears that iodide is critical for the development of pigments in corals, corallimorpharia, and anemones, both the golden brown of the zooxanthellae and the green and red colours of accessory pigments. Alf Nilsen observed a link between coral bleaching and subsequent poor health with iodide deficiency. Corals that appear to be suffering from light shock can be cured, that is, their ability to adapt to the light can be restored, with iodide additions. What this indicates is that in an iodide deficient system, corals and anemones cannot easily adapt to changes in light.

Corallimorpharian "Mushroom anemones" expand and become most colourful when iodine and other trace elements are routinely replenished. J.C. Delbeek.

We have also seen greater expansion of mushroom anemones after a routine of iodine additions has been established, and we have noticed that iodine is useful for treating infections which occasionally affect zoanthid anemones. It may be administered directly over the colonial anemones (pers. obs.) and over *Xenia* (J. Burleson, pers. comm.), but not all invertebrates tolerate direct exposure. Stony corals do not appreciate direct administration of iodine, and will display their displeasure by shriveling up and exuding mucus and mesenterial filaments. Too much added to the tank at once may injure the fish and invertebrates, (pers. obs., and M. Paletta, pers. comm.), and can stimulate the growth of both desirable and undesirable algae (A. Nilsen, pers. comm.). Please see Synergistic effect of Phosphate and Trace Elements in chapter 9 for additional information on algae stimulation. So, a word of caution is in order regarding the dosage. Iodine must not be over-dosed.

The use of activated carbon media, though beneficial in our opinion, may significantly remove iodine from the water. This is not a problem since the plants and invertebrates also remove iodine, and the iodine is easily replenished with supplemental additions. Some other trace elements are also added to reef aquariums with beneficial results. These include iron, molybdenum, barium, and lithium. The exact functions of which in corals are not completely understood.

Iron

The addition of chelated iron to marine aquariums is beneficial to both plants and photosynthetic invertebrates. Only solutions made from ferric citrate or chelated with EDTA remain stable at the pH of seawater (Spotte, 1979).

Additions of iron to reef aquariums achieve a number of qualitative and quantitative effects. For the corals, anemones and other photosynthetic animals, it appears to benefit their photosynthetic partners primarily, and the result is improved colouring, growth and expansion in the host animal. The free-living algae and seaweeds also benefit, and their growth is enhanced through the addition of iron. For this reason we caution the aquarist that the addition of iron or trace element solutions containing iron should be done slowly, to achieve the benefits without the risk of encouraging excessive growth of undesirable algae. Please see Synergistic effect of Phosphate and Trace Elements in chapter 9 for additional information.

Molybdenum

The addition of molybdenum to aquariums has become popular in the past few years in North America, after reports of its use in aquariums by Peter Wilkens. Some recent manufacturers of molybdenum containing trace element supplements claim that it benefits stony corals and prevents separation of the tissue from the skeleton. In fact, no such function of molybdenum has been demonstrated. The value of molybdenum for aquariums is its use by various forms of bacteria, including cyanobacteria, nitrifying and denitrifying bacteria (P. Wilkens, pers. comm.; Howarth and Cole, 1985). Thus molybdenum is useful in both reef aquariums and fish only aquariums, for biological filtration. As with other trace elements it must be added sparingly, in this case to avoid stimulating the growth of cyanobacteria. Please see Synergistic effect of Phosphate and Trace Elements in chapter 9 for additional information.

Barium and Lithium

These two elements are deposited by stony corals in the formation of the skeleton (Livingston and Thompson, 1971; Pingitore et al., 1989), and they are included in some commercially available trace element solutions. It is not known exactly what function they might serve.

Other Trace Elements

Commercially available trace element supplements contain a variety of other elements, such as copper, zinc, selenium, manganese, etc. These elements are important for metabolic processes and pigment formation in the invertebrates and plants, and may serve numerous functions that have not been identified. Water changes help replenish the small quantities of these elements, and use of trace element supplements helps to maintain levels between water changes. We really don't know how much these elements are depleted from the water with time by the invertebrates, plants, microorganisms, and filtration methods. Water changes may further benefit the aquarium by removing excess accumulation of certain elements added with food and supplements.

Chapter Nine

Control of Nutrients and Undesirable Algae in Closed Systems

This is the topic that is the biggest stumbling block for everyone. We're told that the reef is a "nutrient poor" environment. Compared to our aquariums it is. It's hard to call the reef one environment, since there are many different zones of different reefs in different localities around the world. (see chapter 1, Reef Zonation). In general, the further out from shore, the further out in the open ocean that a reef occurs, the more nutrient poor it will be. Land run-off and the waste from birds and other animals associated with the land nutrify the water closest to shore. The natural flow of water and the sweeping action of waves also tends to nutrify the near-shore waters by depositing there the decaying plant and animal debris broken and transported from the outer reefs.

For this reason, reefs occurring close to shore or around islands near a large continent tend to be fairly nutrient enriched. These coastal reef, reef flat, and lagoon environments just happen to be where many of the corals and colonial anemones (zoanthids and corallimorphs) we keep commonly occur, which improves their suitability for our closed aquarium environments.

The decision about what kind of marine environment you wish to create depends heavily on the quality of the water with respect to dissolved nutrients. There is no moral judgment to be passed here. Optimal results for one environment are not optimal for another, that is all. For example, you can decide to maintain a tank full of many species of algae. The water in this aquarium may appear relatively nutrient poor because of the water purifying or "scrubbing" ability of the algae, but the algae themselves may contain large quantities of nutrients. These can be released back to the system by the algae both before and after they decompose.

Caulerpa can quickly overrun an aquarium unless it is kept under control through regular pruning. J.C. Delbeek.

The primary focus on nutrient control for home aquarists is management of undesirable algae. The plague of filamentous algae and cyanobacteria, the "slime" or "grease" algae, is something that takes all the pleasure out of reef keeping, and it imperils the creatures we keep. The plant nutrients that we must control include phosphate, nitrate, and silicate.

Controlling Nitrate

Much has been written about the negative affect of nitrate on the health of fish and invertebrates, but most of the statements have been based on anecdotal information and the generally accepted belief that nitrate is toxic. Little actual documentation of nitrate toxicity exists, and public aquariums as well as home aquariums have demonstrated quite plainly that fish and invertebrates not only tolerate high nitrate levels, they can thrive in water with high nitrate. It is likely that any harmful affects noted are a result of nitrate accumulation's affect on the pH. When nitrate forms, nitric acid is also formed, which causes the pH to fall. This pH drop affects invertebrates' ability to deposit calcium carbonate, because both the carbonate buffer is depleted and the pH is not ideal for calcification. Reef aquariums with live rock tend to have a naturally achieved upper limit for nitrate accumulation since some nitrate is denitrified within the rock, and the plants and photosynthetic invertebrates also remove nitrate and its source, ammonium. Many reef aquariums have naturally low nitrate levels (less than 1ppm as nitrate ion) despite good fish populations and regular feeding (inputs). Still others have persistent or chronic nitrate in the water despite low fish population and almost no feeding at all. Where this nitrate comes from and how it can be controlled is a subject with many facets.

Sources of Nitrate

Feeding

The majority of nitrate in most aquariums comes from the breakdown of food added to the aquarium to feed the fish and invertebrates. Restricting the inputs of food can effectively reduce the nitrate accumulation, but this practice should not be done to the detriment of the fishes' health. We have seen anorexic fish resulting from the psychology of their owners who fear too much food added to the tank will harm it. Please don't neglect your fish this way, it is cruel! Many tanks produce enough food (algae, amphipods, copepods, worms, crustacean larvae) to feed the fish indefinitely, so that little food inputs are necessary at all. Most of the photosynthetic invertebrates will survive, grow, and reproduce with no additions of food whatsoever. When the nitrate accumulation is managed by techniques we are about to describe, feeding can be frequent, and it improves the growth of the invertebrates. It also inhibits the fishes' tendency to sample your favorite invertebrates in their search for a bite to eat!

Liquid Foods

Liquid invertebrate foods benefit non-photosynthetic invertebrates primarily. Filter-feeders capture the tiny particles of food. They also may absorb certain organic compounds from these foods. Photosynthetic invertebrates that do not feed on particulate food (i.e. *Xenia*) may likewise benefit by absorbing specific organic compounds, including some amino acids. Other photosynthetic corals (i.e. Caribbean gorgonians) do capture particles of food and occasional feedings do boost their growth. Nevertheless, we must emphasize that these liquid food additions must be administered conservatively, for they are hardly necessary at all. Natural production of plankton, detritus, and bacteria in reef aquariums mostly eliminates the need to supplement filter-feeder diets. Ahermatypic corals do need supplemental feeding, however, and we suggest that these be fed directly, by means of a long pipette. Liquid foods should only be administered this way, so that very little wasted food is dispersed in the water.

Vitamins

Vitamin additions are another potential source of nitrate. Not all vitamins contain nitrogen, but some do. This is a minor source of nitrogen when compared with food additions, but we wish to emphasize that additions of vitamins, though clearly beneficial to some organisms, should be made conservatively.

Tap-Water, Make-Up Water

Another potential input of nitrate to the aquarium is make-up or top-off water used to replenish evaporation. This daily addition of water can provide significant quantities of nitrate to the "closed" system aquarium when the tap-water, well-water, or other source used is tainted with nitrate. Not all tap-water or well-water contains nitrate, but many urban and rural areas (especially agricultural areas) may have it in the water, often in high concentrations. The use of a water purification system for one's top-off water (and drinking water!) is a good means of insuring the water's purity. Reverse osmosis and deionization systems are readily available now through numerous aquarium industry companies.

Cyanobacteria

Cyanobacteria, the blue-green, "smear" or "grease" algae are a potential source of nitrate to the closed system aquarium since they have the ability to fix atmospheric nitrogen dissolved in the water and incorporate it in their tissue. As herbivores graze these algae, the nitrogen which has been fixed is converted to ammonia, and thus enters the pathway to nitrate (see chapter 2).

Methods of Nitrate Control

Protein Skimming

Protein skimming removes some amino acids from the water along with the various carbohydrates and lipids that make up the majority of the fraction removed. By removing amino acids from the water the skimmer eliminates a significant quantity of nitrogen that would be mineralized by nitrifying bacteria and converted to nitrate. Thus it works at the source in lowering nitrate accumulation. Skimmers also remove organically bound nitrogenous material primarily from the animals but also from plants, which collects with plant exudates and accumulates in the water (see carbon cycling, chapter 2). In this manner too, a potential source of nitrate is removed before it can be mineralized.

Denitrifying Zones

Nitrate is denitrified by facultative anaerobic bacteria that live in anaerobic zones within the aquarium. These occupy the cores of the live rock and the bottom substrate. Their proximity to aerobic layers means that nitrate, produced by nitrification there, is immediately available to these facultative anaerobes that convert nitrate into nitrogen gas and nitrous oxide, which blow off into the atmosphere. The best zone for denitrification is the sand on the bottom of the aquarium (see chapter 7, for more details about sand management).

Denitrification in the sand is capable of eliminating all nitrate produced in the aquarium. This benefit is lost when aquariums are set-up without any substrate on the bottom. Such aquariums have denitrification taking place within the rocks, which may be enough, but often it does not entirely eliminate nitrate accumulation.

Denitrification occurs in the rocks and bottom substrate by natural processes that also occur on reefs. Detritus collecting in the porous structure of the rocks and in the sand is consumed by numerous detritivores such as worms, crustaceans, and microorganisms including protozoans and heterotrophic bacteria. They excrete organic compounds liberated by digestion of the detritus, and these organic compounds feed the facultative anaerobes which consume nitrate as a source of oxygen, and convert it into nitrogen gas and nitrous oxide (see chapter 2).

Denitrifying filters capture this process outside the aquarium. In a denitrifying filter facultative anaerobes are encouraged to grow. Slow flow of water in the chamber allows all the oxygen to be consumed, and an organic supplement such as lactate or methanol is added as food for the bacteria. The system works, and there are many variations on the theme, some with automatic food dosing. Removing the process from the aquarium and putting it in a filter makes for a less than natural solution to nitrate control. It works, but not as simply, in our opinion, as natural processes which occur within the rocks and sand without the need for manual additions or automatic gadgets.

There are also some filter media on the market that are supposed to remove nitrate. In our experience, their ability to do so is limited compared to denitrification processes that naturally take place within a complete ecosystem aquarium. We do not dispute that these media can work, but we feel that they are unnecessary considering the natural means by which nitrate is easily managed.

Algae Filtration

The use of algae grown in a separate aquarium, in an algal turf scrubber, or within the display aquarium can reduce nitrate accumulation by removing ammonium before it is mineralized, and by removing nitrate from the water directly. Some aquarists like the appearance of macroalgae in the aquarium, and it can be both aesthetically pleasing and functional. It not only purifies the water, it provides food and refuge to numerous tiny crustaceans, and food to some fish as well. We caution that vigorous algae growth is harmful to stony corals that can become shaded, smothered, or injured by algae piercing through their tissues. If your goal is to grow lots of coral, the algae must be kept in check through the use of herbivores. Algae also exude substances into the water that are harmful to some corals in the closed system environment. The use of protein skimming or activated carbon effectively removes these substances. See Algal Turf Scrubber filtration, in chapter 5.

Herbivores

Various herbivorous snails, sea urchins, crustaceans, and fish aid in the reduction of nitrate levels in the aquarium. Their grazing habits encourage the algae in the aquarium to remain in a highly productive state, thereby removing ammonium and nitrate from the water. As they graze, they consume some of the nitrogen trapped by these plants, and their fecal pellets can be trapped and removed from the system through settling or mechanical filtration, or utilized as food by deposit feeders, filter feeders, or scavengers such as serpent stars.

Controlling Phosphate

Phosphorous is an essential component of living tissue, being the source of energy for the cell. Therefore, the notion that one can eliminate phosphate from the aquarium is false, unless one empties its contents completely and scrubs the glass clean, leaving a bare, dry aquarium, which isn't much fun to look at. Phosphorous in its many forms will always be there, in living tissue, in the substrate, in dissolved organic and inorganic forms, and in particulate detritus. Please refer to chapter 2, for additional information about phosphorous and its cycling in nature.

Though phosphorous is essential for life, a delicate balance between enough, and too much, exists in closed systems. Inputs must be matched by outputs and assimilation or precipitation. If phosphate is allowed to accumulate in usable organic forms it can trigger overwhelming blooms of algae. Inorganic phosphate (orthophosphate) has limited occurrence in the water since it is quickly consumed by plants, and readily blows off into the atmosphere as aerosol from the bubbling action in protein skimmers or airlifts (Spotte, 1979). It also readily combines with organic compounds or precipitates out of the water onto the calcareous substrate. The high pH in the vicinity of calcium hydroxide addition to the water also has a side benefit of precipitating phosphate there with calcium. Chronic levels of inorganic phosphate in the water typically come from stored organic forms being broken down to release the inorganic phosphate, from sources that leach phosphate, from polluted tap water used as freshwater top-off, or as a result of heavy feeding.

Fish and invertebrate foods added to the aquarium are the most significant phosphate input for most systems. The fish must be fed in most small aquaria, though larger closed ecosystems produce sufficient algae and little crustaceans to provide enough, or more than enough, food for the fish. Still, most aquarists enjoy feeding their aquariums at least once in a while, and this is fine provided one doesn't feed too much. In this regard, we wish to point out again that liquid invertebrate foods that benefit non-photosynthetic organisms primarily, (don't confuse with liquid trace element supplements that are essential) are to be fed only very sparingly or not at all. Most of the cloudy water ends up as pollution that will be mineralized by bacteria, instead of food benefiting target organisms. It is best to administer such foods by means of a pipette or eyedropper to spot-feed the target organism, thereby avoiding the practice of filling the whole tank with the cloudy solution.

Protein skimming removes inorganic phosphate from the water because the bubbling action blows the phosphate off into the air. Protein skimming also removes organic sources of phosphate very efficiently, depositing them in the collection cup along with the other components skimmed from the water. Thus a protein skimmer is a most important instrument in the control of phosphate in closed system aquariums.

Another potential input of phosphate to the aquarium is make-up or top-off water used to replenish evaporation. This daily addition of water can provide significant quantities of phosphate to the "closed" system aquarium. Not all tap-water or well-water contains phosphate, but many urban and rural areas have phosphate in the water in both organic and inorganic forms. If you are using this water to make kalkwasser then the high pH of this solution may result in the rapid precipitation of any remaining phosphate, in the form of calcium phosphate. Nevertheless, the use of a water purification system for one's top-off water is a good means of insuring the water's purity. Reverse osmosis and deionization systems are readily available now through numerous aquarium industry companies.

We mentioned earlier under the topic of chemical filtration that some types of activated carbon can leach phosphate into the water. We suggest that aquarists routinely check newly purchased batches of activated carbon with a phosphate test kit as a precaution (see the section on activated carbon in chapter 5).

One additional input of phosphate comes from airborne dust. This can be especially significant for aquariums located outdoors, and is more significant for aquariums without covers of course. In most indoor home aquariums the phosphate input from airborne dust is slight. Covers, mechanical filtration, and protein skimming help to counter this type of input for aquariums located where there is a lot of airborne dust.

Exports of Phosphate

The use of herbivorous snails is a valuable aid in the export of phosphorous from the aquarium. Algae growing on the rocks and walls of the aquarium remove phosphate from the water and the snails graze the algae. Their little fecal pellets contain some phosphate then, and the removal of these pellets through mechanical filtration or siphoning of settled material is a way to remove the phosphate from the system.

Mechanical filtration is not essential, but it does greatly assist in the export of particulate organic phosphate. Without mechanical filtration or other removal systems such as settling filters, the particulate material must be broken down by bacteria and microorganisms, or consumed by various filter feeders and detritivores. Their consumption of detritus plays a key role in the denitrification process, (see previous section on nitrate control).

The detritivores also play a key role in substrate management. Gobies sift the sand for detritus and the small crustaceans that feed on detritus. The little crustaceans multiply well in thick substrates, as do numerous species of polychaete worms that feed on particulate matter. These organisms all play a role in the reduction of available phosphate in the system, and prevent the sand from becoming a dead muck zone where detritus merely accumulates.

Another means of phosphate control is effective, but not natural. There are numerous brands of phosphate removing filter media on the market, and they all work reasonably well. We wish to emphasize that proper aquarium technique prevents the accumulation of detrimental levels of phosphate. The routine addition of calcium hydroxide solution precipitates phosphate from the water. Combined with protein skimming and proper aquarium husbandry the phosphate level is easy to control without dependency on phosphate removing media. Nevertheless, these media do have some use in emergencies. (see subject heading Dinoflagellates in chapter 10).

Synergistic Effect of Phosphate and Trace Elements

The trace elements iron, iodine, and molybdenum are important for marine algae, and it has been inferred that this means they might stimulate algae growth. When there is sufficient phosphate available in the aquarium, either inorganic or organic, these elements may enhance the growth of algae in the same manner as illumination, but it would be incorrect to say that they cause algae to grow. The availability of phosphate can be compared to the power button of a radio. The intensity of light is the volume dial, and the sound emanating from the speakers is the growth of algae.

When the power button is on, the higher you turn the volume, the louder the sound which comes out of the speakers. When the power is off, no sound comes out, and the volume dial has no effect. These trace elements work like the volume dial, and do not cause a problem when there is no significant availability of phosphate. Please remember that typical phosphate test kits do not measure dissolved organic phosphate, nor particulate organic phosphate, and these are potential sources of phosphate for algae. When starting to add trace elements it is best to begin slowly, and progress to the recommended dosage. If it seems that algae growth is enhanced by your additions, cut back. Then work on reducing the dissolved organic and particulate organic substances which might be accumulating in your tank. Employ better circulation, more protein skimming or activated carbon. Always test the new carbon to be sure that it is not a source of phosphate.

Silicate

Silicate is a minor component of seawater that is an important nutrient for many invertebrates and the golden-brown algae called diatoms. Silicates are most commonly imported to the system with freshwater used for topping-off evaporation. While some silicate is essential, excess silicate can produce undesirable blooms of diatoms that coat the glass and rocks very rapidly. With herbivores, especially snails, silicate can be effectively exported from the system via fecal pellets, and diatom growth essentially eliminated. There is some debate about whether or not it is prudent to maintain silicate levels well below natural seawater values (2 to 3 mg/L). From an aesthetic point of view, silicate (and diatoms) be damned in the aquarium, and their absence is no loss for the clear view and ease of maintenance afforded. From an ecological standpoint, however, silicate is important in the formation of sponge skeletons and serves other functions for different invertebrates and plants. The herbivorous snails, filter feeders, and some fish also gain nutrition from diatoms. Water changes typically supply sufficient silicate to maintain at least some growth of diatoms and other organisms that need silicate. Tiny siliceous sponges regularly reproduce throughout our systems, in the plumbing and on the rocks and glass, not apparently harmed by any deficiency. Adding small quantities of silicate to the system in the form of a supplement risks the stimulation of too much diatom growth, in our experience. If the tap-water contains silicate, or if one adds silicate as a supplement, herbivores can manage the excess growth of diatoms on the reef, if one doesn't mind having to clean the glass more often.

Problem Algae

The presence of problematic filamentous and slime algae in the aquarium results from a combination of factors, including the introduction of the algae with live rock, the lack of herbivory by herbivores, and the availability of limiting nutrients, especially in organic form. We have already discussed means of controlling the nutrients that limit algae growth. Here we present a description of some of the most common problematic species, and our recommendations for natural ways to control them. See chapter 10 for additional information.

Filamentous Algae

Several different types of filamentous "hair" algae or "thread" algae plague reef aquariums, and they can be a problem in localized areas of natural reefs as well. Every aquarist encounters them at some time. Beginners may find the management of hair algae so frustrating that they give up and get out of the hobby. We hope that by identifying the different problem species and the best manner to control them, we can prevent such frustration.

Bryopsis

Another kind of filamentous algae, *Bryopsis,* is very similar to *Derbesia,* except that the tips of some of the filaments have feather-like shape. The colour is dark green, often with an iridescent blue sheen. Both *Derbesia* and *Bryopsis* can occur simultaneously, growing together as tangled masses. While *Derbesia* is readily eaten by fish, *Bryopsis* is tougher to chew and apparently not too delicious. Tangs will avoid it unless they are very hungry. Sea urchins will eat it, however. In established reef aquaria, *Bryopsis* usually occurs as isolated tufts on particular rocks, or on the glass. Like *Enteromorpha,* it commonly blooms explosively in the first several weeks after the aquarium has been set-up, causing much alarm to the aquarist not familiar with this common occurrence. The growth subsides, but unlike *Enteromorpha* which disappears, *Bryopsis* usually remains present in a few areas in the tank, unless one makes an effort to eradicate it. Removal by hand helps, but the algae grows back. Allowing turfs to grow unchecked is risky. They release spores that settle in new locations, creating new "fires" to put out. The photosynthetic Lettuce Sea Slug, *Tridachia* sp., from Florida and the Caribbean, may eat *Bryopsis,* and is a good control. The bluish, reef-inhabiting variety of *Tridachia crispata* (which is probably a separate species) feeds on *Bryopsis,* while the green variety from seagrass

Above: An example of an herbivorous nudibranch, *Tridachia crispata.* Some forms of this nudibranch will eat *Bryopsis.* S.W. Michael.

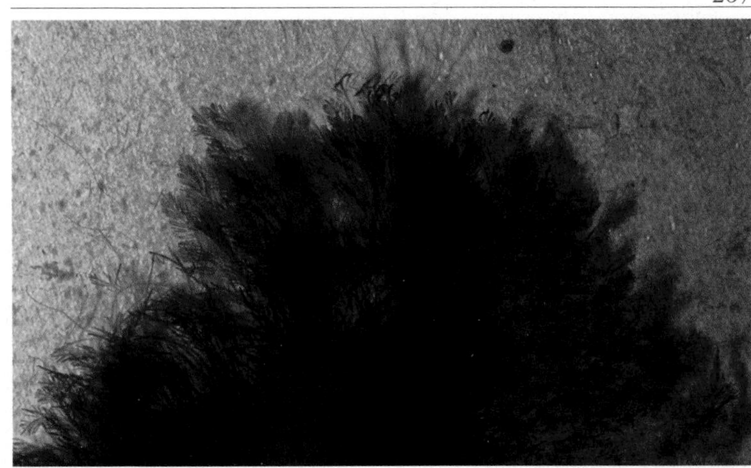

Bryopsis forms thick turfs with feather-like tips. J.C. Delbeek.

beds does not, usually. If the *Bryopsis* grows only from a particular small rock, just remove the rock from the aquarium. The rock can be held separately in the dark. When the algae has died and disappeared, zoanthid anemones and soft corals can be attached to the rock, covering the areas where the algae might have returned.

Cladophora

Cladophora spp. form tangled filaments that are usually lighter green than either *Derbesia* or *Bryopsis,* and much coarser, with a consistency like polyester filter floss. Snails cannot control it when the growth is thick. They are only able to graze fine turfs. Small tangs have difficulty chewing *Cladophora,* but larger tangs plow through it, devouring it quickly. They are the best control. Sea Urchins will also eat *Cladophora.* Give the herbivores a head start by removing large growths by hand. The similar *Cladophoropsis* forms thicker, tougher filaments, and is more difficult to control.

Derbesia

The most common problematic hair algae is *Derbesia* sp. It forms dark green strands that are soft and trap a lot of detritus. It is the "hair algae" most often encountered in reef aquariums and marine fish aquariums. It is easily controlled with tangs and herbivorous snails when the nutrient and organic load on the tank is properly managed. The tangs eat the long filaments, and the snails mow down the short turfs. Elimination of this algae can be achieved when there are sufficient numbers of herbivores. In aquariums too small for tangs, the hair algae must first be cropped by hand, and then the snails can be added to finish it off. *Derbesia* has an alternation of generations, with a form called the *Halicystis* stage that looks like a green bubble, similar to *Valonia* spp.

An infestation of *Derbesia* "hair" algae. J.C. Delbeek.

Caulerpa and *Derbesia* blooming in a nutrient rich aquarium. J.C. Delbeek.

Enteromorpha

Enteromorpha forms soft, tube-like hairs, light green in colour. It is common in the first few weeks after the tank has been established settling all over the glass, and is typically one of the first algae to settle on new substrates in the natural environment. In the natural environment it is an early successional species, occurring on new, bared substrates, being replaced by later growth, and a characteristic omnipresent species in high nutrient areas, especiall intertidally on rocks along the shore. *Enteromorpha* may bloom in a reef aquarium initially, but usually disappears after several weeks, and thereafter occurs only in very small patches or not at all. It worries beginning aquarists by its sudden and dramatic appearance when it occurs, but it can easily be controlled with herbivorous fish (i.e. tangs) and it eventually subsides naturally.

Red Hair Algae

Members of the genera *Ceramium, Anotrichum,* and *Griffithsia* form red turfs, and several other algae also form either permanent red turfs, or temporary ones, as part of an alternation of generations of form. These algae are easy to control with herbivorous snails such as *Astraea tectum* and *Turbo* spp. These algae can become a problem only when there are few snails to graze them. Herbivorous fish will also crop them down eagerly, but do not eliminate them as efficiently as the snails. Apparently red turfs taste pretty good.

"Slime" Algae

"Slime", "grease", "smear", or "sheet" algae are usually types of cyanobacteria, known also as "blue-green algae". They form slimy or gooey mats or sheets that are typically red to maroon, green, or black. Pale brown slime algae are usually either diatoms or dinoflagellates. However, there are brown cyanobacteria too. Slime algae mats often trap the oxygen produced during photosynthesis, the bubbles shining like little silver pearls.

Cyanobacteria

Cyanobacteria are important to reef ecosystems in nature because of their ability to fix nitrogen (see chapter 2). The most common problematic types are red or maroon in color, and they can rapidly cover every surface in the aquarium under the right conditions. The

Cyanobacteria mat on an aquarium substrate. J.C. Delbeek.

right conditions in closed system aquariums involve high levels of dissolved organic compounds in the water. Removal of these dissolved organic compounds is achieved with protein skimming, as described earlier, and in chapter 5. The protein skimmer must adequately remove these organic nutrients to control cyanobacteria growth. Snails are also helpful for controlling cyanobacteria, as are herbivorous fish. We wish to emphasize that the best cure is good protein skimming. A little bit of skimming may not solve the problem, but enough skimming will do the trick. We also wish to discourage the use of antibiotics (i.e. erythromycin) or other algicides. Such remedies are not cures! They do kill the cyanobacteria, but do not prevent it from coming back. One must attack a problem at the root cause (i.e. accumulation of dissolved organic compounds, in this case), to solve it. Cyanobacteria are always present somewhere even in the most successful reef aquariums. It is only when they bloom that they become a problem. One can usually find them in clear plumbing and in the overflow chamber, both places that have high exposure to dissolved organic compounds. They are also common around the bases of soft corals, particularly *Sarcophyton* spp., where they apparently receive some kind of organic food. The "black band disease" of corals, in addition, involves cyanobacteria. See chapter 10 for details about that condition and how to control it.

Diatoms

Diatoms form golden-brown films on the glass and substrate. They are easily controlled by limiting silicate, as mentioned already, and with herbivorous snails. Diatoms are an important part of an herbivorous snail's complete diet.

Dinoflagellates

Dinoflagellates occasionally bloom in reef aquaria, and they can be toxic to invertebrates and fish. They form nearly colourless to rust-brown gelatinous mats and films that trap oxygen bubbles. They can also be present in large numbers in the water column and on the surface of the water during a bloom. They coat bare surfaces so quickly that it is futile to siphon them off. The blooms can be

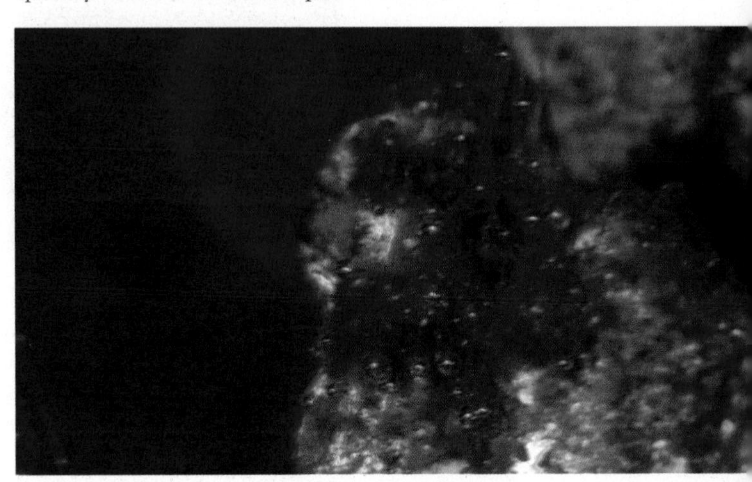

Dinoflagellate bloom. Note trapped oxygen bubbles. J.C. Delbeek.

persistent and maddening! From Peter Wilkens we have learned of a cure that really works. Elevating the pH to 8.4 or a little higher via kalkwasser additions effectively causes the bloom to crash within about a week, usually. See chapter 10 for a complete description and additional recommendations for control.

"Bubble algae" (*Valonia* sp.)

The dreaded "bubble algae" belong to the genus *Valonia*. They form clusters of spherical or oblong green bubble-like structures, actually single cells with many nuclei. They are quite attractive when few in number, like green sapphires, but that's the allure and the trick they play. They can quickly achieve epidemic proportion in the aquarium if left to their devices. They reproduce at an alarming rate, by forming from the nuclei in the cytoplasm thousands of complete miniature bubbles that brood within each "parent" bubble. When the old cell wall tears, these baby bubbles are released, and they settle all over the substrate. The problem is that they encroach on living coral tissue and cause it to recede (see chapter 10). Control after a bloom has already occurred is difficult. Prevention is best. Remove individual bubbles by grabbing them

Valonia sp. algae crowding out zoanthid polyps. J.C. Delbeek.

with a needle-nose pliers or small scissors. Twist and yank them off, and remove them from the aquarium. Don't worry too much about popping them (which does release the tiny brooded bubbles if the algae is in the reproductive stage), but avoid it if you can. If the bubbles are already in plague proportions, disassemble the reef structure and pull them off outside the aquarium. After the rock is reassembled in the aquarium, if the surfaces are covered by anemones, soft corals, and stony corals, there will be less available surface for *Valonia* to grow, and any new bubbles can be seen and easily removed as a general maintenance procedure every few weeks. In time, if purple coralline algae are encouraged to grow vigorously and coat the bare surfaces of the rocks, *Valonia* will become rare and maintenance will be less critical. *Valonia* seldom bloom in aquariums with heavy growth of coralline algae.

Herbivores

There are numerous herbivorous fish that can be utilized for control of algae in the reef aquarium.

The most efficient herbivorous fish are tangs, particularly members of the genus *Zebrasoma* that have a long snout. The Yellow tang from Hawaii, *Zebrasoma flavescens* is the most readily available. The Sailfin tangs, *Zebrasoma veliferum* from the Pacific, and *Z. desjardinii* from the Indo-Pacific and Red Sea are also excellent grazers, as are the *Scopas* tang, *Z. scopas,* and the Red Sea Purple tang *Z. xanthurus*. All of these eat filamentous algae and are capable of grazing even tough algae, like *Cladophora* and *Cladophoropsis*.

The Red Sea Purple Tang, *Zebrasoma xanthurus.* J.C. Delbeek.

Large specimens of *Zebrasoma desjardinii* have been reported to feed on *Valonia* and other problematic algae. J.C. Delbeek.

The Yellow Tang, *Zebrasoma flavescens*, is an excellent herbivore in the reef aquarium. J.C. Delbeek.

Tangs of the genus *Ctenochaetus*, such as this Chevron tang, *Ctenochaetus hawaiiensis*, are excellent grazers of microalgae, but beware that they are very prone to *Amyloodinium*. J.C. Delbeek.

Ctenochaetus tangs, such as the Kole, *C. striatus,* and the Chevron tang, *C. hawaiiensis*, have broad, fat lips that they bang and flap across the rocks constantly, scraping up short algal turfs, slimy algal mats, and detritus. They effectively polish the rock, like snails, with their brush-like teeth.

Beware that all tangs are especially prone to fish diseases such as "Ich" and *Amyloodinium*. Yellow tangs are prone to "black spot" disease. See Moe 1992, Blasiola, 1992, and computer programs listed in appendix C for details about treatment of fish disease.

Some types of blennies are excellent herbivores. *Salarias fasciatus,* from the Pacific, and the Red-lipped Blenny, *Ophioblennius atlanticus,* from Florida and the Caribbean, have lips very similar to the *Ctenochaetus* tangs, and they eat in the same manner, though it is a bit more comical to watch one of these graze as it seems to bang its entire head against the rocks. Red-lipped Blennies are very territorial and nippy, however, and they can be bothersome especially in small aquariums. Not all blennies eat algae. Some are carnivorous, and others that have the same kind of brush-lipped face as the algae eaters actually feed exclusively on coral polyps (i.e. *Exalias brevis*). Be careful with blennies! (see also chapter 10, poor expansion).

A few types of gobies also eat algae. Members of the genus *Amblygobius,* in addition to sifting sand, graze filamentous algae from the rocks. *Amblygobius rainfordi* and *A. phalaena* may starve in an aquarium that is completely devoid of filamentous algae. These gobies are quite shy, and do not do well in an aquarium with aggressive tankmates (see Delbeek and Michael, 1993).

Amblygobius rainfordi are timid fish that feed on filamentous red and green algae. J.C. Delbeek.

Amblygobius phalaena.
J.C. Delbeek.

Pygmy angelfish of the genus *Centropyge* make attractive additions to any aquarium, and they are good herbivores as well. Unfortunately, they do not always restrict their diet to algae. They may begin to feed on corals and clams suddenly, and without mercy, much to the aquarist's chagrin. In our experience, *C. argi* seldom bothers corals. Popular species such as *C. loriculus*, the Flame angel, and *C. acanthops,* the African Flameback angel, have proven unpredictable. Sometimes they bother corals and clams, sometimes they don't, and sometimes they don't bother them for years, but change their habit suddenly.

We should warn you that all of these herbivorous fish are inclined to mistake soft corals and sometimes even stony corals for algae on occasion, nibbling on the polyps, but usually do no permanent harm. The *Centropyge* angels, however, often develop an appetite for coral and tridacnid clams, so their inclusion in a reef aquarium is a little risky (see chapter 10).

Herbivorous Snails

Several species of herbivorous snails are available to aquarists, from the genera *Astraea, Turbo, Trochus, Nerita, Cerithium* and *Calliostoma*. Other herbivorous molluscs include limpets, chitons, and abalones. Some herbivorous snails grow large and have an undesirable attribute of knocking things over in their search for food. Some don't live very long, or are easily preyed upon. Ideally the rate of attrition of the snails is low, and only a few will need to be added per year, but sometimes fish and other predators will cause heavy losses. A few types readily reproduce in small aquariums, and all types do reproduce successfully in really large

The African Flameback angel, *Centropyge acanthops*. J.C. Delbeek.

aquariums. Herbivorous snails are a true blessing from the reefs. They make the management of algae really simple, and they are more efficient than fish at removing algae from crevices in the rocks. Still, the best way to manage the growth of algae with herbivores is to use both fish and snails. The fish graze the longer growths, and the snails polish off the fine turfs.

Astraea tectum and *Astraea caelata* (Trochacea) are excellent snails from Florida and the Caribbean. They are long-lived, and do not grow very large, so they aren't too bad about knocking things over. Use about one snail per gallon for the best results in small reef aquariums. In large aquariums, 2000 L (500 gal.) or more, fewer snails may be needed as the surface of the rock exposed to light does not increase as dramatically as the water volume.

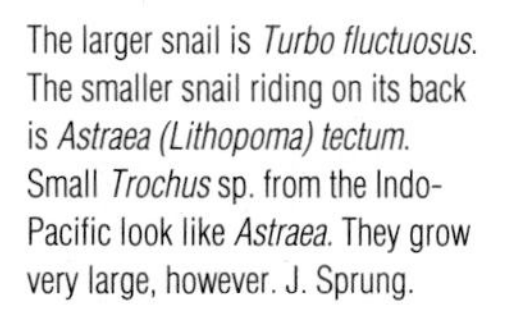

The larger snail is *Turbo fluctuosus*. The smaller snail riding on its back is *Astraea (Lithopoma) tectum*. Small *Trochus* sp. from the Indo-Pacific look like *Astraea*. They grow very large, however. J. Sprung.

Cerith snails shown here are common on live rock from the Caribbean. They feed on diatoms primarily, but larger ones feed on some turf algae. Some species live on sand intertidally, migrating with the tide. J. Sprung.

Nerites are common intertidal snails that feed on diatoms, cyanobacteria, and turf algae. Some species must crawl out of the water to breathe, and are unsuitable for aquariums without lids. They wander too far. The two attractive species shown here are good for reef aquariums. J. Sprung.

Turbo species (Trochacea) are very popular for algae control. They grow fairly large, about 8cm (3 in.), and have a voracious appetite, so fewer are needed. About two snails per 60 liters (15 gallons) is sufficient. Because of their large size they are less desirable in small reef aquariums, less than 120 liters (30 gallons), since they can knock things over. Some species aren't hardy in captivity, possibly as a result of temperature or shortage of food.

Trochus species (Trochacea) are similar to *Astraea* in appearance. They are also excellent consumers of filamentous algae! *Trochus niloticus*, which has been depleted from some locations by pacific islanders who eat them and use the shells for a variety of ornamental purposes, is now being farm raised by the MMDC in Palau. Initially reared as a means of algae control in grow-out raceways for tridacnid clams, the snails are now being sold to the aquarium industry. These are the only farm raised snails available. All others are taken from rocky coasts and seagrass meadows. In the future it is likely that other Trochacea such as *Turbo* and *Astraea* will be farm raised for aquarists. Although *Trochus niloticus* is a fine algae eater, it does grow very large, ultimately too large for small aquariums. Fortunately small juveniles are offered for sale. These snails are long lived.

Limpets, which look like a Chinese hat, also are introduced with the rocks, and some species such as *Diodora*, the keyhole limpet, will reproduce in the aquarium. They are all good herbivores, but sometimes they will eat both soft and stony corals (see chapter 10).

Nerite snails, *Nerita* spp., are usually restricted to the intertidal region, while most other herbivorous snails occur in both intertidal and subtidal regions. Some nerites are unsuitable for aquariums since they crawl out, leave a slimy trail across the floor, and go "crunch" underfoot. Some nerites do adapt well to aquariums, occasionally going above the water line, but usually remaining in the aquarium and eating algae. Nerites are best employed as a supplemental variety of herbivorous snail, in addition to the primary ones such as *Turbo, Trochus,* and *Astraea*.

Another supplemental variety of snail, *Cerithium litteratum* is usually introduced to the aquarium with live rock. It remains quite small and is long-lived, but it is not very efficient at consuming filamentous algae, preferring diatoms and cyanobacteria.

Stomatella varia, a tiny herbivorous snail from Indonesia. J.C. Delbeek.

A small herbivorous snail, *Stomatella varia,* that looks like an abalone, is common on live rocks from Indonesia. These are actually related to *Turbo* and *Astraea*, family Trochacea, despite their appearance. They readily reproduce in the aquarium, like limpets. They do not harm corals, remain small, and are excellent, voracious herbivores. Though they cannot tackle dense turfs of filamentous algae, they are good for controlling diatoms.

Amphipods

Amphipods are laterally compressed, tiny crustaceans that look like fleas. They occur abundantly on live rock and amongst algae. Most are omnivorous, though many are herbivores. They are secretive by day to avoid predation by fish. If you want to see amphipods in your reef tank, take a look at night using a flashlight You will see them scurrying about the rocks with their little arched backs. A high population of amphipods grazing at night can dramatically hinder the growth of filamentous algae, diatoms, and cyanobacteria. Many fish feed on amphipods, which are a natural, perpetual food source. Some fish, i.e. mandarinfish (*Synchiropus* spp.), subsist on amphipods primarily, and are capable of depleting the population in a small aquarium so much that they slowly starve to death. In a large aquarium with plenty of area for the amphipods to multiply, mandarins never run out of food. Sand and gravel bottoms provide additional refuge to amphipods, and encourage the development of a large population. See the topic "refugium" associated with algal turf scrubbing, chapter 5.

Crabs

Crabs are mostly omnivorous, but there are several primarily herbivorous species that can be utilized in a reef aquarium. We discourage their use in small reef aquariums (less than 400 liters; 100 gal.) because the crabs generally grow large enough to present a threat to fish, which they can capture at night. The two types of herbivorous crabs commonly available are the "Sally Lightfoot" crab, *Percnon gibbesi*, and the various "hardback" or "coral" crabs, *Mithrax* spp. Though these crabs are especially good at cropping down short turfs of filamentous algae, we recommend the use of snails and fish as herbivores instead, to avoid the possibility that hungry crabs will eat fish or invertebrates (see chapter 10 for additional information about crabs).

Hermit crabs

There are some species of hermit crabs that remain small and present little risk to the fish. Though omnivorous, these small

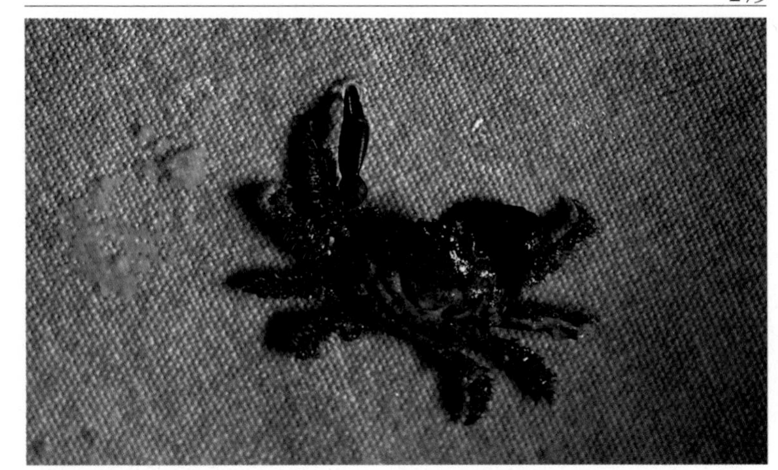

Mithrax sculptus, an herbivorous crab from the Caribbean. J.C. Delbeek.

Mithrax crabs are good herbivores for large aquariums. *Mithrax sculptus* only grows to about 6.3 cm (2.5 in.), but is capable of capturing small fish. Some *Mithrax* species grow very large, and are therefore unsuitable for most aquariums. We have heard reports of a coral crab (*Mithrax* sp. ?) from Indonesia that eats *Valonia* (bubble algae). While we have not observed this crab, it is possible that it will be recognized as an important means of controlling this and other problem algae.

hermit crabs feed primarily on algae, both filamentous algae and cyanobacteria. They also stir the sand, a desirable attribute. They can be a bit clumsy, and their habit of walking over corals can be irritating, causing a temporary contraction of the polyps, but they are harmless. The so-called "Red-legged hermit crabs" *Paguristes cadenati, Calcinus tibicen,* and *Phimochirus operculatus* are ideal, and remain small (<1 in.). Avoid large hermit crabs that may crush, taste, spindle or mutilate anything within reach. Hermit crabs require small snail shells to live in and must be supplied with them to move into as they grow.

Sea Urchins

Urchins make excellent herbivores if your only goal is to eliminate algae. If you want to keep your corals where you placed them, sea

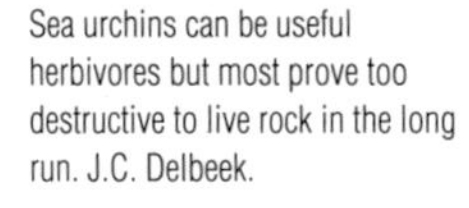

Sea urchins can be useful herbivores but most prove too destructive to live rock in the long run. J.C. Delbeek.

urchins' habit of bulldozing and demolition won't make you happy. Urchins grind away at the rock as they graze, which means they also are removing coralline algae and clearing bare substrate for the re-growth of filamentous algae. It makes sense that they should do that because the behavior insures the continued presence of their food source. In really large aquariums with rapid growth of coralline algae, the effect of a few sea urchins on coralline algae abundance is negligible. In small aquariums their effect is noticeable. There are many exquisitely beautiful types, but by far the least clumsy are the longspine ones, *Diadema* spp. These are effective natural controls for the problem algae *Valonia, Bryopsis,* and *Cladophoropsis.*

Physical Means of Control

In addition to control by limiting nutrients and using herbivores, one can intervene physically in the eradication of undesirable algae. *Derbesia* is easily removed with a siphon or with a toothbrush, such removal affording the herbivores a head start. A wooden stick or toothpick can also be used to remove algae growing on invertebrates. The stick is twirled slowly to draw the algae off. After physically removing algae such as *Cladophora, Cladophoropsis,* and *Bryopsis* from the rock, re-growth in the same spot can be prevented by placing a live coral specimen over the spot, using monofilament line to hold it in place (see chapter 7). Zoanthid anemones and soft corals, which attach very quickly to the rock, are especially effective for this purpose.

Temperature

Temperature has a slight connection with the control of problem algae. In our experience, problem algae in aquariums are more prevalent at higher temperatures. Temperatures above 25 °C (78 °F) tend to promote more rapid growth, while temperatures below 24 °C (75 °F) tend to slow the growth of algae. In the natural environment, some algae blooms occur when the water becomes cooler in winter months, but most algae blooms occur during the summer, and don't subside until the temperature drops again in the autumn.

To conclude, we again wish to emphasize that protein skimming is your number one aid to preventing excessive algae growth, via nutrient limitation. It is not enough to have a protein skimmer, it must be of sufficient size for the aquarium, and it must be operating properly. Any skimmer will skim, producing volumes of foam and dark liquid. To really work well for the aquarium, the skimmer must have the column filled densely with tiny bubbles, and the tank volume capacity should be passed through the skimmer(s) at least once per hour or more (see chapter 5 for more detail). Herbivores are as important as skimming, and having combinations of different herbivores along with proper skimming is the key to solving plagues of algae. Finally, when coralline algae are encouraged to grow rapidly, by maintaining a high pH and carbonate hardness (8.4 and 12 dKH respectively), filamentous and slime algae are virtually non-existent, and the growth of stony corals also proceeds rapidly. The addition of kalkwasser, therefore, has a side benefit of tending to eliminate problem algae indirectly.

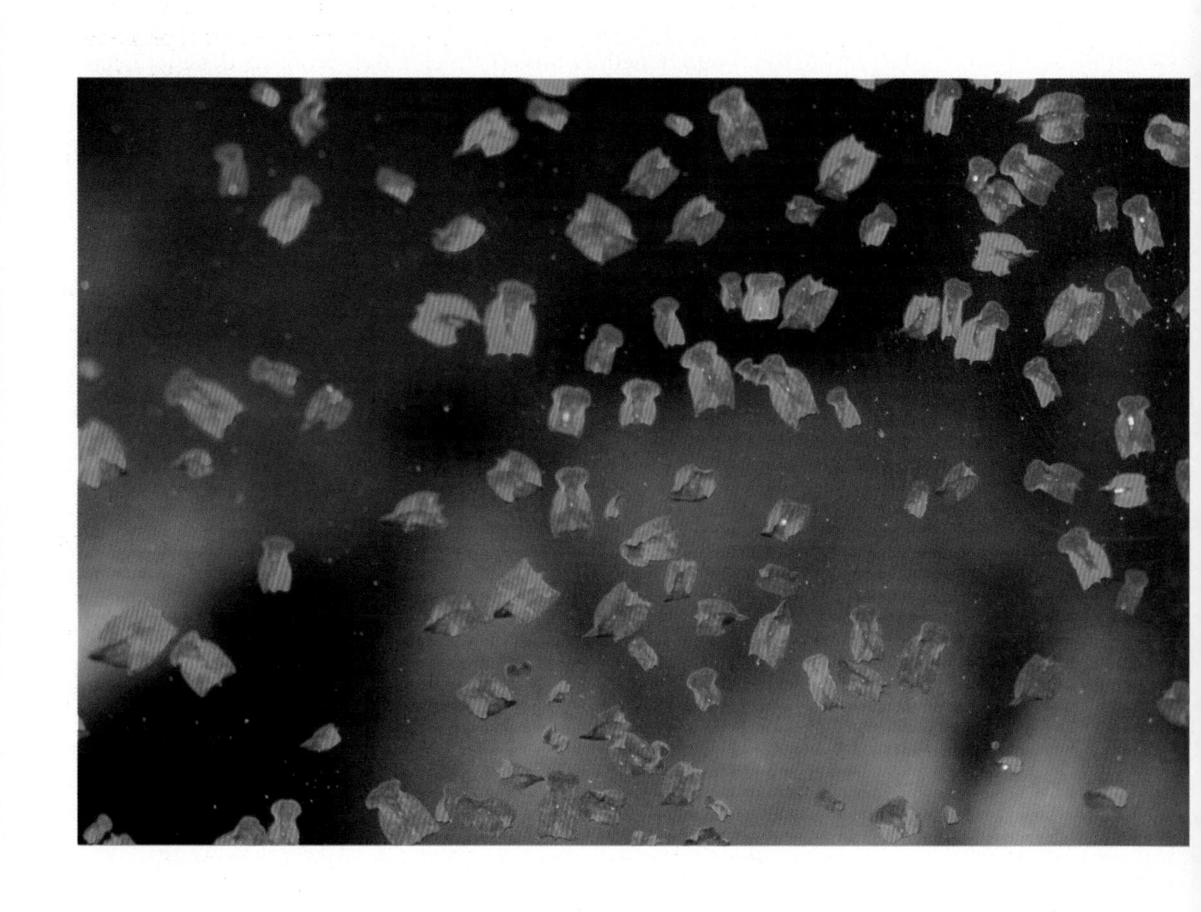

Chapter Ten

Diseases, Parasites, Pests and Commensals of Corals and Clams

Many unwanted "hitchhikers" often come along for the ride with the corals and clams we keep. Typically, these interlopers go unnoticed and the decline of the specimen is attributed to some other cause. Failure to recognize and remove potential problem organisms can result in serious damage that is often lethal. The pests include but are not limited to, snails, nudibranchs, crabs, shrimp, polychaete worms, flatworms, anemones and algae. The vast majority of these are introduced along with newly purchased corals or clams. Therefore it is a wise precaution to quarantine any new arrivals in a separate aquarium. Here they can be closely examined for a few days to determine if they are carrying any parasitic organisms such as nudibranchs, snails, crabs or flatworms. At the very least, any new arrival should be closely examined before being purchased and before placing it in the display aquarium. Once invertebrate pests are introduced into the main aquarium, they can be extremely difficult to remove completely, and can cause considerable damage to your animal collection.

Another area of concern is infections caused by various bacteria, viruses and protozoans, which are normally present on corals and clams. In some cases it is not until the coral or clam is placed under some form of stress, that these microorganisms become a problem. Again, quarantining any new arrivals would be prudent. This can be as complex as keeping the specimen in a separate aquarium with its own filtration system, to keeping it in a floating mesh container in the display aquarium. A separate quarantine aquarium is the best method since bacterial and protozoal pests could easily pass through a mesh container into the aquarium. Mesh containers are best used when larger pests such as flatworms, nudibranchs or crabs are suspected.

Red planaria, *Convolutriloba retrogemma*. J.C. Delbeek.

Bacteria and Viruses

Very little research has been done on the types of bacteria and viruses that can afflict corals. This is unfortunate as it is becoming increasingly clear that in areas such as the Caribbean these organisms are potentially more devastating to coral reefs than the mechanical effects of storms and humans (Shinn, 1989). The research on bacteria and viruses affecting bivalves, especially mussels and oysters, is much better. This is solely due to the fact that they represent billions of dollars in cash crops. Until recently, very few studies had been done on pathogens affecting tridacnid clams (see Norton et al., 1993a; Norton et al., 1993b and; Sutton and Garrick, 1993). Hopefully, more research funds will be forthcoming for work on both coral and tridacnid clam diseases as their potential impact on local and national economies becomes better appreciated.

Certain coral diseases such as White Paste, White Band and Black Band Disease (see trouble-shooting section) may be caused by pathogens that are opportunistic and only become plentiful when some other stress has occurred. For example, physical damage caused by mechanical injury or predation may expose the underlying tissue and skeleton to infectious agents. Increased water temperature, poor water quality, excessive UV light, coral stings and lack of trace elements can also weaken corals, making them more susceptible to infections.

In tridacnid clams, a large number of bacteria, both pathogenic and non-pathogenic, have been isolated (see Humphrey et al., 1987). Among these, *Vibrio alginolyticus* and *V. anguillarum* have been associated with mortalities in larval and adult oysters (Humphrey, 1988). However, these organisms have commonly been isolated from healthy clams too. Bacterial infections have also been implicated in mass mortalities of larval cultured tridacnid clams in Australia (see Humphrey, 1988). Antibiotic treatments with streptomycin, neomycin, penicillin, and/or rifampin have been tried, and combinations of streptomycin and neomycin proved most effective for increasing survival of tridacna clam larvae (Fitt et al., 1992). (This combination may also prove helpful for treating corals.) At the present time no viral agents have been isolated from tridacnid clams (Humphrey, 1988).

The biggest problem with bacterial and viral diseases is that they are very difficult to identify and treat. Many treatments can severely stress the other inhabitants and are therefore of limited use in the reef aquarium. In the next section we will offer some treatments that have proven effective in many cases of bacterial coral infection. The study of bacterial and viral diseases is definitely an area where the captive maintenance of corals and clams can play a useful role.

Phylum: Protozoa

Subphylum: Ciliophora
Common Names: Ciliates, Protozoans

Like bacteria and viruses, many of the protozoans found in corals and clams occur naturally. In aquariums their numbers only increase in response to a decrease in health of the specimen. It is therefore possible that many of these protozoans are opportunistic feeders, not true parasites.

In corals, protozoal infections often take on the appearance of a "brown jelly". The most commonly encountered protozoa in these infections belong to the family Philasteridae such as *Helicostoma nonatum,* but others such as *Euplotes* spp. may be involved too.

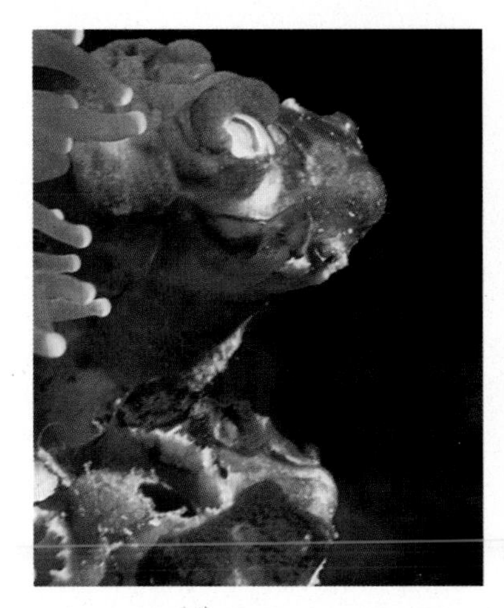

"Brown jelly" *Helicostoma* infection on a *Euphyllia glabrescens.*
J.C. Delbeek.

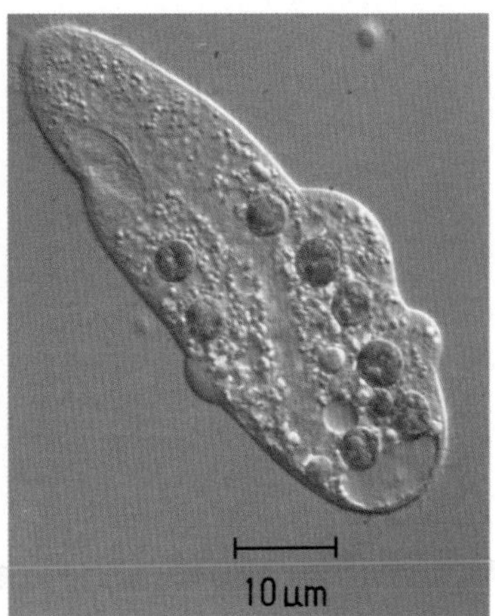

Helicostoma sp. protozoan. Note the ingested zooxanthellae.
J. Yaiullo.

Some of these protozoa may feed directly on the coral tissue, while others may feed on bacteria associated with the open wounds. This is explained more thoroughly in the troubleshooting section of this chapter.

In tridacnid clams, protozoa from the genus *Perkinsus* have been found in specimens collected throughout the Great Barrier Reef. However, their role in clam mortalities remains unknown. Additionally, an unidentified ciliated protozoan has been found that invades the mantle and ingests zooxanthellae (see Humphrey, 1988). We have no experience with protozoal infections in tridacnid clams and they do not appear to be very common in aquaria.

Most protozoan infections in corals are the result of some sort of physical injury. Once the tissue has been damaged, the protozoans can quickly appear on both hard and soft corals (see "brown jelly" infection in the troubleshooting section). Protozoan infections can also occur as a result of the same adverse conditions mentioned above for bacterial infections. In the case of the Atlantic hard coral, *Acropora cervicornis* (Staghorn Coral), it was found initially that no matter how healthy a specimen was when collected, they always succumbed to protozoan infections in captivity. It was not until additions of a strontium chloride solution were begun that this coral survived for more than a few months (J. Sprung, pers. obs.). With strontium additions this particular species is now easily kept in properly maintained closed systems, and grows very rapidly. This is an excellent example of the importance of an inorganic element to overall coral health.

Phylum: Cnidaria
Class: Anthozoa
Subclass: Zoantharia
Order: Actiniaria
Genus: *Aiptasia* spp.
Common Name: Glassrose Anemones

Aiptasia spp. anemones are probably one of the most bothersome pests in the reef aquarium. In some cases they can proliferate so quickly that they can completely cover an entire aquarium in only a few months. In aquariums that receive regular feedings, they can multiply even faster. Some aquarists consider this is a desirable thing and they proudly present these aquariums as "reef tanks",

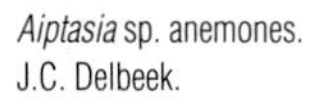

Aiptasia sp. anemones.
J.C. Delbeek.

however, we do not share the same view. We have even seen dead coral rocks covered with *Aiptasia* offered for sale as "anemone rock". Do NOT buy such rocks! These small anemones can easily sting and irritate corals and clams to death and should not be allowed to gain a foothold in the aquarium. *Aiptasia* are usually introduced along with live rock when setting up a new aquarium. They can also arrive on pieces of rock with corals on them, or with zoanthid and mushroom anemone colonies.

Removal: There are several techniques that can be employed to eliminate these pests from the aquarium but they can be divided into two basic categories: biological and mechanical. Biological controls consist of using natural predators to eliminate the anemones. One drawback to this method is that these organisms

Aiptasia anemones stinging a colony of star polyps, *Pachyclavularia* sp. S.W. Michael.

may eat not only the *Aiptasia* but also any other cnidarians in the aquarium such as zoanthids. For this reason, it is prudent to introduce these predators to a newly established aquarium before adding any other corals. This will cause them to search the live rock and eat any *Aiptasia* they find. Once all the *Aiptasia* have been eaten, the predators can be removed and corals introduced.

Several species of butterflyfish have been used in this manner with varying degrees of success. The best species for this purpose is the Raccoon Butterflyfish, *Chaetodon lunula* (M. Awai in Sprung, 1990). Klein's Butterfly, *Chaetodon kleinii*, will also eat *Aiptasia* (Fosså and Nilsen, 1993c), as will the Caribbean Banded Butterfly, *Chaetodon striatus*. Be warned, however, that these fish may eat almost any coral or anemone. The Copperband Butterflyfish,

Chelmon rostratus, has been used with varying degrees of success and has the added benefit of not bothering other corals, although they will eat annelids such as feather duster worms. Certain shrim can also be used to control *Aiptasia* but they will also eat other corals like zoanthids. Some have suggested using the Peppermint shrimp, *Rhynchocinetes uritai* (Tullock, 1991). These shrimp will, however, also eat zoanthids such as Yellow Polyp, *Parazoanthus gracilis.* Furthermore, these shrimp tend to eat only juvenile *Aiptasia,* leaving behind the reproductive adults. Certain species o anemone-eating nudibranch such as those belonging to the genus *Spurilla* (e.g. the Caribbean nudibranch, *Spurilla neapolitana*) ca also be used successfully in newly set-up aquariums that contain no other corals or anemones. It is also possible that the Caribbean nudibranch *Dondice occidentalis* will eat *Aiptasia,* and it may not bother other anemones or corals.

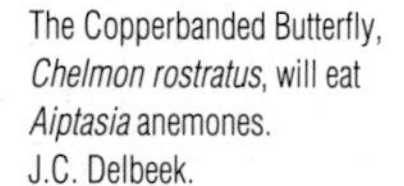

The Copperbanded Butterfly, *Chelmon rostratus,* will eat *Aiptasia* anemones. J.C. Delbeek.

Mechanical removal can be used at any time before or after other animals have been added to the aquarium but may involve the removal of the rock the anemones are attached to. One of the dangers in mechanical removal is that *Aiptasia* have amazing regenerative powers. If even a small portion of tissue is left behind it can regrow into a new individual. If the tissue becomes lacerate into several small pieces, each piece has the potential to develop into a new anemone. Therefore it is important that all tissue be removed. Mechanical removal can be quite difficult and time consuming since most *Aiptasia* are anchored in holes and crevices in the rock. Even when lightly touched they can retract into these holes with amazing speed!

One method of mechanical removal involves placing the rock with the anemones into a separate bucket. Then, using a knitting needle (or equivalent), simply grind the anemone out of the hole. Scrape as forcefully as you can to remove all the tissue. A strong stream of saltwater can then be used to rinse out any loose tissue. It is also possible to fill the hole with freshwater, to destroy any remaining tissue. This will kill any other marine organisms in the hole too, but the loss is usually minimal. Some authors have recommended removing the rock into boiling water and then scrubbing it clean (Haywood and Wells, 1989). This technique should only be used on dead rock and NEVER on live rock or rock covered with coralline algae!

Another method involves the use of an hypodermic needle and various solutions. The needle is used to inject a solution into the body of the anemone, killing it quickly. When injecting the anemone, the needle must pierce the body wall without passing completely through the anemone. The solution must be injected so that it fills the body cavity. One of the problems involved with this method is that the anemones must be easily accessible, and let's face it, they don't just sit there and take it, they rapidly retreat into the rock . Another problem is that some of these solutions can be detrimental to other tank inhabitants so only a few anemones can be treated each day. A supersaturated solution of calcium hydroxide (2 grams in 100 millilitres) has been used in this manner with excellent results without any noticeable side affects to the rest of the aquarium (see Hemdal, 1992). Approximately 1-2 mL is drawn into a 5 mL syringe with a wide-bore needle. The solution should be kept agitated to prevent clogging of the needle by gently twirling the syringe between the fingers. A 5% solution of hydrochloric acid injected by syringe has been recommended also (Wilkens and Birkholz, 1986) as has boiling water (P. Wilkens, pers. comm.). The boiling water can be effective even without penetrating the anemone, but be careful not to burn yourself! Aquarists have also successfully killed these anemones by injecting them with a small amount of dilute copper sulfate solution, or by inserting a copper wire into them and leaving it there for several minutes. We caution that while copper is an important trace element for invertebrates, excess copper is highly toxic to them. Therefore we do not encourage the novice to use copper to control *Aiptasia,* though it can be done safely by the experienced aquarist.

Phylum: Platyhelminthes

Class: Turbellaria
Common Names: Flatworms, Planaria

Flatworms come in many shapes and sizes. Some are parasites, commensals or opportunistic feeders of corals while others are predators and scavengers feeding mainly on algae, detritus, dead organisms, copepods or diatoms. These small worms are dorso-ventrally flattened and normally have a rounded head with a forked tail. In the aquarium, the non-parasitic forms are commonly encountered in the first few months when a reef aquarium with live rock is setup. The parasitic, opportunistic and commensal forms are primarily introduced with new coral specimens.

Many species of flatworms can reproduce both sexually and asexually and for this reason all it takes is one specimen to infect an aquarium. Their regenerative powers are phenomenal. If a single individual is cut into several pieces, each piece can develop into a new adult.

Flatworms have a variety of feeding mechanisms. Some species envelop and trap prey using their skin folds and a sticky slime. Others pierce the body wall of their prey using their protrusible pharynx and proteolytic enzymes, and suck out the contents (Barnes, 1974).

Non-Parasitic Forms

There are approximately 14 different non-parasitic forms that have been seen in aquariums (Beul, 1987). Some of these feed on small copepods, others feed on diatoms and others act as scavengers (Wilkens and Birkholz, 1986; Beul, 1987). The most commonly encountered flatworm in the reef aquarium are the small, semi-transparent, whitish ones belonging to the suborder Maricola. These are usually found in newly setup aquariums with live rock. They have a length of 5-10 mm (0.25 in.) with a rounded anterior end and fork-shaped rear end. Mainly active at night, they are usually found crawling along the glass or rock but they can swim short distances when disturbed. These small worms should not cause any undue alarm as they are actually quite helpful and will usually disappear within a few months as their food supply diminishes. If they do not, it could be that you are overfeeding the aquarium (Wilkens and Birkholz, 1986). Colourful, iridescent flatworms that feed on diatoms and other microalgae sometimes proliferate in aquariums with strong illumination and heavy growth of algae on the glass.

Reduction of the algae growth through limitation of plant nutrients usually causes the flatworm population to decline.

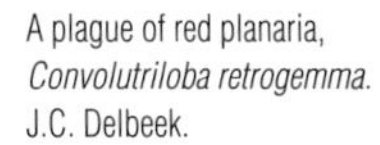

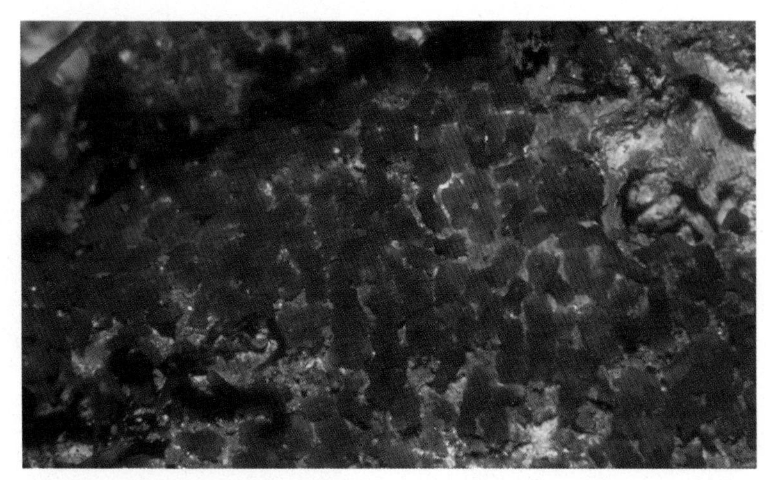

A plague of red planaria, *Convolutriloba retrogemma*. J.C. Delbeek.

Another non-parasitic form (*Convolutriloba retrogemma*) is rust-brown in colour and is an opportunistic feeder on photosynthetic corals (Wilkens and Birkholz, 1986; Lange and Kaiser, 1991). They grow to a length of 5 mm (0.25 in.) with two, and sometimes three, pointed projections on the rear (Beul, 1987). It is believed that these flatworms are not true parasites and only attack damaged or infected corals (Wilkens and Birkholz, 1986). They feed on the damaged tissues and ingest the zooxanthellae. Their colour comes from the fact they incorporate the zooxanthellae in their body wall and use them much as the coral did. For this reason they can multiply asexually very quickly in brightly lit aquaria and can quickly overrun an entire aquarium. These flatworms are usually found in aquariums that have Indonesian live rock, or they can be acquired by purchasing specimens from a dealer whose tanks see a high turnover of specimens imported from Indonesia.

Parasitic Forms

Parasitic flatworms are usually distinguished by the fact that they are always found on their hosts. They are generally light grey to brown in colour and many species have a stripe down the back. Although many types of corals can be affected, mushroom anemones (*Discosoma* sp., *Actinodiscus* sp.) are the most often infected. Large polyped hard corals can also be infected, especially Elegance Coral (*Catalaphyllia jardinei*) and Bubble coral (*Plerogyra sinuosa*). Soft corals that can be infested include *Sinularia, Cladiella, Litophyton* and *Sarcophyton* (Wilkens, 1990).

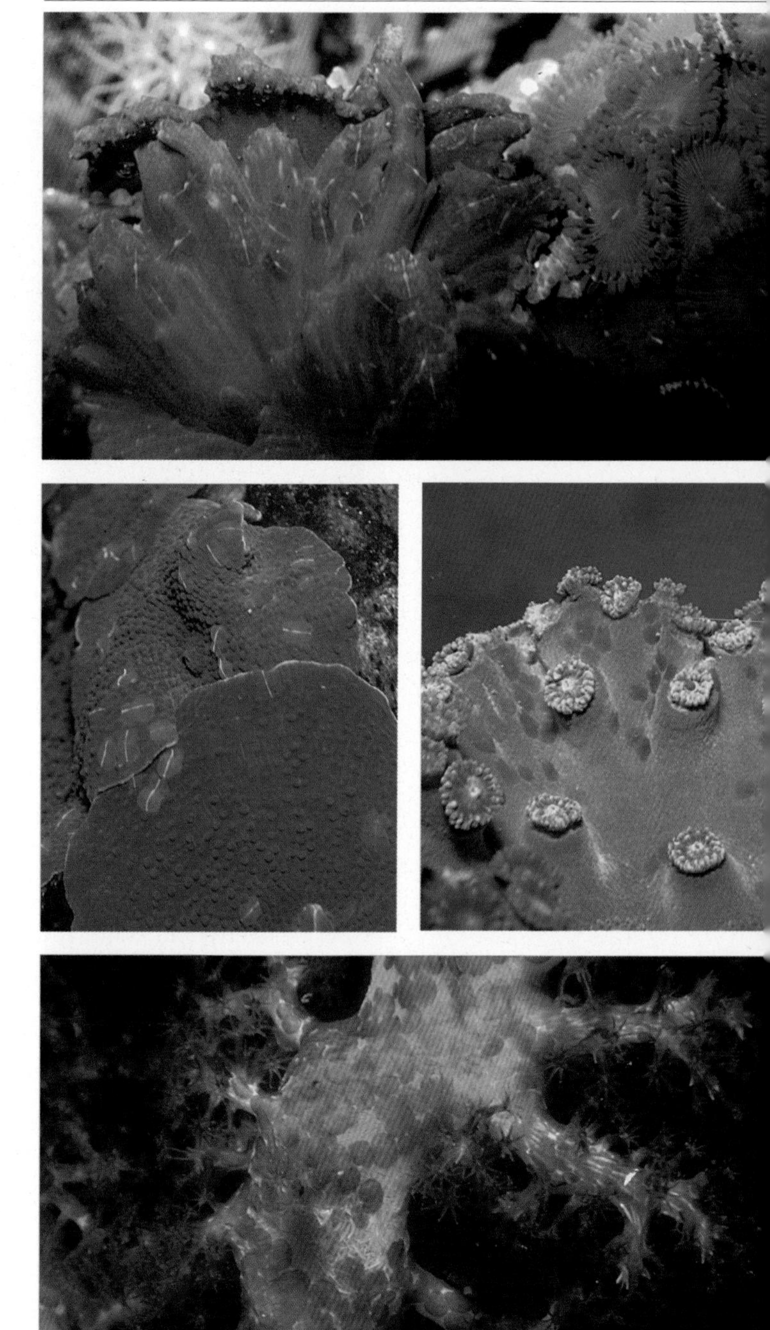

Planaria infesting a specimen of *Sinularia dura*. Note that the stripes on the planarians look like the mouths of the zoanthid anemones on the right. J. Sprung.

Mushroom anemones infested with planaria. J.C. Delbeek.

Planaria can also affect stony corals such as this *Turbinaria peltata*. J.C. Delbeek.

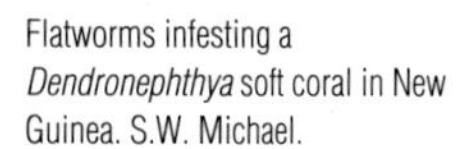

Flatworms infesting a *Dendronephthya* soft coral in New Guinea. S.W. Michael.

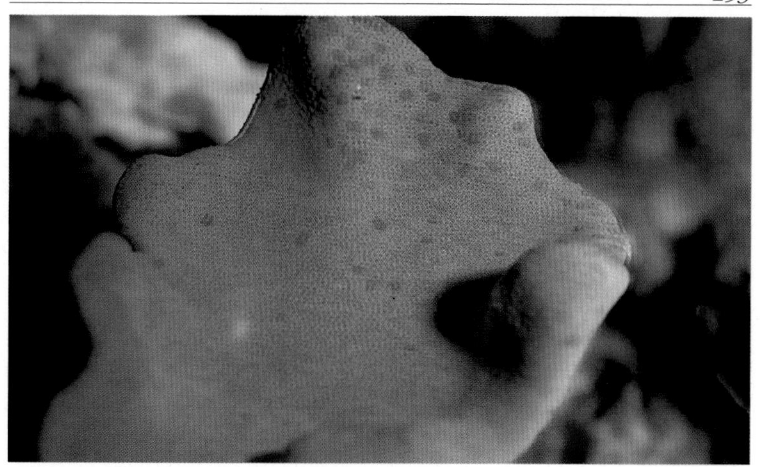

Planaria can affect soft corals such as this *Sarcophyton*. L.N. Dekker.

If these flatworms are allowed to multiply they can completely cover the coral, resulting in its death. However, it is unclear if the coral dies from the flatworms or from lack of light. In some cases, the coral does not appear to be harmed by the infestation. It may be that the flatworms are merely feeding on the detritus and bacteria trapped in the body slime of the coral and not on the actual tissue (Wilkens, 1990). If this is true, then these species are commensals and not true parasites.

Removal: There are a number of methods that can be used to remove harmful flatworms from the aquarium. The best method is of course prevention. Check any purchases very closely for any sign of flatworms and remove them before placing the specimen in the aquarium. If possible, isolate the coral in a separate aquarium for a few days where you can closely observe it. Since the worms are quickly killed by freshwater, a brief (5-10 second) freshwater dip can be employed to prevent their introduction into the aquarium. In freshwater the worms immediately lose their grip and fall off (see additional information below).

No matter how careful an aquarist is, flatworms can still be introduced unnoticed. It is important to remove them as quickly as possible before they multiply. Failure to do so can result in a serious plague that will prove to be extremely difficult to control later.

Physical removal is possible when only a few corals are affected. For example, mushroom anemone colonies can be removed from the tank and vigorously shaken in a separate bucket of seawater. This will remove the majority of flatworms. Any remaining worms can then be brushed off with a soft paintbrush and siphoned out (Wilkens, 1990).

Some species (e.g. *C. retrogemma*) are attracted to light and this can be used to remove them from the aquarium. Turn off all the aquarium lights and shine a single light onto a spot in the aquarium. The flatworms will soon begin to collect at this point. It is now a simple matter of siphoning the flatworms from this area with a small hose (Wilkens, 1990).

If only a certain coral is affected it is possible to remove the coral and give it a short (5-10 second) freshwater dip (Wilkens and Birkholz, 1986; Beul, 1987; Ruiter, 1987a). This will effectively dislodge the flatworms from the coral. Make sure the freshwater is at the same temperature and pH as the aquarium, to avoid chemically shocking the coral. Not all corals will accept a freshwater dip. Such dips work best on mushroom anemones and stony corals, and the tougher soft corals *Lobophytum, Sarcophyton,* and *Sinularia*.

Water chemistry changes can also be used to control flatworms. It has been found that increasing the specific gravity above 1.022 can rid a tank of flatworms (Wilkens and Birkholz, 1986; Ruiter, 1987a; Wilkens, 1990). However, mushroom anemones do not react well to such a change (Ruiter, 1987a). High pH values (8.4-8.6) can also reduce flatworm populations (Wilkens and Birkholz, 1986).

Natural predators of flatworms must exist or else they would multiply unchecked in the wild. The problem is that very few aquarists have found effective predators, and not every individual of a species will eat flatworms. For example, certain wrasse species are reported to eat flatworms. These include the Six-Lined Wrasse (*Pseudocheilinus hexataenia*) and the Yellow Wrasse (*Halichoeres chrysus*) (Wilkens, 1990). A Leopard Wrasse, *Macropharyngodon varialvus* has been reported to eat flatworms on mushroom anemones and *Sarcophyton* soft corals, and it is possible that other members of this genus may do so too (DeVries, 1987; B. Carlson, pers. comm.). Wrasses of the genus *Anampses* may also feed on flatworms, and might be worth trying (B. Carlson, pers. comm.). The Spotted Mandarin, *Synchiropus ocellatus*, has been recommended by Wilkens (1990) but it is not always effective. We have seen evidence that the Goldenheaded Sleeper goby, *V. strigatus* may eat the rust-coloured flatworm, *C. retrogemma*. Other bottom sifting gobies have also been observed eating them (Joe Yaiullo, pers.

Macropharyngodon wrasses may be planaria eaters. S.W. Michael.

Anampses wrasses may also be planaria eaters. S.W. Michael.

comm.). Starfish from the genus *Nardoa* and snails of the genus *Thais* have also been found to feed on planaria, but only when the population of flatworms has not increased greatly (Wilkens, 1990).

Synchiropus ocellatus, the spotted mandarin, sometimes feeds on planaria. J. Sprung.

Lange and Kaiser (1991) found that Tetra's Marin-Oomed™ was useful in eliminating the rust-coloured flatworms, *C. retrogemma*, from an aquarium that contained only leather corals of the genus *Sarcophyton*. When used as directed the flatworms burst, releasing a red-coloured toxin into the water. This toxin is deadly to fish and must be removed immediately either by water changes or the use of large quantities of activated carbon. We recommend that this technique be attempted on aquariums that contain only *Sarcophyton*, and only after all other methods have been exhausted. Other, safer, chemical methods for the removal of flatworms will no doubt appear within the next few years.

Phylum: Mollusca
Class: Gastropoda
Common Name: Snails and Sea Slugs

The are large numbers of snail species found in coral reef areas. Some are harmless herbivores, others feed on small live fish, certain ones act as omnivorous scavengers and there are numerous types that feed on the corals and clams that we keep in our aquariums. These are often inadvertently included in shipments of live rock, corals and clams. If these snails are not removed immediately they can quickly rid your tank of clams, corals, gorgonians, anemones and zoanthids. Fortunately most of these snails are easily removed from the aquarium. The biggest problem is recognizing a predatory snail or realizing that a sick coral is the victim of snail predation. Once the offending snail is discovered, it is a simple matter to remove it from the aquarium. As with so many coral and clam predators, prevention is the key. Examine new specimens very carefully before and after purchase for any suspect snails.

Subclass: Prosobranchia
Order: Archaeogastropoda
Superfamily: Fissurellacea
Common Name: Keyhole Limpets

A limpet feeding on a stony coral.
J. Sprung.

Keyhole Limpets are primitive gastropods with a cap-like or depressed cone- shaped shell that is adapted for close adherence to the substrate (Gosner,1971), and a single apical hole from which the siphon emerges. These small snails are usually introduced into the aquarium along with live rock. Although these snails are primarily herbivores, we have observed them feeding on both stony and soft corals.

Order: Mesoagastropoda
Family: Architectonicidae
Common Names: Box Snails, Sundial Snails

One of the most commonly encountered coral predators is the Checkered Box snail, *Heliacus areola.* These snails are often found nestled between the polyps of a zoanthid colony on which they feed at night. They outstretch their extendible proboscis and make a small hole near the base of the polyp from which they proceed to suck out the contents. The polyp eventually shrivels up and falls

Heliacus snail feeding on zoanthids.
S.W. Michael.

off the rock (Ruiter, 1987b). Other members of this family such as *Philippia* and *Architectonica,* are known to eat coral polyps and anemones (Ates, 1987).

Family: Coralliophilidae
Common Name: *Rapa Rapa* Snail

Leather corals are often preyed upon by snails. For example snails of the family Coralliophilidae such as *Rapa rapa* are often associated with *Sarcophyton, Sinularia* and *Lobophytum* leather corals (Wilkens and Birkholz, 1986). These cream-coloured snails have an onion-like appearance and can reach a length of 50-90 mm (2-4 in.). They are often found near the base of the coral and have the ability to bore a hole into the side of the animal, moving into the body column from where they proceed to hollow out the entire specimen (Wilkens and Birkholz, 1986). Similar snails of the genus *Coralliophila* feed on stony coral tissue. They have a round shell either greyish in color or covered with algae, and the animal is yellow or orange.

Family: Cymatiidae
Common Name: None

Cymatium muricinum has shown to be predatory on tridacnid clams, and is particularly a problem for clam farms that have ocean based grow-out (Perron, et al., 1985). This snail is found throughout the Indo-Pacific and in the western Atlantic, so its introduction to the aquarium with live rock is a possibility.
The snail's larvae are believed to settle directly on a tridacnid, and

then undergo metamorphosis. At this point they enter through the clam's byssal opening and lodge between the shell and mantle (Perron et al., 1985). The clam appears to be little affected as the snail begins to feed on the juices of the mantle. Eventually the clam will react by closing its valves. The clam may even try to enclose the snail inside a pearl-like blister (Perron et al., 1985). Eventually the clam exhibits wide gaping, and death shortly follows. The grown snail then leaves the dead clam and moves on to another victim. Larger snails, 25-50 mm (1-2 in.), position themselves at the bottom of the clam, next to the byssal opening. They extend their proboscis into the opening and feed on the tissue inside (Perron et al., 1985). Larger clams are generally immune to predation by these snails as their great weight presses the byssal opening down onto the substrate so firmly that the snails cannot reach it (Perron et al., 1985). *Hippopus* spp. are generally immune as their byssal opening is too small to allow the snails to enter.

Family: Cypraeidae

Common Name: Cowrie

Cowries are often offered for sale in aquarium shops. Although the majority of these are herbivores, certain species can feed on corals. Cowries tend also to be rather destructive in that they can easily topple over live rock and coral as they move about the aquarium.

A common cowrie *Cypraea* (*spadicea*?). S.W. Michael.

Family: Ovulidae

Common Name: Egg Cowrie

Closely related, and sometimes mistaken for true cowries, are Ovulidae snails. Many of these are predators of soft corals and gorgonians. The unidentified ovulid illustrated came attached on a colony of *Lemnalia* on which it was feeding. The well-known Caribbean Flamingo Tongue, *Cyphoma gibbosum*, is an example

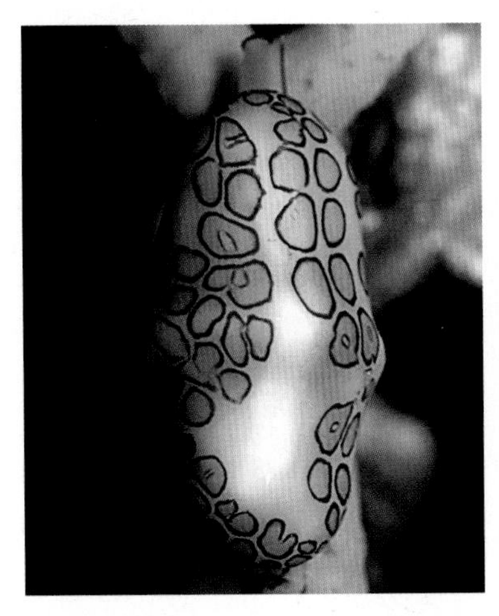

Cyphoma gibbosum feeding on a Caribbean gorgonian, *Plexaurella* sp. J. Sprung.

An unidentified ovulid snail feeding on *Lemnalia* sp. imported from Indonesia. J. Sprung.

of an ovulid that eats mainly gorgonians, particularly sea fans. There are, however, many more, lesser known species that prey exclusively on soft corals such as *Volva brevirostris*, *Prionovolva* spp. and *Simnia loebbeckeana*. Some of these are so specific that each species of Pacific gorgonian has its own species of predatory ovulid snail. Many of these snails blend in very well with their prey, often exhibiting the same colours and surface textures on their mantles. The cover of the August 1991 issue of Freshwater and Marine Aquarium magazine shows an excellent example of a gorgonian-eating ovulid photographed by Alex Kerstitch. Wilkens and Birkholz (1986) go into more detail on these animals.

One snail that is often offered for sale as the Egg Cowry, is in fact an ovulid snail, *Ovula ovum*. This snail has a snow white shell covered by a jet black mantle with white spots. This species feeds exclusively on soft corals such as *Xenia* and especially, *Sarcophyton,* and should not be purchased (Rudman, 1984; Coll and Sammarco, 1986). Even though these soft corals contain various toxic chemical compounds, *O. ovum* can apparently transform them into less toxic compounds without any ill effects (Coll and Sammarco, 1986).

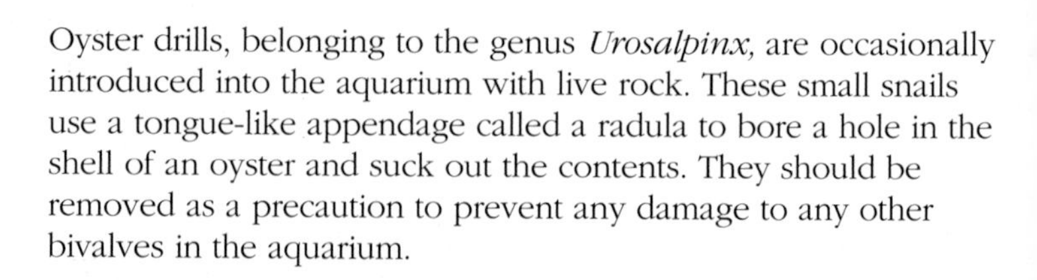

Order: Neogastropoda

Family: Muricidae

Common Name: Oyster Drills

Above:Two unidentified muricidae? snails feeding on *Zoanthus* sp. colonial anemones from Indonesia. J. Sprung.

Oyster drills, belonging to the genus *Urosalpinx,* are occasionally introduced into the aquarium with live rock. These small snails use a tongue-like appendage called a radula to bore a hole in the shell of an oyster and suck out the contents. They should be removed as a precaution to prevent any damage to any other bivalves in the aquarium.

Muricidae snails also affect corals. *Drupella cornus* feeds on acroporid and pocilloporid corals in the Indo-Pacific (Schuhmacher, 1991), and a similar appearing snail has also been found feeding on zoanthid anemones from Indonesia, see photo.

Subclass: Opisthobranchia

Order: Nudibranchia

Common Names: Nudibranchs, Sea Slugs

Nudibranchs, or sea slugs, are found in all the world's oceans. They exhibit a wide range of body forms, colours and feeding preferences. It is these feeding preferences that we are most concerned with. Most nudibranch species are quite specific in what they will eat, often feeding only on one species of prey, while others have more catholic diets, feeding on several different species within a family. Most nudibranchs only have a maximum life span of 1-2 years so they are not long lived in the wild. Given that their food type may be impossible to provide and that they are extremely short lived it is best not to purchase these animals, regardless of their beauty. Unless one knows the exact food item a particular nudibranch requires, and can supply it easily, we do not recommend keeping these invertebrates; they are better left on the reef. Nudibranchs are occasionally introduced with live rock or corals. In this case it is important to know what type it is and determine if it can harm your corals or clams. If you are unsure of the type, it is always a wise precaution to remove the animal before it can cause any damage. More often, the aquarist does not notice the nudibranch until it has already begun to cause damage to the other tank inhabitants. Nudibranchs are easily removed by simply siphoning them out of the aquarium or removing them with a pair of tweezers. Exceptions occur with certain species that can reproduce in the aquarium and can reach plague proportions if left unchecked (Achterkamp, 1988). There are four suborders within the order Nudibranchia, each of which has specific morphological characteristics that help us to differentiate between them.

Suborder: Doridacea

The Doridacea contain over 25 families and is the most familiar form of nudibranch to aquarists. They are characterized by having two prominent rhinophores; horn-like projections on the head that are involved in chemoreception. Dorids also have a dorso-

Above: *Phyllida* nudibranchs can release toxins into the water. S.W. Michael.

A dorid nudibranch on the left and a small aeolid nudibranch on the right. S.W. Michael.

ventrally flattened body where the edge of the back extends over the foot. At the rear of the animal, on the dorsal surface, feathery gills are located in a ring surrounding the anus. These nudibranchs feed primarily on sponges although some species also feed on bryozoans and ascidians (sea squirts). As a consequence they do not pose any threat to corals or clams in the aquarium. However, Wilkens (1990) reports that some dorids will prey upon stony corals, so the aquarist must consider this a possibility. Many of them are very specific in the type of sponge or ascidian that they will eat, and do very poorly in the aquarium. One potential danger with some genera in this group is the possession of poison secreting glands for defense. In at least one reported case the dorid nudibranch, *Phyllidia varicosa*, caused the death of fish and invertebrates kept in the same container (Johannes, 1963).

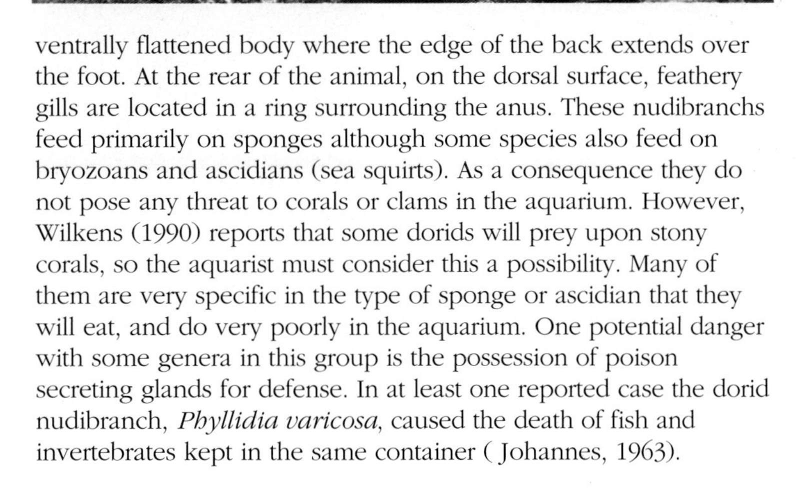

Suborder: Aeolidacea

Aeolidacea are the second largest suborder with over 20 families. These nudibranchs also have rhinophores but they do not have feathery gills. Instead they have prominent oral tentacles around the mouth and numerous projections on the back called cerata.

Right: An unidentified aeolid nudibranch found feeding on a *Cladiella* sp. soft coral. J.C. Delbeek.

Above: Another unidentified aeolid nudibranch associated with "Green Star Polyp" (*Pachyclavularia viridis*). The grey cerata with pale yellow tips make it look like a small sea anemone. It is a commensal that apparently causes no harm, feeding on mucus(?) from the host. It readily reproduces in the aquarium. J. Sprung.

Right: A well-camouflaged *Phyllodesmium* sp. nudibranch within a *Xenia* colony. Note the polyp-like appearance of the cerata. J.C. Delbeek.

Far right: *Phyllodesmium* sp. nudibranch removed from the *Xenia* colony. Compare the "polyps" in this photo with the previous one. J.C. Delbeek.

Cerata are extensions of the body into which the midgut extends. In the Aeolidacea, the tips of the cerata contain tiny sacs called cnidosacs, where the nudibranch can store nematocysts from their prey to use for their own defense. These cerata can also store zooxanthellae ingested from the prey, and can be used by the nudibranch to make food.

Aeolidacea species can feed on bryozoans, sea anemones and soft corals. The genus *Phyllodesmium* feed predominately on soft corals such as *Sarcophyton, Anthelia, Xenia, Clavularia* and *Pachyclavularia*. The ingested zooxanthellae are stored in wide, flat cerata and are used by the nudibranch to manufacture food. Other examples include the genus *Spurilla,* which feed mainly on sea anemones, and *Phestilla melanobrachia,* which feeds on

Tubastrea coral (Rudman, 1984; Coleman, 1989). Often a clue to the type of prey eaten can be seen by looking at the shape of the cerata as they will often mimic the shape of the coral polyps preyed upon (see Wilkens and Birkholz, 1986).

Suborder: Dendronotacea

The Dendronotacea, contain 10 families and are characterized by having rhinophores that can be retracted into rhinophoral sheaths. Members of this group also have cerata that are often highly branched in appearance. The most commonly encountered genus in the aquarium is *Dendronotus*. These nudibranchs are usually 1-6 cm (0.4-2.4 in.) in length and have a milky white body covered with highly branched cerata. They prey on just about every type of soft coral but are most often associated with *Sinularia, Cladiella* and *Sarcophyton*. As with most nudibranchs, they are most active at night. During the day they will usually gather at the base of the coral, right next to the substrate, with all their cerata and rhinophores retracted, making them difficult to detect. At night they make their way along the coral, feeding as they go. Tell-tale signs of damage by this nudibranch include small white or yellow scraped areas or holes on the body column and top, loss of colour, shortening of polyps and a failure to expand (Wilkens and Birkholz, 1986; Achterkamp, 1988; J.C. Delbeek, pers. obs.). It is extremely important that the aquarist carefully check any new corals for this pest. Although they can be easily removed with tweezers or a siphon, they can sometimes occur in groups of 60 or more on a single coral. They also have the ability to multiply in the aquarium without a larval stage (Wilkens and Birkholz, 1986; Achterkamp, 1988).

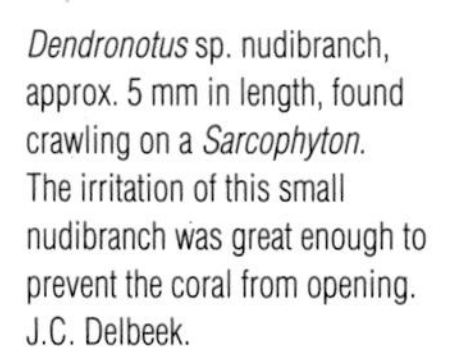
Dendronotus sp. nudibranch, approx. 5 mm in length, found crawling on a *Sarcophyton*. The irritation of this small nudibranch was great enough to prevent the coral from opening. J.C. Delbeek.

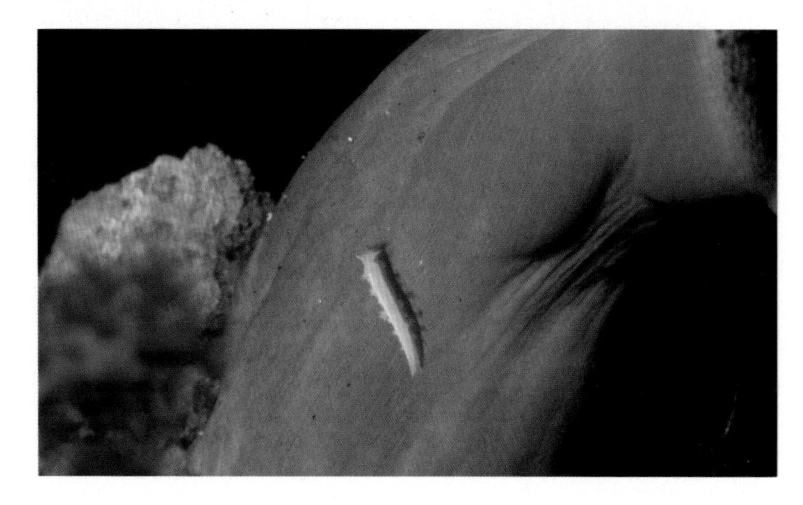

The genus *Tritonia* are also predatory nudibranchs occasionally found on newly imported Pacific gorgonians. These tiny nudibranchs can quickly devour an entire colony and should be removed as soon as they are detected. Other soft coral eating genera include *Tritonopsilla* and *Marioniopsis* (Coleman, 1989). *Tritoniopsis frydis* occurs in the Caribbean and is commonly found on gorgonians from this region.

Tritonia sp.(?) nudibranch feeding on a gorgonian, *Pseudopterogorgia* sp. S.W. Michael.

Tritoniopsis sp.(?) that feeds on leather corals, *Sarcophyton* spp. This specimen was imported with a yellow leather coral from Tonga. It was discovered after it had consumed half of its host's base practically overnight! It layed yellow coils of eggs on the rock near the base of the leather coral. This nudibranch is difficult to detect as it is nearly invisible on yellow leather corals, and remains hidden at the basal attachment to the rock during the day. D. Ramirez.

The white "strings" coiled around this gorgonian branch contain the eggs of a *Tritonia* sp. nudibranch. S.W. Michael.

Removal: Once Dendronotacea gain a foothold in the aquarium they can be very difficult to eradicate. One method that can be used to make this nudibranch easier to detect, involves shutting down the water circulation in the aquarium for several hours. The decrease in oxygen levels in the aquarium often cause the nudibranchs to move about, making them easier to see and remove (Achterkamp, 1988). Of course, if you are using one, the external filtration system should be kept running by cycling water through it. Do not shut down an external biological filtration system for any length of time, lest anaerobic conditions build-up. Another means of catching these nudibranchs is to temporarily fill the aquarium with fine air bubbles by means of a venturi or injecting air to the suction side of a circulatory pump. The millions of tiny air bubbles irritate the nudibranchs and cause them to move frantically about the aquarium, making them easy to detect and remove (Achterkamp, 1988). Please keep in mind that both of these techniques can also be detrimental to other aquarium inhabitants. In our experience these predatory nudibranchs are easier to detect at night when they are most active.

Suborder: Arminacea

The fourth suborder, Arminacea, contains only 9 families. These nudibranchs have no rhinophoral sheaths, lack a first pair of oral tentacles and some species have cerata. They feed mainly on cnidarians such as corals, gorgonians and sea pens (Coleman, 1989).

Order: Pyramidellacea

Family: Pyramidellidae

There are a wide variety of snails that feed on clams and oysters. The majority of these belong to the family Pyramidellidae, of which there are at least 1000 species in the Pacific alone (Cumming, 1988). These snails are generally quite small reaching a maximum length of 2-4 mm (0.08-0.16 in.), and look like small grains of rice. The majority of what is known about these snails comes from the species that are commonly found feeding on oysters and clams in commercial operations. Very little is known about how many species may affect tridacnid clams. *Tathrella iredalei* and *Pyrgiscus* sp. are two species that have been isolated in Australian tridacnid clam farms (Cumming, 1988). J. Sprung observed a small snail which fed on stony corals, using a long stylet to suck up fluids as in other pyramidellidae. It apparently causes no harm to the coral as a single parasite, but if allowed to multiply within the aquarium it could become a pest.

The typical feeding location of *Tathrella* snails, along the upper shell edge of a *Tridacna crocea*. J.C. Delbeek.

Studies of *Pyrgiscus* have shown them to be very serious pests of tridacnids in aquaculture systems. They can exhibit an extremely rapid population growth when in land-based seawater tanks or in trays raised above the substrate in the wild (Cumming, 1988).

The protrusible proboscis of one of these *Tathrella* sp. snails is evident as it pierces the mantle of a *T. derasa* clam. Note the egg mass on the shell of the clam (far right). The upside-down vase like object above and to the right of the snails is a harmless common syconoid sponge that often multiplies in reef aquariums. S.W. Michael.

Reproduction of these simultaneous hermaphrodites is quite rapid, with a 2.5 mm (0.1 in.) snail capable of producing 2-3 egg masses/day, each with up to 120 eggs (Cumming, 1988). The eggs are held in a jelly-like mass, often with several egg masses close together on the clam shell. The eggs hatch after 10 days and the larvae are capable of feeding within 3 days of hatching (Cumming, 1988). The young tend to remain on the clam they hatch out on. As a result, all it takes is two of these snails to cause a rapid population growth on a single clam.

Pyramidellid snails feed mainly at night. During the day they stay near the base of the clam or, in species with large scutes such as *Tridacna squamosa*, between the scutes, out of direct sunlight. At night they migrate up towards the lip of the shell where they extend their proboscis and, using their needle-like stylus, poke a hole into the mantle of the clam. They then suck out fluid contents from the mantle. When left unchecked, large numbers of these snails can easily kill a clam within days or months, depending on the size of the clam.

Removal: Interestingly, these snails are relatively rare in the wild, which indicates that some form of natural biological control must be in place. One such natural predator is the portunid crab *Thalamita sima*. Small specimens of these crabs (1-1.5 cm; 0.4-0.6 in.) are often found in clam beds and have been used to control pyramidellid snails in aquaculture projects with some success, however, they have also been found to feed on small (4 mm; 1.6 in.) tridacnid clams (Cumming, 1988). Certain wrasse species are more promising predators. Members of the genus *Halichoeres*, specifically

For those polychaetes that invade the tissues of soft corals there is little that can be done to remove them. The only effective method is to cut off the infected portions and try to propagate the healthy sections that remain.

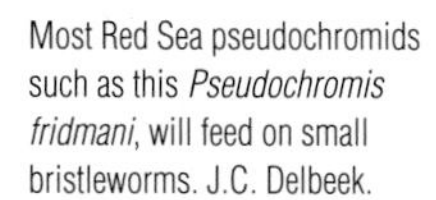

Most Red Sea pseudochromids such as this *Pseudochromis fridmani*, will feed on small bristleworms. J.C. Delbeek.

Phylum: Arthropoda

Class: Crustacea

Crustaceans include many of the invertebrates that we commonly encounter in the marine aquarium such as amphipods, barnacles, copepods, crabs, lobsters, Mantis shrimp and true shrimp. Very little is known about the food, feeding methods or behaviour of reef crustaceans (Bruce, 1984). A large number of crustaceans live symbiotically with other reef inhabitants such as corals, gorgonians, sponges, sea squirts, clams, snails, starfish, sea cucumbers, sea urchins and anemones. In many of these relationships it is unclear whether the shrimp is a parasite, predator or scavenger (Bruce, 1984). In some cases the shrimp merely uses the host as a refuge from which to foray for food. A large number of these crustaceans can be destructive in the reef aquarium but we will only deal with the more commonly encountered ones.

Subclass: Cirripedia

Common Name: Barnacles

Most reef barnacles belong to the order Thoracica and many of them live in association with sponges and corals (Bruce, 1984). These barnacles are often encountered by the aquarist as they are

A harmless barnacle commonly found on *Euphyllia glabrescens*. J.C. Delbeek.

commonly found growing on *Euphyllia* corals, especially *E. glabrescens*. They do not bother the coral, but merely live in association with it, filter feeding from the water column. There are, however, parasitic barnacles belonging to the order Ascothoracica that can be either external or internal parasites. The external forms are commonly found on Sea Lilies and Brittle Stars, while the internal forms can infect soft corals (Barnes, 1974).

Subclass: Malacostraca

Superorder: Hoplocarida
Order: Stomatopoda
Common Names: Mantis Shrimp, Thumb Splitters

Mantis shrimp are the bane of almost every reef aquarist. We all dread hearing that tell-tale snapping sound coming from deep within the rock-work of our aquaria. These shrimp are voracious predators feeding on small fish, shrimp, worms, clams and crabs. They have an elongated body with two well-developed eyes located on short stalks, and a pair of powerful pincers held folded underneath the body; similar to the terrestrial praying mantis. There are numerous species ranging in size from 5 cm to over 30 cm (2-12 in.) in length, many of them brilliantly coloured. The larger species tend to live in burrows in the sand, feeding on passing fish. The smaller species belonging to the genus *Gonodactylus,* are commonly found in coral rock, and these are the ones we commonly encounter in our aquariums.

A commonly encountered live rock-dwelling Mantis shrimp, *Gonodactylus* sp. S.W. Michael.

Mantis shrimp are extremely wary and are very difficult to eliminate from the aquarium, but they should be removed at all cost if you value your fish and invertebrate collection. These shrimp are quite intelligent and if you fail to catch them with one method you will have to resort to another because you will never get them the second time. Baited traps, placed close to the mantis' home work sometimes. A trap with a quick release door is a must. These can easily be constructed using a small, rigid plastic box and lid. Weigh the box with a rock to make it sink. Wrap an elastic band over the lid and prop it open with a piece of rigid plastic pipe (rigid airline tubing works well). Tie some transparent line to this post and lead it out of the aquarium. Bait the trap with fresh shrimp, clam or fish and wait. These crustaceans are most active at night so the trap should be set just as the lights go out. If not, your fish will eat most of the food. It would be best to place the trap into the aquarium a week or two before you bait it. This will allow the tank inhabitants time to adapt to its presence. You can also bait the trap each night and train the shrimp to come for the food. Then, when you are ready to trap out the shrimp, wait till it enters the trap and pull sharply on the line. This will pull out the plastic post and cause the trap to shut quickly. Make sure the elastic bands are fresh or else they will not close the trap quickly enough. This trap can also be used very effectively to remove small fish from the aquarium.

If you know exactly which rock the shrimp is living in, it is a simple matter to remove the rock to a separate bucket. Of course this is only simple when the rock is easy to move! Be aware that the rock may have a backdoor where the shrimp can escape, so make sure you have a large enough net to hold underneath the rock before you pull it out of the aquarium. Be very careful when handling the rock and make sure you are wearing a thick pair of gloves, these shrimp did not earn the name "Thumb Splitters" for nothing! Mantis Shrimp are fascinating animals and once removed from the main tank they can be placed in their own separate little aquarium, where you can safely watch their behaviour.

There are some fish species that will eat these shrimp but they are very aggressive species that may not be suitable for all aquariums. Specifically the Australian Dottyback, *Labracinus lineatus,* is reported to feed on Mantis shrimp but is a very nasty fish (S. Michael, pers. comm.). Roger Bull related to us a rather interesting technique for eradicating mantis shrimp from a newly set up reef aquarium. An octopus will quickly hunt down any crustaceans like Mantis Shrimp or crabs hiding in the rocks. Once they have been eliminated, the octopus can be removed. This technique is meant for new set-ups only, before the fish are introduced, because octopus can feed on fish (see also chapter 7, aquascaping).

Another method that can work to eliminate mantis shrimp is to use a sharp, pointed end scissors to injure or slice the mantis shrimp in two. The shrimp can be encouraged to come out by placing some bait near the entrance of the hole or by prodding the hole with a stick (Mantis shrimps are quite aggressive, and will attack an object inserted down the hole into which they have retreated). If you can hold the open scissor blades just around the opening of the hole, and can manage the beads of sweat running down your brow, when the shrimp pops its head out of the hole far enough, you might be able to pop its head off with a quick slice of the scissor. This is not a job for the faint-hearted. It is also possible to skewer a shrimp with a long sharp needle forcefully inserted into the hole into which it has retreated, or through nearby holes in the same rock. We wish to point out that these rather violent techniques are no more cruel than the methods used by predators of these powerful shrimp.

True Shrimp

Superorder: Eucarida

Order: Decapoda
Suborder: Natantia

Family: Hippolytidae

Shrimp of the genus *Saron* are occasionally offered for sale to aquarists. When kept in fish only aquariums, they make very

Saron marmoratus (female).
A.J. Nilsen.

interesting additions. They are also easy to sex and should prove to be a good breeding challenge. However, they can wreak havoc in a reef aquarium with corals and clams.

The Marble shrimp, *Saron marmoratus*, so named due to their green mottled appearance (although they can alter their colour to blend in with the background), is a large shrimp, reaching over 9 cm (3.6 in.) in length. They usually occur singly or in pairs. The males are easily identified due to their extremely large foreclaws. The females have short, hairy foreclaws held lower beneath the body (Achterkamp, 1985a).

Marble shrimp are shy and only venture about the aquarium at night. It is during these forays that the shrimp can attack tridacnid clams, corals, mushroom anemones and small polyped corals such as zoanthids (Achterkamp, 1985a; Wilkens, 1990). We do not recommend that you keep these shrimp in a reef tank but if you wish to do so you should keep a close eye on the health of your corals and clams.

The Buffalo shrimp, *S. inermis*, is a smaller version of *S. marmoratus*, reaching a maximum length of 5 cm (2 in.) (Achterkamp, 1989). The males of this species also have elongated foreclaws, but the difference between males and females is not as pronounced as in *S. marmoratus.* This species is also shy and will spend most of it's time hidden. At night, however, they will venture out to feed. These shrimp will eat most small polyped corals, especially zoanthids and parazoanthids such as *Parazoanthus gracilis* (Achterkamp, 1989). Also included in this

Lysmata amboinensis, the common cleaner shrimp. J.C. Delbeek.

family is the common cleaner shrimp, *Lysmata amboinensis.* Although not a coral eater, when hungry they will not hesitate to wade into a coral and remove food from it or even tear open the body cavity to remove food (Wilkens, 1990; J.C. Delbeek, pers. obs.) The related *Lysmata wurdemanni* from the Caribbean, also commonly called a Peppermint Shrimp, does not feed on corals but will steal food from them. It is a cleaner shrimp that will pick parasites from fishes, and it is most useful for the generation of plankton within the aquarium. Large numbers of these shrimp can be housed in the same aquarium, and they readily spawn, producing a continuous supply of planktonic larvae. Do not confuse this beneficial shrimp with *Rhynchocinetes* spp.

Family: Rhynchocinetidae
Common Names: Dancing or Peppermint shrimp

The commonly sold, *Rhynchocinetes uritai*, is another potential coral eater. These small, 4 cm, shrimp are usually found in harems of one male with 4 to 6 females. The males are easily recognized by their enlarged foreclaws. This species can produce a nearly constant supply of planktonic larvae that can serve as a food supply for other aquarium inhabitants. In Holland, this species has been raised to adulthood using rotifers, protozoans and baby brine shrimp (Achterkamp, 1985b). In an aquarium with large active fish these shrimp will remain hidden during the day. At night they move out across the reef to feed.

Unfortunately, these shrimp will feed on most corals in the aquarium including stony corals, mushroom anemones and

Rhynchocinetes uritai, the Dancing or Peppermint Shrimp, is a voracious predator of anthozoans. J.C. Delbeek.

especially zoanthids (Achterkamp, 1985b; Wilkens, 1990). Although they can be utilized to eat juvenile *Aiptasia* anemones (see Tullock, 1991), they will not restrict their diets to just this one food item and will eventually attack other cnidarians. Note: though they have similar appearance and the same common name, do not confuse this species with *Lysmata wurdemanni*, a cleaning shrimp which does not feed on cnidarians.

Family: Stenopodidae
Common Name: Banded Coral Shrimp

Shrimp in this family include the Banded Coral shrimp, *Stenopus hispidus*. These shrimp are commonly referred to as cleaner shrimp but they tend to clean mainly larger fish and it is not unheard of for them to catch and eat smaller fish. These shrimp can also destroy corals and anemones by ripping them open to remove ingested food. We have seen these shrimp kept in reef tanks but the aquarist should be made aware of the potential for trouble.

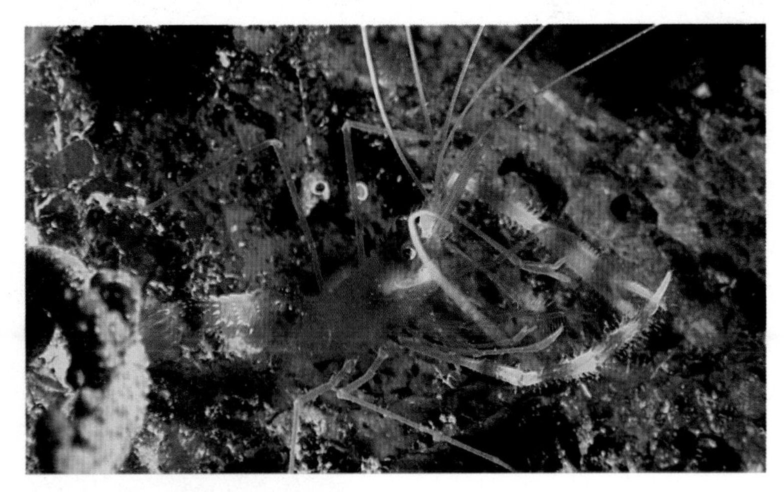

The Banded Coral Shrimp, *Stenopus hispidus*. J.C. Delbeek.

Suborder: Reptantia

Section: Macrura
Common Name: Lobsters

Most small lobsters offered for sale belong to the genera *Enoplometopus* (Reef lobsters) and *Panulirus* (Spiny lobsters). These lobsters are omnivorous scavengers and can be quite destructive in the reef aquarium. We do not recommend them for an aquarium with corals and clams, as they will undoubtedly feed upon many of the tank inhabitants, including small fish and shrimp.

Crabs

Most crabs are not to be trusted in the reef aquarium. When small they usually do not present much of a problem but as they grow they can become more and more destructive. There are so many varieties of crabs that it is difficult for the lay person to identify which ones may be dangerous, therefore the safest course would be to remove any crabs found. Crabs are divided into two main sections, Anomura and Brachyura. Anomura are between crabs and lobsters in appearance, having characteristics of both, and their abdomen is folded underneath the body. There are approximately 13 families. Brachyura are the so- called "true crabs" and compose over 45 families.

Section: Anomura

Hermit and Porcelain Crabs

Most hermit crabs are harmless when small, but as they grow they can become quite destructive. Some may attack living snails and some will pester corals Certain species remain small and can be used to help control algae. Porcelain crabs are usually harmless and most such as the common Anemone crab, *Neopetrolisthes maculatus*, are filter feeders.

The common Anemone crab, *Neopetrolisthes maculatus*, is a filter feeder. J.C. Delbeek.

A baited plastic cup easily traps this hungry crab. S.W. Michael.

glass and place it vertically against the rock, closest to the crab's home, just before the lights go out. When the crab climbs down into the glass, the smooth sides will prevent it from climbing back out and you can remove it easily the next day (Nooyen, 1990).

As we described for mantis shrimps, a sharp, pointed-end scissors can be used to extract crabs. The crabs can move quickly, but aren't as swift as mantis shrimp, so scissors are quite effective for catching them. The pointed end allows the scissors to be inserted in between rocks, and the force of a scissors closing can slice through the crab's shell, disabling the poor creature. The scissors can also provide a strong grip, allowing easy extraction of the crab from the rocks. The crab held between the blades can be turned or pulled until it lets go.

Problem Algae

Ostreobium

Ostreobium are green algae that live within the skeleton of living corals, just below the tissue layer. They are also free-living in rocks and shells made of calcium carbonate. *Ostreobium* bores through the calcium as it grows. Under nutrient poor conditions it causes no harm, but sufficient availability of nitrate and/or phosphate can cause it to grow more rapidly than usual, allowing it to encroach upon the newly deposited calcium where it can interfere with the deposition of new skeleton. This can cause the live coral tissue to recede from a localized area. Limiting nitrate and phosphate levels is the only known control.

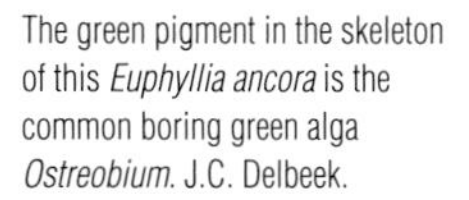

The green pigment in the skeleton of this *Euphyllia ancora* is the common boring green alga *Ostreobium*. J.C. Delbeek.

Valonia and Other Encroaching Algae

Any bare spots on a coral head or the rock that it is attached to, are sites where many species of algae can gain a foothold and shade the adjacent living coral tissue. The algae not only shade the coral tissue, but may also leach compounds that damage it. In the case of *Valonia* spp. and *Ventricaria* spp., "Green Bubble" algae, the growth of the plant may physically push the coral tissue away.

Valonia sp. algae encroaching upon living tissue of this *Goniastrea palauensis*. J. Sprung.

In dense filamentous species that trap detritus, the elevated phosphate in the microenvironment resulting from the breakdown of this detritus can poison coral tissue or prevent calcification (growth) in the area. *Cladophoropsis, Cladophora, Derbesia* and some cyanobacteria are the most common forms of algae that can cause this problem (see chapter 9 for photos).

The safest algae are the calcareous red coralline species, though even these can be a problem, particularly the articulated varieties such as *Jania* spp. and *Amphiroa* spp. that can encroach living coral tissue and cause harm.

Removal: Control of algae is achieved through the use of herbivores and the limitation of the nutrients nitrogen and phosphorous. Algae may be physically removed by hand, with a toothbrush, or siphoned off. See chapter 9 for more information on undesirable algae control.

Dinoflagellates

Recently, a number of hobbyists have reported the appearance of a mysterious slimy material coating both living and non-living surfaces in the aquarium. This material has the consistency of nasal mucus and it can be dark brown, light tan or nearly colourless. It usually develops as a light coating but soon entraps air and forms long strings that float up into the water. In some cases, microscopic examination of this material has shown it to be composed of the dinoflagellate, *Gamberdiscus toxicus*, that causes ciguaterra disease in fish, but other dinoflagellates have been found as well, such as *Ostreopsis* sp. (M. Foust and J. Norris, pers. comm.). Sometimes these dinoflagellate blooms are toxic to aquarium inhabitants. Herbivorous snails may roll over, stop eating and die. Tangs that eat the dinoflagellates may stop feeding and starve to death. Sea urchins that eat them may lose their spines and die. Water changes, salinity changes and siphoning do not seem to help since these organisms multiply so quickly. Ultraviolet sterilization has been indicated as a possible means of control, but not a cure (P. Wilkens, pers. comm.). Ozone has proved to be ineffective in the control of this plague in our experience. Oddly this plague seems to disappear almost as suddenly as it appears.

One effective control is photoperiod. Try leaving your lights off for a full day, thereafter maintaining a photoperiod of only four hours per day until the bloom dies. This is not in itself a cure, but keeps the bloom at a manageable level.

Dinoflagellates have much of the same nutritional requirements as other microalgae, and once introduced into the aquarium, they can subsist on extremely low levels of phosphate. This plague is difficult to control, though we have found that elevated pH is the most reliable cure (see below). The use of a phosphate absorbing medium has resulted in their eradication in certain cases but in our experience it does not always help. If you have a bloom, you should reduce additions of trace elements. Trace elements don't cause algae problems, but they can enhance growth when a problem already exists. After the bloom subsides, you can resume adding the proper dosage of trace elements.

Other types of algae may bloom simultaneously with the dinoflagellates, including diatoms, cyanobacteria, and green algae, all forming mucous-like gelatinous strings (M. Foust and J. Norris, pers. comm.). It is not known whether they contribute to the toxicity.

If you have a pH meter that is accurately calibrated, and you maintain the pH at 8.4 to 8.5 during the day, and above 8.2 at night through the controlled dosed addition of a saturated solution of calcium hydroxide, you might just tackle this problem. Such additions can precipitate phosphate from the water, and the high pH further seems to discourage the growth of undesirable algae, including dinoflagellates. Peter Wilkens made this suggestion to us, and our own experience confirms his observations. This seems to be the best solution at present (see chapter 9 for more information and photo).

Trouble-Shooting Ailing Corals and Tridacnid Clams

"Brown Jelly"

Disintegration of coral tissue sometimes proceeds as a front or band of what looks like "brown jelly". Behind the front, hard coral show a white, denuded skeleton; in front is healthy tissue. The front can move several inches per day and may rapidly consume an entire coral head. Sometimes a large portion of the head can "turn to jelly" when it is injured. In soft corals such as *Xenia* spp., entire colonies can be totally consumed within a day or two. If this brown matter is examined under the microscope, one finds that it is teaming with protozoans that appear to be filled with consumed zooxanthellae. The protozoan *Helicostoma nonatum* has commonly been identified with this condition but there are other protozoans typically present too (Wilkens and Birkholz, 1986; Beul, 1987). One can also find small worms feeding on the coral tissue in a "brown jelly" infection. It is suspected that several organisms work together to destroy coral tissue in these kinds of infections. Current theory is that bacteria or physical injury usually creates the initial insult. This begins a chain of events allowing protozoans, worms, and even copepods to do their dirty work. It

"Brown jelly" infection consumed a healthy colony of *Xenia* within a day. J.C. Delbeek.

may simply be that these organisms are opportunistic and ingest dead or damaged tissue. It is possible that this condition is related to the so-called "white band" disease reported on coral reefs. There the strong surge action would remove the excess "jelly" so that only the line of destruction would be apparent.

This condition most commonly attacks species of the stony coral genus *Euphyllia*, particularly *E. ancora* and *E. glabrescens*, but it is also common on newly imported (damaged) *Goniopora* species, and *Acropora* species. It is not limited to stony corals. This infection is also a common problem with the colonial anemones (*Zoanthus* and *Palythoa* species), and many soft corals, particularly *Anthelia*, *Cladiella*, *Xenia* and corals belonging to the family Nephtheidae. It is particularly common when the temperature of the aquarium exceeds 27 °C (80 °F), which is stressful to the animals.

Treatment: Siphon off the mass of "brown jelly" using a small diameter hose such as airline tubing. For stony corals, dip the entire colony in a freshwater bath for about one minute, and place the specimen in a holding tank with clean seawater and good water circulation (B. Carlson, pers. comm.). The osmotic shock from the freshwater dip effectively kills the protozoans and bacteria, affording the coral a chance to heal. Be sure that the temperature of the freshwater is about the same as the aquarium. It is not necessary to adjust the pH of the freshwater. Powerful scissors may then be used to cut off the dead skeleton immediately after the dip or later, after the coral has healed. Usually it is best to cut the denuded skeleton off immediately, since the rotting tissue can further stress the living tissue and start the infection again.

Unfortunately, freshwater dips are not usually effective with soft corals. For these corals, siphon off the "jelly" and any dead tissue, then cut off the affected branches using sharp scissors. Sometimes it may be beneficial to apply an antibiotic such as Chloramphenicol, directly onto affected areas. DO NOT treat the whole tank with a dose of antibiotic as this could damage the resident populations of beneficial bacteria. Just apply a small amount directly onto the affected areas of the coral. The antibiotic is prepared by mixing the contents of a capsule (or crushing one tablet to powder) into about a quarter teaspoon of tank water in a cup and applying the paste with an eyedropper. After applying the antibiotic and allowing a couple of hours to pass, direct a strong current over the coral for several days to flush away the protozoans and encourage healing. Beware that Chloramphenicol and other antibiotics are very toxic to humans! Wear gloves and a dust mask as precaution when handling them. Corals may also be removed from the display tank and treated with antibiotics in a separate aquarium. Streptomycin and neomycin, shown to be effective in larval tridacnid clam rearing experiments (Fitt et al., 1992), may prove beneficial in the treatment of bacterial infection in corals. One drop of Lugol's iodine solution diluted with aquarium water in an eyedropper can be directly applied over affected zoanthid anemones, and some soft corals (see chapter 8).

Once the affected corals have healed, the protozoans seem to disappear and do not usually affect healthy corals in the same aquarium. Still, be careful not to allow much of the "brown jelly" to slough off into the water. If clumps of the "jelly" land on soft corals, zoanthids, or corallimorpharians where there is little water movement, it may affect them. This condition can spread and affect healthy corals, but usually it does not. Water currents break the jelly up and flush the protozoans off, minimizing the risk. As mentioned earlier, such infections are more common, and more readily contagious at high temperatures, above 27 °C (80 °F) (J. Sprung, pers. obs.). When the corals are stressed, their resistance to the protozoans is weakened. Corals are much healthier at cooler temperatures, ideally about 23 °C (74 °F). If, after the above treatment, there are still some protozoans visible in the aquarium, it may be possible to use biological means to remove the few remaining ones. Fish such as the various Mandarin dragonets species (*Synchiropus* spp.) and small wrasses like the Six-Lined Wrasse (*Pseudocheilinus hexataenia*) will feed on any exposed protozoans (Wilkens, 1990).

White Film

A white film that looks like tissue paper draped over the coral is a sign of anaerobic decay. This is the same white film that can be seen on fouling live rock and dying sponges. A tell-tale sign is that the specimen will smell of hydrogen sulfide (rotten eggs) when removed from the water.

Treatment: Gently lift the coral out of the water and swish it back and forth in a bucket of seawater. Determine by smelling the coral and observing the tissue where the anaerobic fouling is occurring. Sometimes it is merely an attached sponge that has died, which is all that needs to be removed, and the rest of the coral is unharmed. If a portion of the coral is affected, cut it off, and direct a good circulation of water over it. Keep it elevated off the bottom. This

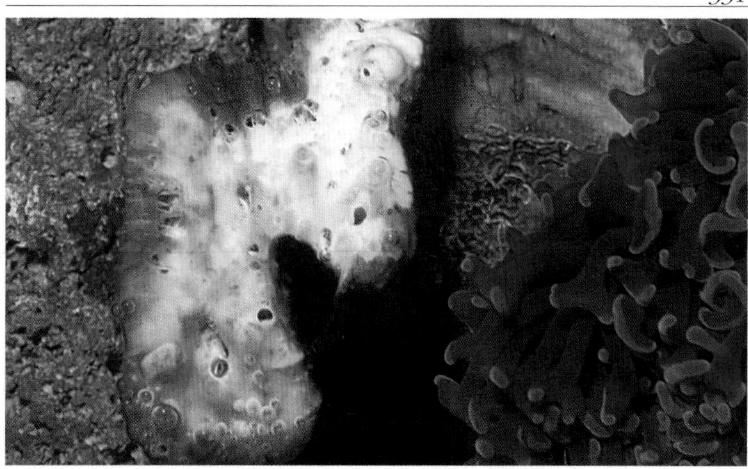

"White film" condition on a *Euphyllia ancora.* J.C. Delbeek.

can be accomplished by placing it on live rock or by making a short stand out of a piece of plastic pipe. If most of the coral appears damaged, one can try to save it by directing a strong current over the specimen, but if the film returns, the coral is dead and should be removed from the aquarium to prevent further fouling of the water.

White Paste

When a stony coral does not open up and appears to be engulfed by a pasty white substance, it is an indication of another kind of infection. A stressed coral secretes excess mucus to protect itself from an irritant in the water or the attack of a neighbouring coral. This excess mucus, which is teaming with bacteria, can suffocate the coral when there is insufficient water flow. In aquaria, this condition most often affects *Catalaphyllia jardinei, Euphyllia* spp. and *Goniopora* spp.

Treatment: Move the coral to another location in the aquarium. Flush the affected area with a directed current of seawater. Apply Chloramphenicol to the area as outlined above. Perform a water change if an irritant in the water is suspected and add a fresh amount of phosphate-free activated carbon. Check the coral frequently to be sure the coral is not being stung by other corals or anemones.

Acontia Filaments

Sometimes a stressed coral may appear to be covered with fine white strings. These are its digestive filaments that can be ejected externally for defensive purposes (see photo chapter 3). Acontia are used to attack neighboring corals and where they contact another species, tissue disintegration is sure to follow. The release of terpenoid compounds by soft corals can also stimulate acontia release in stony corals, as can the presence of other noxious compounds in the water. Supersaturation of oxygen, as can occur from too strong illumination or the administration of pure oxygen into a pressurized "oxygen reactor", will also stimulate the ejection of acontia filaments. In the case of Corallimorpharia (mushroom anemones), rough handling can stimulate the release of acontia. In this case, however, the filaments are typically reabsorbed by the coral in a few hours; no harm done.

Treatment: Allow enough space between corals so that their tissues cannot come in contact. Use barriers if relocation is not possible. Perform a water change if a noxious chemical is suspected and add a fresh amount of phosphate-free activated carbon to remove any remaining toxins. Move the coral into a shadier region if over-illumination is suspected. Adjust the protein skimmer to maintain it's peak performance, so that noxious compounds don't accumulate.

Excess Mucus

If a stony coral appears to be emitting a lot of clear slimy mucus, it has been physically injured, disturbed by a noxious chemical in the water and/or stung by another coral. It is normal for the coral to shed some mucus all the time, but if the polyps are closed and the mucus is thick and trailing off the coral, it is an indication that the coral is irritated. Please note that some species of stony corals normally release more mucus than others. For example, *Turbinaria peltata* will produce copious amounts of slime when slightly handled, and this is not necessarily an indication that it has been greatly stressed.

Release of some clear mucus from around the mantle and upper surface of tridacnid clams, often with attached air bubbles, is normal. This is a means of getting rid of excess carbon from photosynthesis. Excessive mucus production, which may clog mechanical filters, is a sign of irritation from something in the water. It can also be that a neighboring coral is irritating the clam. John Burleson (pers. comm. 1988) observed that a species of *Xenia* irritated a *Tridacna squamosa*, causing it to produce copious mucus, which quickly clogged the mechanical filters in his aquarium.

Treatment: Administer a stronger flow of water and avoid handling the coral too much. Perform water changes if the condition does not improve and add a fresh quantity of phosphate-free activated carbon

Brown Mucus Under Tridacnid Clams

Clear, brownish, thick gelatinous mucus around the byssal opening of tridacnid clams is a harmless condition. It appears to serve some protective function, preventing contact with potentially irritating substances (i.e. coral mucus), and blocking predation by worms, snails and crustaceans.

Packets of Zooxanthellae Emitted from the Mouth

The appearance of dark brown strings or pellets emitted from the mouth of a coral or the exhalent siphon of a tridacnid clam is an indication that it is regulating the population of zooxanthellae in its tissues. It is normal for them to do this periodically, especially when they are first placed in the aquarium or moved. The symbiotic relationship between these organisms and zooxanthellae depends on a variety of environmental factors, most important of which are temperature, light intensity and spectrum. Changes in these parameters create the need to readjust the zooxanthellae population. Too great a change (i.e. too much or too little light, or a photoperiod that is either too long or too short) can shock the system, causing the coral or clam to expel all of its zooxanthellae, a condition known as bleaching (see next item).

Brown strings of zooxanthellae being released by a colony of *Turbinaria*. J.C. Delbeek.

Bleaching

Loss of the symbiotic zooxanthellae makes the coral or clam look white, pale yellow or pale green, as if it had been dipped in bleach. The tissue is still alive, but now it is transparent. Bear in mind that some corals are naturally quite pale brown or yellow, especially if they come from a shallow, heavily illuminated zone. Corals and clams that have undergone bleaching appear less healthy than they should, generally exhibiting poor tissue expansion. In some cases there is no actual reduction in zooxanthellae number but a decrease in the pigment content of the zooxanthellae. This is related to changes in light intensity, whereas actual zooxanthellae expulsion accompanies changes in temperature (Hoegh-Guldberg and Smith, 1989). Bleaching can occur suddenly if the coral has been exposed too long to water that is too hot (over 30 °C (86 °F) for most species) or too cold (below 19 °C (66 °F)) for most species), or if the light field has changed too radically (Hoegh-Guldberg and Smith, 1989). Bleaching can also occur slowly as a result of inadequate or too much light. In a study of the stony corals *Stylophora pistillata* and *Seriatopora hystrix,* Hoegh-Guldberg and Smith (1989) found that after 7 hours exposure at 30-32 °C (86-90 °F), zooxanthellae expulsion rates were 1000X greater than control corals, and remained high even after they were returned to the normal temperature of 29 °C (85 °F). Furthermore, oxygen metabolism was found to be abnormal for up to 4 days after the exposure and evidence suggestive of recovery did not occur until after a total of 19 days.

Bleaching is evident in the polyps along the growing edge of this *Favia.* This coral is recovering from injury, and new tissue has begun to regrow into the empty calices. J. Sprung.

Coral bleaching may also be a method of adaptation for the corals (Buddemeier and Fautin, 1993). Different species of zooxanthellae may tolerate different environmental extremes with different species of coral partners. Bleaching may afford corals the opportunity to repopulate their tissues with different species of zooxanthellae that are better suited to symbiosis under a wider range of environmental parameters (temperature, light, etc.). Corals may be playing "musical chairs" with zooxanthellae. We speculate, therefore, that a high diversity of corals in an aquarium (and therefore a high diversity of zooxanthellae species) may afford better adaptation to environmental extremes, and better success with the individual specimens.

Coral bleaching, leading to death in hard corals, can also occur as a result of the excessive use of activated carbon and the subsequent reduction of trace elements, especially iodide (A. Nilsen, pers. comm.). This condition is particularly common in stony corals, but can also occur in clams and soft corals. The loss of iodide, when not replaced, can cause rapid bleaching and death in these corals (A. Nilsen, pers. comm.; J.C. Delbeek, pers. obs.). As we discussed in chapter 6., the iodide may help the coral by detoxifying excess active oxygen that develops within the tissues as a result of photosynthesis (R. Buddemeier, pers. comm.).

Treatment: If the coral is in dim light, move it up slowly over a period of weeks, thereby increasing the light intensity it receives. If the coral is bleaching from too much light, move it down or over to a slightly shadier region, and be sure that an iodide solution is being added regularly to the aquarium. Iodide appears to assist in the adaptation of corals to strong illumination (A. Nilsen, pers. comm.). To prevent coral bleaching, avoid temperature extremes, be careful when moving corals to greater light intensity, and when adding new lighting fixtures or when replacing old lamps. The ability to raise the fixture or shield it in some way is a useful precaution to have available when replacing old lamps with new ones. Otherwise, the sudden increase in intensity may cause coral bleaching in some of your specimens. Unless accompanied by some other stress factors such as poor water quality, infections or tissue damage caused by excessive UV exposure, most corals can completely recover from a bleaching episode.

Tissue Recession (White Band Disease)

The retreat of living tissue off of the skeleton is a malady that can have numerous causes. Injury can cause loss of tissue by physically abrading it or tearing it off, thus weakening the coral, exposing it to infection. Stings by neighboring corals with sweeper tentacles can also cause rapid tissue recession as can predation by fish and invertebrates. Most problematic tissue recession is of a much slower nature, and the incidence may be the result of several factors combined.

Tissue recession often occurs when the coral stops depositing calcium. The tissue may continue to grow, but the skeleton shows no new deposition and the tissue recedes like gums pulling away from teeth. Often the skeleton is greenish in color, which indicates the presence of *Ostreobium*, an alga that may contribute to this condition (see previous section).

The deposition of skeleton for coral growth is dependent on sufficient calcium being in the water. When the coral does not have a pool of at least 400 mg/L to draw calcium from, its growth may be slowed. If the calcium level is below 250 mg/L it may not grow at all, and tissue recession can occur (see addition of calcium, chapter 8).

The carbonate hardness of the water is also critical for calcification. It is suspected that external sources of carbonate are not directly utilized in skeleton formation, but that the CO_2 from coral metabolism is used to form carbonate within coral tissue. Nevertheless, when the carbonate hardness declines below 7 dKH, calcification is slowed. The ideal level is about 7-12 dKH. The ocean around coral reefs is typically 7-8 dKH. It is difficult to maintain the carbonate hardness above 12 dKH in an aquarium. Some authors have recommended levels as high as 20 dKH. This goal results in wasteful, frequent additions of buffers, and is simply unnecessary. With proper, slow additions of kalkwasser (a saturated calcium hydroxide solution), the dKH should remain quite stable and only occasionally should buffers be necessary. The link between carbonate hardness and calcification appears to be related to the pH of the water, which results from a number of factors, including dKH.

The pH value is critical for calcification since it is the elevated pH at the junction between coral tissue and skeleton that allows the shift to precipitation in the reaction of calcium and carbonate. When the pH in the environment is too low, as can happen in a closed system due to the accumulation of organic acids, nitrate and depletion of the buffer (low dKH), the calcification process is halted. The ideal pH for calcification is approximately 8.4. The ideal range lies between 8.0 and 8.45. At pH levels above 8.5 calcification slows (A. Nilsen, pers. comm.).

Certain trace elements are important for coral tissue growth and skeleton deposition. Very little is known about this subject, and aquarists keeping corals are really pioneering this research. Of use are the limited number of scientific papers listing trace elements found in coral skeletons or tissues (see Livingston and Thompson, 1971; Pingitore et al., 1989). However, one must exercise caution in extrapolating such studies into meaningful uses in the aquarium. It is very easy to draw erroneous conclusions from such works. Very little is known even about the biochemical mechanisms of calcification, let alone the role played by various elements. Strontium is the only widely recognized minor element that specifically aids coral and coralline algae growth and inhibits tissue recession. Barium, rubidium and lithium have also been indicated as important for coral growth (P. Wilkens, pers. comm.) (see the section on trace element additions for more discussion of this topic, chapter 8).

Another potential cause for tissue recession is the accumulation of phosphate in the system, or chronic additions of excessive quantities through the make-up water or other sources.

Phosphate can act as a poison to the calcification process in both corals and coralline algae by interfering with their ability to deposit calcium (Simkiss, 1964; Brown et al., 1977). It may bind at the site of calcification and prevent the crystallization of calcium carbonate. This cessation of calcification may cause the coral tissue to recede. Phosphate test kits made for aquarists only measure inorganic phosphates. They do not measure organic phosphates. Organic phosphates may accumulate in the water or in the sediments, and release chronic, low levels of inorganic phosphate through microbial activity. Refer to chapters 2 and 9 on nutrient control and nutrient regulation on the reef for further information about phosphate control.

Certain predators can leave behind white markings on stony corals. The bristle worm, *Hermodice carunculata,* hides by day and feasts on corals and anemones by night. Where it eats living coral tissue, bright white skeleton is exposed, leaving an appearance like white band disease. Other polychaete worms that live within live rocks may feed on corals as well. Limpets also will eat coral tissue, leaving white exposed skeleton. Typically the limpets will eat tissue around the base of a branch and work their way up. They will also consume soft corals. Certain fish like Parrotfish (Scaridae) and large wrasses can also leave white markings on stony corals, though these tend to be more localized. Stings from sweeper tentacles or acontia can also leave white band disease-like markings on the surface of hard corals. Protozoans and other infections that rapidly consume coral tissue proceed as a front of tissue disintegration, behind which the white skeleton is exposed. See heading "Brown Jelly".

Finally, while most of the reasons for tissue recession are linked to water quality, the growth of coral skeleton is intimately linked with photosynthesis, so inadequate light can also result in the inability to lay down new skeleton. This can produce tissue recession, since the coral may continue to form its organic matrix where calcium crystals should be deposited, and in the absence of calcification, the layers of matrix separate the tissue from the old skeleton. This may also be accompanied by loss of zooxanthellae, known as "bleaching" (see above).

UV Burning

Some corals lack or have reduced amounts of UV blocking substances in their tissue and can be damaged by wavelengths of light below 380 nm. Many corals utilize light in the wavelengths between 360 and 380 nm, but too much causes harm. The harm may be direct burning of the tissue, like a sunburn, or it may result from the toxic build-up of oxygen in the tissue. Usually the coral will be contracted in the most heavily illuminated areas, and the shaded or lower-down areas will be open. In soft corals this gives the appearance of "bald spots" on the most heavily illuminated, top portions. In stony corals, the tips of the polyp tentacles may appear shriveled and may have a coating of excess mucus. High levels of UV radiation can also result in zooxanthellae expulsion and bleaching.

This *Cladiella* soft coral exhibits "bald spots" on the branch tips due either to UV burning or build up of oxygen in the tissues from photosynthesis. J. Sprung.

Treatment: Move the coral lower down in the aquarium. Be sure to use a lens with unshielded HQI metal halide lights (e.g. Osram Powerstar™) that blocks all UV wavelengths below approximately 360 nm and most below 380 nm. With globe shielded metal halides and metal arc lamps the glass globe around the element blocks most UV and a separate lens is optional (e.g. Coralife® and Ultralux®). Some forms of fluorescent lamps, particularly H.O. and V.H.O. lamps, produce significant amounts of UV light, necessitating UV shielding for the organisms closest to the surface. UV absorbing acrylic can be used to make a lens which blocks these wavelengths and protects the bulbs from splash. The difference in temperature and humidity on the two sides of acrylic lenses can cause warping, however. UV absorbing tubes are also available that fit around individual fluorescent lamps, but these often melt from the heat produced by H.O. or V.H.O. lamps. The heat is concentrated at the ends of the lamps, so these tubes can be cut short to avoid contact up to six inches from the ends.

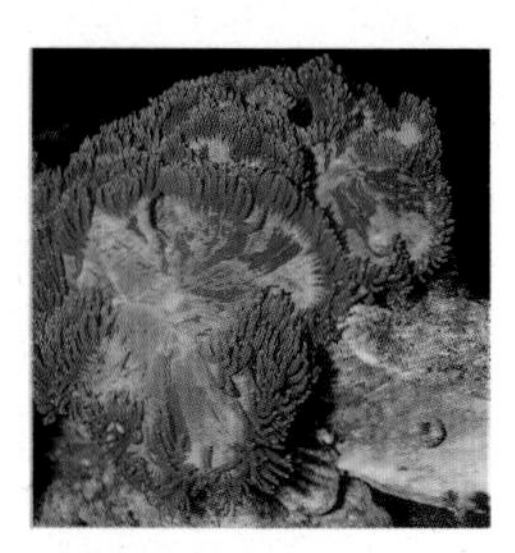

Note the shortened stubby tentacles on the upper portion of this *Catalaphyllia* compared to the more natural, elongated tentacles lower down. This coral may have been over-illuminated. J. Sprung.

Poor Coral and Clam Tissue Expansion

At times a coral or clam will not expand fully, or will remain completely contracted for several days or longer. There can be quite a few reasons for this occurring, some of which are mentioned above; it is mainly a process of eliminating the possibilities.

Certain soft corals and at least one stony coral genus, *Porites,* produce a waxy film that results in the coral being closed for a number of days. This film is eventually shed, but this process can be aided by directing a current across the coral's surface. The film contains waxes that trap bacteria and algae, and basically acts to cleanse the coral of these growths; a natural anti-fouling mechanism. This process is especially common in the leather coral genera *Sarcophyton, Lobophytum* and *Sinularia,* and some Caribbean photosynthetic gorgonians. If the coral remains closed for more than two weeks, then there is most likely another cause.

Many corals will periodically shed mucus or a waxy skin to rid themselves of fouling growths and excess carbon, as this Leather Coral (*Sarcophyton* sp.) is doing. J.C. Delbeek.

Tridacnid clams, stony and soft corals can all exhibit poor expansion when exposed to a pH that is too high or too low. This can occur through the misuse of various calcium additives that may exhaust the alkaline reserve of the aquarium. Check pH and alkalinity, and adjust both as required to maintain the pH between 8.0 and 8.5.

In some cases poor expansion is a result of improper water movement for that particular specimen. Check to make sure that the water current is not too strong or too weak for the affected specimen. Excessive water velocity can cause even a current loving coral like *Sarcophyton* to remain contracted, and too weak a current can cause photosynthetic Caribbean gorgonians to remain closed. See the species descriptions chapter 13 and appendix A for specific water movement recommendations.

As mentioned in Chapter 3, many soft corals can release some pretty nasty chemical weapons called terpenoids. If these have accumulated in the water, or if the affected specimen is downstream from an "offending" coral, prolonged tissue contraction can occur. As a temporary step you can move the affected coral or move the offending coral. In some cases this is enough. However, if a number of corals are affected, then the use of phosphate-free activated carbon, protein skimming, water changes or all three may be necessary to remove the offending compounds from the water. If these steps fail, then the aggressive coral(s) must be removed.

In Chapter 3, it was also pointed out that many stony corals can produce stinging structures such as sweeper tentacles. If these continuously brush against a sensitive neighbour, the specimen may react by remaining contracted. Sweeper tentacles are quite common in *Catalaphyllia, Euphyllia, Favia, Favites, Galaxea, Plerogyra* and many other stony corals. In this case, the corals should be moved so that they cannot reach other corals. In some aquariums this is not possible due to spacing constraints, therefore dividers made of glass, plastic or pieces of live rock can be used to block the tentacles. It is also possible to simply pinch the sweeper tentacles off using your fingers, without any damage occurring to the coral. As a precaution, you should wear a Latex glove to protect yourself from any possible reaction to the stinging cells (nematocysts). In some species, (e.g. Bubble coral, *Plerogyra sinuosa*) sweeper tentacles are more evident at night, so the aquarist must inspect the aquarium at this time.

In crowded aquariums, corals may actually touch each other on a continuous basis. In this case, strongly nettling (stinging) species will cause more sensitive ones to contract. In this case the corals should be separated or barriers should be used. In many cases corals can touch each other without any problems occurring, however, the aquarist should continuously observe any such interactions for signs of aggression.

Sometimes poor expansion can be due to fouling or over-growth by algae, bacteria or mucus. Siphon away the fouling substance or take the coral from the aquarium and carefully remove the fouling material. When returning the coral to the aquarium either place it in the same location and alter the water flow to direct more current over it, or place it in a new location with more current.

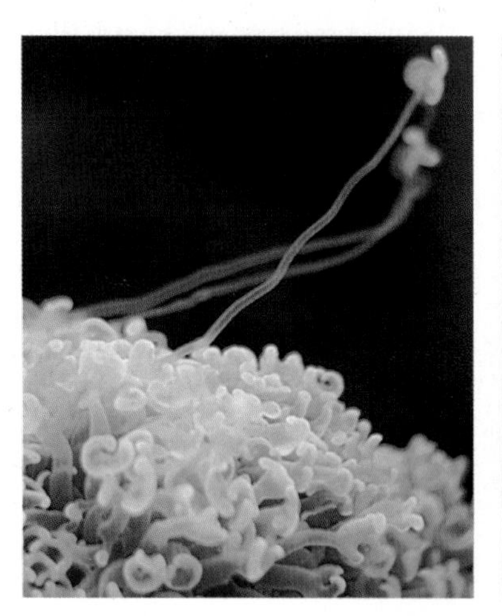

Sweeper tentacles produced by *Euphyllia ancora* can extend several cm. J.C. Delbeek.

Euphyllia divisa stinging the colonial zoanthid *Palythoa caribaeorum.* Note that a tentacle has actually attached itself to the zoanthid. When placing specimens in the aquarium make sure there is enough space between them. J.C. Delbeek.

Corals and clams may not expand fully because they are being irritated by some organism in the tank such as a fish or some sort of parasite or predator. Although most will tolerate the occasional nip from a passing fish, or in the case of hawkfish, clownfish, gobies and blennies, fish actually resting on them, some corals do not. Tridacnid clams and gorgonians are especially sensitive to being bothered by fish. The danger in this is that the coral or clam will not expand fully and will begin to suffer from lack of light.

Pygmy angels *(Centropyge* spp.) are notorious pests of corals and clams, and should be closely watched. The problem with fish is that they are individuals; no two are alike. One aquarist may have a Pygmy angel that never bothers his/her corals, while another aquarist, with the same species of angel, may have nothing but problems with the fish picking at the corals and clams. Short of removing the offending fish, there is very little that can be done in this situation. Fish spend the majority of their time searching for food. This involves exploring their environment and making a number of "exploratory nips" to find something edible. In a large aquarium, the corals and clams may escape such "sampling" for many hours but in a smaller aquarium, the frequency increases greatly. If the fish are well fed, they may make less of these forays and some aquarists have found that increased feeding has helped. Others leave pieces of lettuce tied to rocks or stuck to the glass in several places in the aquarium to keep their fish occupied, thus reducing the frequency of nips at the corals and clams.

If the fish are not the problem then perhaps there are other predators or parasites in the tank such as crabs, shrimp,

Centropyge resplendens may eventually begin feeding on soft corals or tridacnid clams. J.C. Delbeek.

nudibranchs, polychaete worms or snails that may be irritating the coral or clam. If you suspect this may be the case you should closely inspect the specimen for any of the above, especially a few hours after all the lights have gone out. Take a flashlight and tie a piece of red cloth or acetate, around the lens; most aquatic organisms are insensitive to red light. Using this setup, closely examine the coral or clam for strange looking organisms crawling on it. Pay particular attention to the underside. If you see anything suspicious remove it as soon as possible. In many cases the coral will begin to expand again within a few days. Be warned though in the case of polychaete worms, there could be many more still present in the tank and you should still keep a close eye on the affected coral. Some worms are not visible on the outside of a coral since they burrow into the flesh and consume it from within. In this case there is little that can be done short of cutting off the infected portion. When dealing with snails and nudibranchs remember that many of these can leave behind egg masses, either directly on the coral, or on a rock adjacent to the coral. For this reason, keep an eye out for any suspicious looking jelly-like masses and keep a close eye on your corals or clams for a few more weeks after removal of the parasites.

As mentioned above, excessive UV light can also result in poor expansion of tissue. First try moving the coral lower in the aquarium. If this doesn't work, and iodide is being added in sufficient amounts, then it may be necessary to place UV shielding material between the light source and the aquarium. Even light that is too intense can cause corals and clams to not expand fully. This is often the case when new clams or corals are first placed into brightly lit aquariums. In this case the specimen should be moved lower in the aquarium.

Sometimes high temperatures can cause corals to remain contracted or to expand only partially. It is best to keep the aquarium temperature below 27 °C (80 °F) as much as possible. As short term emergency measures, freezer packs or bags of ice cubes can be floated in the sump to lower the tank temperature but you should avoid any sudden drops in temperature. Increasing the evaporation rate will also cause a decrease in water temperature through evaporative cooling (see chapter 8 for temperature control).

Another potential problem could be oxygen poisoning. When deep water corals are placed under bright light, their zooxanthellae produce greater amounts of oxygen from the increased rate of photosynthesis. These increased oxygen levels in the tissues can poison both corals and anemones (Dykens and Shick, 1984; Wilkens and Birkholz, 1986). Shallow water anemones cope through various mechanisms, including the use of enzymes to break down oxygen, withdrawing their tentacles to decrease surface area, covering their body column with gravel to protect it from the sun, and seasonally varying the amount and ratio of chlorophyll in their zooxanthellae to correspond with seasonal changes in light intensity (Dykens and Shick, 1984). Similar behaviour may exist in corals. However, sudden shifts in light intensity do not allow the coral time to adapt, and it quickly expels its zooxanthellae and/or shrivels away from the light (Wilkens and Birkholz, 1986).

Black Band Disease

Black band disease is characterized by a front of black, gooey material (mostly algae) behind which is denuded skeleton and in front of which is healthy tissue. The denuded skeleton is rapidly coated with many species of algae, including the black gooey mats characteristic of the black bands. It affects hard corals mostly, but also some soft corals, including gorgonians and *Sinularia* spp.

Current scientific studies indicate that several organisms cause this disease. While a particular species of cyanobacteria, *Phormidium corallyticum*, has been identified with this condition (see Ruetzler and Santavy, 1983), other species of cyanobacteria and filamentous algae are often present (P. Dustan, pers. comm.), sometimes in the complete absence of *Phormidium*. Also, many bacteria and microorganisms have been isolated from the bands, and these may either be directly involved in the tissue destruction or simply opportunistically consuming damaged tissue, other microorganisms, or cyanobacteria.

Treatment: Siphon the black front away using small diameter airline hose with a rigid airline tip. A teasing needle can be used to lift the black mat and separate it from the coral skeleton with the siphon close behind to catch the debris. The coral can be removed from the display tank and placed in the reservoir without illumination for about a week to allow the tissue a chance to begin healing without competing with new algae growth. Be sure that the coral receives adequate water movement if it is placed in the reservoir. If it is not possible to remove the coral from the display tank, be sure to siphon the band away as described and, if the band returns, apply small quantities of antibiotics such as

Chloramphenicol or Erythromycin directly on the affected area, as described in the treatment of "brown jelly". Freshwater dips also described for treating "brown jelly" kill cyanobacteria as well, and are an effective treatment. If the affected coral takes food, feeding it (sparingly) chopped shrimp or fish will assist healing.

Blackened Areas on Soft Coral

Soft corals may exhibit black, necrotic areas on branches, particularly *Sinularia* spp. and Caribbean photosynthetic gorgonians, especially *Eunicea* spp. This occurs when tissue has suffocated in shipment or when it has been attacked by a microorganism and died. Stinging from other corals can also cause this in gorgonians.

Treatment: Generally it is best just to direct a strong current stream over the coral and allow the necrotic areas to fall off on their own. Usually they do not spread. If they appear to be spreading despite good water flow, then it may be necessary to cut off the affected branches with a sharp scissors. When gorgonians lose tissue and the skeleton is exposed, it is a good site for attachment and growth of algae. If the exposed skeleton is only about half a centimeter or less, the tissue usually grows back rapidly before algae can grow. Larger gaps than this may make it necessary to cut the branch off. It is best to wait and allow the gorgonian a chance to heal to determine if cutting will be necessary. If algae grows on the skeleton and the tissue seems unable to grow there, the branch should be cut as close as possible to living tissue of the main colony. The free branch, if substantial, may then be transplanted in the aquarium after the algae is scraped off its base.

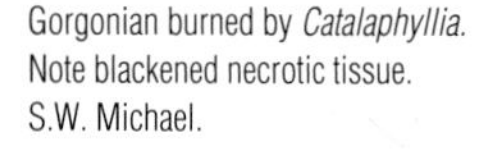

Gorgonian burned by *Catalaphyllia*.
Note blackened necrotic tissue.
S.W. Michael.

A second excurrent siphon has formed between the normal inhalent and exhalent siphons in this *Tridacna derasa*. It probably is the result of an injury that healed without closing up. J. Sprung.

Hole in the Mantle of Tridacnid Clams

Sometimes a tridacnid clam will develop a hole in the center of the mantle, between the inhalent and exhalent siphons. The wound can be fatal, but usually it spontaneously heals, leaving no scar. Sometimes a scar remains or, as in the photo, a second exhalent siphon forms. The causes of a hole developing in the mantle include physical injury, infection (bacterial or protozoan), and light damage. A tridacnid clam located less than eight inches from a metal halide bulb can suffer a burn that causes a hole to open in the mantle. Most metal halide bulbs have a focused ring or "hot spot" of light that can be observed by holding one's hand a few inches below the bulb. The high intensity of heat and light from this ring can burn clams and corals if they are near it. This is not a problem when the bulbs are located more than eight inches above the aquarium. If a clam develops a hole in the mantle, move it to a new location and observe it to make sure it is not being attacked. The wound should heal within a few days.

Broken Hinge in Tridacnid Clams

On rare occasions a tridacnid clam will suffer a break in the brown protein material joining the two shells. If this occurs, re-align the shells and affix a rubber band loosely around them to hold the hinge position while still allowing the shells to part enough for the mantle to extend and receive light. The clam will secrete a new hinge within about two weeks.

Chapter Eleven

Collection and Transportation

Most of the specimens kept by marine aquarists are collected from the wild. This situation is gradually changing, as more species of fish, plants, and invertebrates are being grown in aquariums each year. In fact, while the marine aquarium hobby has seen many periods of growth and subsequent decline over its history, the next decade may well see the greatest renaissance and growth ever, owing to a proliferation in discoveries making it feasible to cultivate numerous species of marinelife, in profitable quantities. It is our aim to encourage this trend, and to discourage any practices that are harmful to coral reefs and the environment. We believe this is the goal of every aquarist. Recognizing this and the fact our hobby now depends on wild collections, we have included this chapter on collecting, not for the purpose of encouraging hobbyists to collect their own, but to serve as a guide for commercial collectors and coral reef researchers, and to promote the proper, legal means of obtaining specimens for those hobbyists who wish to collect some specimens for their own research.

In aquariums in North America, live rock typically comes from Indonesia, Singapore, or Florida, though some comes from the Marshall islands, Micronesia, Mexico, or the Caribbean. Hobbyists in Europe, Australia, or elsewhere have different sources of live rock. In the U.S. the main source of live rock has been Florida, where collectors have been harvesting it in federal waters since the ban on collection from state waters in 1989. The state government has been trying to end the harvest of live rock from federal waters, but may not have the authority to implement such a rule. When the federal government decides to end the harvest of live rock, then aquaculture and overseas imports will be the only sources.

The living corals and anemones we keep are mostly from Indonesia, but some are imported from Singapore, Kenya, the Red Sea, Tonga, and a few localities in the Pacific. The exporters in these areas obtain their specimens from collectors who gather from different island groups or regions where they find particular organisms in abundance. Many species are easily gathered close to shore in polluted or damaged regions. In these areas where there is heavy physical disturbance from human activity or storms, and higher levels of dissolved nutrients, the growth of the soft corals and colonial anemones commonly kept by aquarists is prolific

Dr. Bruce Carlson shows how to harvest coral grown in a reef aquarium. T. Kelly.

(Wilkens and Birkholz 1986). Many species are collected on intertidal flats at low tide when the water is less than one metre deep. Some exporters have transplanted colonies of rare local soft corals and anemones to sites near their facility so that they can have a nearby source for them (P. Wilkens, pers. comm.). This form of coral farming can supply our hobby with many of the popular species.

The *Tridacna* spp. clams available to hobbyists are mostly farm raised, though some wild caught specimens of *T. maxima, T. crocea,* and *T. squamosa* are occasionally available. We expect that in the near future all *Tridacna* clams available to hobbyists will be tank raised.

Before You Collect!

While most aquarists must rely on their local pet shop or a mail-order supplier for their invertebrates, sometimes it is possible for the hobbyist to collect specimens while on holiday. Strict regulations concerning the collection and export/import of marinelife, particularly from coral reefs, limits this activity. Still, it is possible to collect in some areas if one takes the time to contact the appropriate government agencies who issue collecting permits.

The collection of marine invertebrates and fish, whether it be in your backyard or abroad, is often monitored by a permitting system for both the recreational and the commercial collector. Such permit systems allow for the accumulation of valuable collecting data through "trip ticket" reports of the catches and localities, which must be sent in for the permit to remain valid. Permits also are an important means of generating revenue for the regulating agencies. Finally, the permit systems aim to protect "the living resource", endangered species, and to clearly outline special areas where collection is prohibited.

Collection of stony corals and tridacnid clams is usually prohibited except for scientific purposes. When applying for a permit, it is best to make the point that the specimens you wish to collect are for your personal research, not just to decorate your aquarium. The permit may require that you be affiliated with a university or other institution, so it is wise to be prepared with proof of such an affiliation. Also, government agencies are habitually slow with processing permits, so it is essential to initiate the permitting process several months or a year in advance of your trip. In some places recreational collectors are allowed a certain daily "bag limit" of specimens, and no permit is required. Usually, though, even recreational collecting requires a permit of some kind. BE PREPARED!

On the return trip one must be prepared for the other major obstacle to obtaining your own specimens: import/export laws. Usually, having a collecting permit with the species you have collected described with scientific names will be of great assistance to the fish and wildlife examiners or customs agents. Passing through Hawaii as your point of entry is difficult as that state has extremely strict laws about even carrying invertebrates through. You must check with the department of Fish and Wildlife or similar agency for your country to determine what documents you will need to make your return legal and without difficulty. You must also obtain a CITES (Convention on International Trade in Endangered Species of Wild Fauna and Flora) export permit from the country you are exporting the coral from, if the country requires a CITES permit for export of coral. The forms are either free or require a small administrative fee and are usually available through fish and wildlife, or fisheries departments. All stony corals and their by-products such as the coral rock on which soft corals are attached, are covered by CITES, and are listed as Appendix II species, requiring an export permit from the country of origin if the country participates in the CITES system. Make sure that ALL the specimens you collected are listed on the permit and that they are identified by scientific name, to species if possible.

Equipment For The Collecting Adventure

Collecting marine invertebrates is quite easy, requiring only a few tools. A geological hammer with a rock pick tip is most effective for removing specimens with a little of the attached rock (see photos). Alternately, a good stainless steel diving knife can be used for the same purpose, but less effectively. Scissors and other cutting tools are also useful. Finally, you will need special holding and packing equipment, described below.

When collecting marinelife while diving one must use a bag to hold the specimens. A mesh lobster bag works fine for this purpose, but for the best results you must also bring plastic containers with lids and a roll of small plastic bags used to separate the specimens. These can be placed in the lobster bag as well. Individual specimens can be placed inside plastic bags, and the bags can be placed in the plastic containers in the mesh bag. Several of the plastic bags will fit in one plastic container. This technique prevents damage to the pieces from rolling around, abrading, or crushing. During especially rough weather, in strong currents or surge, it may be tricky to manage the bags and containers underwater, but it is well worth the effort. Since damage caused during the initial collection and transport to shore is probably the biggest source of loss of specimens, it is best to plan ahead and use this method of separating and cushioning individual pieces.

Collecting

Corals

Branching stony corals are possible to collect with a gloved hand, as it is easy to find loose heads or branches. It can also be useful to bring along a stainless steel scissor, bone scissors, or gardening clippers to cut branches off a large colony. Be sure to cut branches at least 7.5 cm (3 in.) long for best survival. We wish to emphasize that branch corals such as *Acropora* spp., *Pocillopora* spp., *Porites* spp., and *Stylophora* spp. are fast growing, so that it makes sense to collect small branches rather than whole colonies.

The rubble and gravel bottom found between large coral heads and just behind the reef is a good site to look for loose pieces that have broken off naturally. These are better to collect since they have already healed, are of the right size, and their removal doesn't affect corals on the reef itself. Many branched and solitary species occur loosely in seagrass areas or on soft mud or sand, and may be picked up by hand.

Massive or "head" corals can be tapped off with the geological hammer. Of course one doesn't hammer at the living tissue, but just at the rock around the base until the head pops loose.

One is advised to use gloves while diving on a coral reef. The glove protects the diver from stings and cuts. Gloves offer no benefit to the coral, and make damaging diver contact with the reef more likely. Gloves are abrasive and can injure coral tissue more easily than bare hands. Greater precision of handling collected coral can be realized when gloves are not used. One gloved hand and one ungloved affords flexibility. Some people are especially sensitive to coral stings, however, and should not handle corals without gloves.

A sharp scissors is useful for cutting off branches of soft corals such as *Sinularia* spp. or gorgonians, when the main colony is too large to be collected. The cut should be made swiftly so that it will be clean, not jagged. Such cuts heal rapidly. Usually it is best to collect small colonies attached to a small piece of rock. Often these occur among loose rubble on the bottom, and they can simply be picked up.

Tridacnid Clams

Unlike fishes or corals, tridacnid clams have been exterminated in localized areas due to overharvesting by fishermen who collect them for food and shells. The mariculture of tridacnid clams was begun to re-stock areas where the clams had been eliminated, and to provide a farm raised source of clams for food. A portion of the farm raised clams now goes to the aquarium industry, and profits realized from the demand of aquarists have stimulated interest in producing colourful varieties of all the species. We wish to encourage aquarists to purchase clams that have been raised on farms. Our decision to include information about clam collecting is not intended to encourage their harvest, but to provide the information about the proper techniques of collecting and transporting them for those individuals (i.e. scientists) who might need to collect some wild clams.

Tridacnid clams vary in their ease of capture. *Hippopus hippopus, T. gigas,* and *T. derasa* are usually loosely attached or not attached at all, and are simply lifted off the bottom. *Tridacna maxima* and *T. squamosa* tend to be enmeshed in coral, and attached by their byssus. The safest way to collect these is to lean the clam to one side and cut the byssus threads right next to the site of attachment on the coral with a knife or small scissors. It will not necessarily kill the clam if it is simply pulled off, tearing the byssus free, but this may cause injury. *Tridacna crocea* is the most difficult to collect. Sometimes they can be found in good numbers in relatively small, loose boulders, but usually they are on, or more aptly, IN hardbottom. The dense rock that they bore into must be broken around them with a geological hammer and rock pick. They may then be pulled off, or the byssus threads may be cut with a knife or small scissors.

Do not collect too many specimens! The fewer collected the greater their chance of survival. Soft corals are more difficult to keep alive before shipping than hard corals, and they should be kept separate from them.

The Trip Back to Shore

Once you get back on the boat you may sort your specimens into buckets or live wells, or other temporary holding containers for the ride back to shore. Never place live fish or other invertebrates in the same bucket or holding container with live corals. The exuded mucus and shed nematocysts of corals will kill other invertebrates and fish. If the trip is not long, the corals may be left in the plastic containers inside the mesh bag and unpacked only after you get back to shore. They are safer in bags inside the closed containers than rolling around in buckets of sloshing water on rough seas. Do not leave them out in the sun where they will heat up. Place the whole "goodie bag" inside a live well, or cover them with a wet (with seawater) towel, or bring them below deck.

Holding Your Catch

Hold specimens in buckets or, preferably, shallow trays such as the white plastic refrigerator trays, with air pumps and air stones for water circulation. Strong water circulation is very important. Locate the holding containers in a cool place, ideally cooler than 29 °C (80 °F), where they will receive some sunlight. A small fluorescent lamp may also be used for illumination. It is useful to collect some small shells or coral skeletons from shore to prop the specimens up, and to provide a new foothold for the clams. Shells can also be useful to separate corals, and prevent injury from stinging.

Water changes are necessary about once or twice per day, depending on the number of specimens. If it seems that more frequent water changes are necessary, then too many specimens are in the bucket or some of them are dying. Water changes should be performed as follows: It's best to have extra trays to allow you to rotate the corals to new trays with each water change. The new tray is clean and dry to start, and is filled with freshly collected, clean seawater. Another bucket of clean seawater is placed between the old tray and the new one, and the pieces should be gently swished in this water to rinse them before placing them in the new tray. As each piece is removed from the old tray, take the opportunity to "sniff it" for any fouling. Small sponges or tunicates attached to the coral often die, and can ruin the whole coral or the entire contents of the holding container. Remove any fouling spots with a screwdriver, file, or toothbrush, rinsing the piece thoroughly in the bucket of seawater before placing it in the tray. After all the corals are moved to the new tray, the old one is rinsed in freshwater and dried with a towel to remove slime and bacteria attached on the surfaces.

Tridacnid clams should be kept separately, and they must receive several hours of indirect daylight per day. As mentioned earlier, fishes should also be held separately, as they can be killed by coral mucus. While the corals are best held in shallow trays, to make illumination and water circulation easier, buckets are better for fish to discourage jumping and provide greater volume.

If your holiday will be several days or weeks, and you plan to collect right from the start, a good alternative holding procedure is to find a calm, protected site near shore in relatively deep water where the specimens can be temporarily "planted". Inspect them daily if possible since fish, crabs, molluscs, urchins, and other wandering animals may move your specimens around. Sometimes it's best to forego most collecting until the end of your trip, weather permitting. Soft corals in particular don't hold up well when held for several days in buckets.

For packing you will need to bring several pint to quart size plastic containers with lids, preferably square in shape because they will pack better inside the shipping boxes or luggage, lots of plastic bags, tape, rubberbands, a small knife, and styrofoam boxes, or sheets of styrofoam or foam "peanuts" which can be used to line a cardboard box to insulate the specimens. You cannot bring oxygen cylinders on a plane, so you will need to locate a source of oxygen for packing at your collecting destination, or simply use air to pack the specimens. You might plan to stop and pack the invertebrates at a welding shop. Oxygen can be obtained from welding shops, which can be found in all but the most remote locations. Fire Stations or hospitals are another possible source of oxygen.

Many corals and other invertebrates can be shipped home "dry". Dr. Ed Bronikowski, who is now working at the Tampa Aquarium in Florida, presented a paper several years ago to the AAZPA about a successful method for shipping corals without water. Dr. Bruce Carlson has also used this method successfully for years, and has shared the technique with us (see also Carlson, 1987).

The term "dry" is slightly misleading as the specimens aren't actually dry, but they aren't submerged in water either. The technique is used most often for branched stony corals such as *Acropora, Pocillopora,* and *Stylophora,* and it also works well for fleshy corals such as *Lobophyllia* and *Euphyllia,* for example, but these should be completely contracted before packing. Sea anemones are also best shipped this way, and they too should be deflated before packing to minimize water in the bag. Other invertebrates that may be shipped by the "dry" method include corallimorpharians (mushroom anemones) and zoanthid anemones. Certain soft corals can be shipped dry too, but some may need to be shipped in water. Those that can be shipped dry include gorgonians, star polyps, and the leather corals (*Sarcophyton, Lobophytum,* and *Sinularia* spp.). Those that are typically shipped in water, but would probably fare well shipped dry, include most *Anthelia, Cladiella* spp. and *Xenia.* It is possible that these may be shipped dry if they are first allowed to shed the profuse mucus created when they are handled. Tridacnid clams are best shipped one specimen per bag in water with oxygen. Dry shipment of tridacnid clams has been tried, and it works for short periods, but packing them with water and oxygen gives lower mortality in long term transit.

The dry shipping technique is as follows:

Wrap the specimens up loosely in strips of plastic or with "bubble wrap". Plastic strips roughly 1 to 2 cm broad can be cut from heavy duty fish shipping bags using a pocket knife. If the plastic is from too thin a material (i.e. plastic garbage bags), it will be difficult to handle and will collapse into a useless mass when wet. Note: The plastic strips or bubblewrap should be clean. Old plastic bags may have salt or bacteria on them, and should be avoided. Also avoid re-using the plastic strips. If you must re-use them, or if you only have used bags to make them, then first thoroughly rinse the plastic with freshwater and allow it to air dry before following this packing procedure.

The plastic will cushion the coral and maintain a level of moisture around it. Place the coral in a square plastic container just large enough to fit it, dunk the container in seawater to fill it, then drain out all but about a tablespoon of the water. Snap the lid on the container, and lift up an edge to inject oxygen inside. Then quickly snap the lid shut and wrap the top with a strip of duct tape to prevent the lid from popping off in flight when the external air pressure drops. Pack the containers as usual inside a styrofoam-lined box. With almost no water, the box weighs very little.

Before shipping by the dry method, it is beneficial to perform steps that will cause the coral (both soft and stony) to shed off excess mucus which might otherwise suffocate it during transit, or cause the proliferation of bacteria that could kill the coral during transit or shortly thereafter. "Cleaning" the corals this way may also prevent them from poisoning themselves with their own toxic products.

Chapter Twelve

The Identification and Care of Tridacnid Clams

Live tridacnid clams have proven to be very popular in the reef aquarium trade. They are beautifully coloured, hardy, grow rapidly and require little care. Of course there are a few points to consider that will enhance the enjoyment of these animals, and ensure that these clams will live and grow for many years.

Purchasing Tridacnid Clams

When tridacnid clams are first imported, they generally exhibit a behaviour called "gaping". This means that the shell is fully open, the mantle is poorly extended and the inhalant siphon widely stretched. As this condition eventually passes the inhalant siphon will gape less, the shell will not be open as much and the mantle will begin to expand. If, however, the clam is kept under unsuitable light conditions, is damaged or unhealthy, this "gaping" behaviour will continue and the mantle will begin to pull inwards, shriveling and tearing between the siphons. The inhalant siphon in healthy clams can open wide at times but "gaping" results in a very wide opening and it is constantly held this way when the clam is unhealthy.

Examine the mantle closely, it should be colourful everywhere, there shouldn't be any white or clear areas; check also for rips and tears. Be aware that *T. gigas* may have some clear areas near the centre of the mantle, which is normal. In healthy clams the mantle should be extended over the edge of the shell and not be pulled inwards. If there are some colourless areas these could be the result of lack of light, predation or disease. In the former case the clam should recover quickly when placed in better conditions.

Watch how the animal reacts to an external stimulus, a healthy specimen will respond by closing it's shell with some force (Achterkamp, 1987b). Newly imported specimens tend to react rather sluggishly but as they regain their strength, their reactions will improve.

Check to make sure the byssus gland is undamaged; there should not be any torn or loose tissue hanging from beneath the clam. There may be some byssal strands visible, but there should not be any solid tissue hanging loose. Unfortunately, byssal gland damage is not always readily visible and the clam will appear fine for 1 or 2 weeks, then it will suddenly die for no apparent reason. However,

Closeup view of the mantle of a particularly beautiful *Tridacna crocea*. J.C. Delbeek.

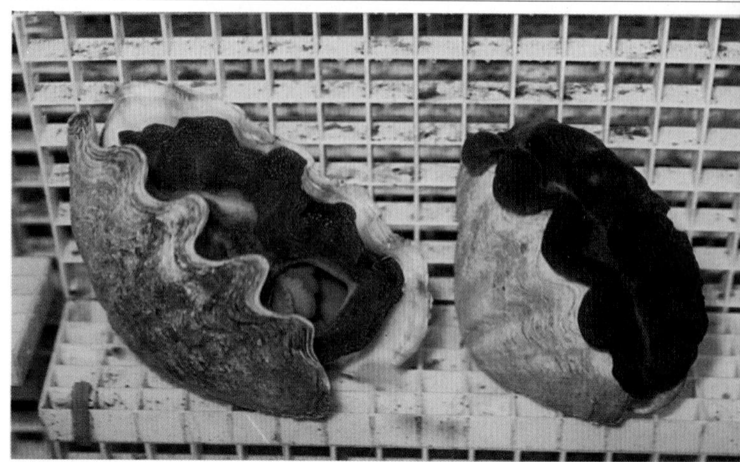

Two *Tridacna crocea*; the one on the left exhibits gaping. J. Sprung.

byssal gland damage is not always fatal and we have collected and purchased several damaged clams with little loss.

If the clam has already attached itself to the substrate before being purchased, great care should be taken when detaching it. Lift the shell up gently and insert a sharp knife, razor or scissors and cut the threads as close as possible to the substrate. DO NOT cut close to the shell or you could easily cut into the extended byssal gland. If the clam is attached to a small rock or some pebbles, there is no real need to separate them. Some species such as *T. crocea* and *T. maxima* are very sensitive to being handled in this manner and they are best left attached, if at all possible.

Associated Organisms

Certain species, such as wild specimens of *T. maxima* and *T. squamosa,* may have extensive growths of encrusting organisms on their shells. Check these growths to make sure there are no dead or necrotic areas that might foul your aquarium.

There are various parasitic snails that can be imported along with the clam. These will usually be visible as small rice grain-sized, cream coloured spots near the base of the shell or, at night, along the upper edge of the shell. If the clam is still attached to a rock check carefully underneath the clam by gently lifting it a short distance off of the rock and looking underneath for any small (2-5 mm long) snails. Check any potential purchases for these parasites and remove all of them. If you have the facilities, it would be wise to hold the clam separately before introducing it to your main aquarium to make sure all the snails were removed. Check also for small, jelly-like masses on the shell. These are the egg masses of these snails and should be removed too. Do not confuse these with the jelly-like mass that some clams may excrete around their byssus opening. For a more detailed description of clam parasites and diseases see Chapter 10.

Occasionally symbiotic shrimp of the family Palaemondidae *(Anchistus, Conchodytes* and *Paranchistus)* or small crabs, will be visible through the inhalant siphon in larger clams (Rosewater, 1965). These crustaceans live inside the clam and do not harm it. What they eat and what they do for the clam is unclear. They might defend the clam from potential predators or parasites, while using the clam as shelter (see "Pea Crabs", chapter 10).

Light

We have successfully kept tridacnid clams under both fluorescent (R.O., H.O. and V.H.O.) and metal halide (H.Q.I.) lighting systems. However, the more light you give them the better they will do. Also, with brighter light you can place them lower in the aquarium. To really appreciate the colours of these animals, they should be viewed from a downward angle. Lower light intensity systems will require that the clams be placed closer to the surface, and therefore the brilliant colours will not be as apparent to the viewer observing through the front glass. Recent trends in lighting have seen open aquariums with suspended lighting coupled with lower tank stands. This allows the tank to be viewed from above, an ideal position to see the full colours of these clams.

Usually those clams with brown mantles do not require as much light as those with blue mantles and are generally considered easier to keep in the aquarium. The blue forms typically live in shallow-water, the blue pigment acting as a light filter. Therefore, they require substantial quantities of light to maintain their bright colours. There are also differences in the light requirements of the various species. Both *T. crocea* and *T. maxima* require the most amount of light, followed by *T. gigas, T. derasa, T. squamosa* and the *Hippopus* spp.

Although tridacnids require strong lighting, the smaller the specimen the less light that is required. This is due to the fact that as a clam grows, its mantle thickens and the number of zooxanthellae increases. This causes the deeper lying zooxanthellae to receive less light. As a result, smaller clams (<4.0 cm) require less light to maintain optimum growth rates, while larger clams require more light (Fisher et al., 1985).

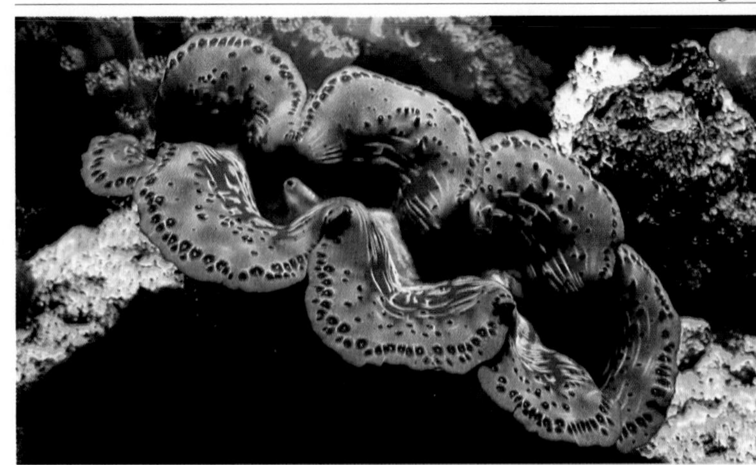

Tridacna maxima illuminated from above. J.C. Delbeek.

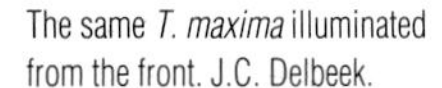

The same *T. maxima* illuminated from the front. J.C. Delbeek.

Placement

Since these animals love light, one is often tempted to place them close to the surface. Be warned, however, that they can close their shells with some force and can expel a surprising amount of water out of their siphons and out of the tank! In some cases, they can also trap small, slow moving fish that rest on their mantle, such as mandarins, gobies, hawkfish or blennies (Achterkamp, 1987a).

When first placing the clam into the aquarium make sure that the byssal opening is placed flush against the substrate and the mantle is facing directly upwards. Try to place the clam on an horizontal surface. If the surface is inclined, make sure the byssal opening is on the lower portion of the substrate. An easy way to remember this is to place the clam so that the inhalant siphon (which lies above the byssal opening) is on the lowest portion of the slope. The reason for this is that as the clam grows it will place greater strain on its byssal gland. If the gland is on the upper portion of the slope, the weight of the clam could gradually pull the gland out. If the incline is too great, the clam will not receive enough light, so do not place it on a steep substrate. Adult *Hippopus* spp. tend to sit more on their hinge than on the byssal opening, therefore they should be placed so that the majority of their mantle is facing upwards. Juveniles will attach themselves to rocks with byssus threads as in other tridacnid clams. Tridacnid clams do not like a strong current, especially *T. crocea*, so do not place them in positions where they would receive strong, direct water currents.

For the first few days, until the clam has firmly attached itself, it may fall over a number of times. During this period it might be wise to place a few small rocks around the clam to help it stay upright. Do not place the clam between large rocks or inside small holes, as this can prevent the clam from opening fully. If the clam falls over do not leave it in this position. It should be righted as soon as possible. Another way to place a clam is to create a custom base and allow it to become attached first. This base can be made by hollowing out a depression in a piece of live rock with a Dremel Moto-tool™ or similar grinding tool. Use the clam's base to determine the fit and modify as necessary. The clam can then be seated upright in it's customized base and placed on the bottom of the aquarium in a brightly illuminated spot. This temporary location makes it easier to retrieve the clam in case it falls out of the base. Once the clam has firmly attached itself you can then place it and it's attached base in the desired location. One can also use half a clam shell for making a base.

Place the clam well away from any aggressive coral or anemone species. If you see sections of the mantle pulled away or shriveled on the same side as a coral or anemone, suspect it of irritating the clam and move the coral, anemone or clam as soon as possible. If a clam does not appear to be doing well, it should be moved immediately! Do not hesitate, or it will quickly die.

Retailers and wholesalers can best display these animals by placing them on small cups or pots filled with crushed coral. This way they will only attach to the gravel grains, making them easy to remove when sold. The pots also help to keep them upright and positioned properly under a light source. To really improve sales and make

An efficient method for dealers to display tridacnid clams. The pot is filled with crushed coral, preventing the clam from becoming permanently attached. The gravel should be rinsed periodically to prevent worms from taking up residence. S. W. Michael.

An ideal invertebrate holding system. The corals and clams are easily illuminated in this shallow tank design. Customers can appreciate the beautiful colours because they can view the clams and corals from above through the open top, which also allows easy access. Note mangrove seedlings. J.C. Delbeek.

customers happy, the dealer may hang a mirror above the aquarium, at about a 45 degree angle, so that the customer can see the true colors of the clams as they appear when the light is reflected back, without actually having to get up over the aquarium.

Pests and Predators

Small tridacnid clams are heavily preyed upon in the wild by a variety of organisms. Since the vast majority of clams available for the home aquarium are juveniles, one must be especially cautious with them. Various species of fish (e.g. triggerfish, large wrasses and puffers), crabs, lobsters, shrimp, polychaetes, octopi, and snails can prey upon tridacnid clams (Heslinga et al., 1990). See chapter 10 for details on snail predators.

Saron marmoratus feeding on a tridacnid clam. A.J. Nilsen.

Among the fish, certain wrasse species (Family Labridae) are bad tankmates with giant clams. Species such as *Coris aygula* (Twin Spot Wrasse) and *Gomphosus varius* (Bird Wrasse) have been known to attack and devour juvenile clams in the aquarium. Sometimes they are simply eaten from above or they are knocked over and eaten through the soft, unprotected, byssal opening. Any large wrasse species should be closely watched when introduced into a tank with tridacnid clams.

Other fish do not so much prey upon the clams, as irritate them. Fish that are constantly grazing such as *Centropyge* spp. (Pygmy Angels), *Ctenochaetus* spp. and *Acanthurus* spp. tangs, and *Ecsenius* spp. (Blennies) will occasional nip at a clam in passing. Sometimes small pieces of tissue are removed but the biggest problem is that the clam will expand less and less, and will eventually expand so little that it will not receive enough light and slowly die (see also failure to expand, chapter 10).

Large crabs are another serious predator of clams. Crabs will eat clams and will usually kill them shortly after the clams are placed in the tank, before they can attach to the substrate (Carlson, 1991). They either attack the clam through the byssal opening, or larger crabs simply crack the shells open.

Certain species of shrimp can also prey on clams. For example large shrimp such as *Saron marmoratus* (Marble Shrimp) and *Saron* sp. (Buffalo Shrimp) will attack clams at night. On rare occasions even the Common Cleaner shrimp, *Lysmata amboinensis*, has been known to attack clams, especially injured ones, when hungry.

Parasitic snails are sometimes imported with tridacnid clams. Any new specimens should be closely examined and any snails removed immediately (see previous section this chapter and chapter 10).

Overgrowths and fouling by algae represent another problem for clams. When the algae begins to grow over the lip of the shell, the mantle may become irritated and it will not expand as much. Macroalgae, such as *Caulerpa,* can irritate the clam from underneath if allowed to grow under the byssal opening. In this case the clam will produce copious mucus from below that surrounds the base. Such mucus production is a normal means of protection for the clam against algae, stinging corals or predators. The mucus is thick, clear, and often with brown patches (see

brown jelly from clams in the troubleshooting section, chapter 10)
The effect of noxious by-products of soft corals (e.g. *Xenia* spp.) can also cause clams to produce copious amounts of clear mucus that can quickly clog prefilters (J. Burleson, pers. comm.).

Small anemones belonging to the genus *Aiptasia* can be a real problem in some reef aquariums. If these small anemones are allowed to grow on tridacnid clams, they can reach underneath th mantle and sting the clam. This will result in the mantle pulling away and the clam will eventually die.

Finally there are certain polychaete worms such as the larger *Nereis* spp. and *Eunice* spp., which can prey upon tridacnids. They are usually active at night and feed on the clam from below, through the byssal opening or by boring a hole through the shell.

For more information on the diseases, parasites and pests of tridacnid clams, and how to deal with them, see Chapter 10.

Care

With proper lighting and careful attention that they are not being irritated or fed upon by other organisms, tridacnid clams require little else in the way of care. As discussed in Chapter 4, these clams receive the majority of their nutrition from their zooxanthellae, so n additional feeding is required. Some people believe that tridacnids should be fed, based on the assumption that they are filter feeders like other clams, but on every occasion that we have attempted to feed them, the clams have closed forcefully, expelling the food like someone who is choking. The only food item they may accept is a dilute suspension of live phytoplankton or a yeast . However, the effort required to feed these items is not worth it in our opinion. Tridacnid clams have been grown successfully in both culture systems and home aquaria for many years without any supplement feedings. Of course bacteria, organic and inorganic compounds are always present in the water in closed systems, and these may be consumed or absorbed by the clams.

To ensure that your clam will grow it must be provided with other necessary ingredients. Calcium is the main building block for clar and should be present in the water at levels of at least 280 mg/L fc growth to occur. However, more rapid, natural growth is seen when the calcium ion concentration is in the range of 400-480 mg/L. Strontium is incorporated in the shell along with calcium and should also be provided for optimum growth. The addition of iodide to the aquarium will also enhance growth and colour in clams (J.C. Delbeek, pers. obs.). See Chapter 8 for a discussion of element additions and formulae.

Species Accounts

Scientific Name: *Tridacna crocea* (Lamarck, 1819)

Common Names: Boring or Crocus Clam, Crocea clam

Colour: *Tridacna crocea* is probably the most colourful member of this genus. Colours can include various mixtures of blue, purple, yellow, green, brown, gold, and orange in a variety of patterns. The mantle usually has numerous iridescent blue, yellow or green blotches, small spots or lines.

Tridacna crocea. J.C. Delbeek.

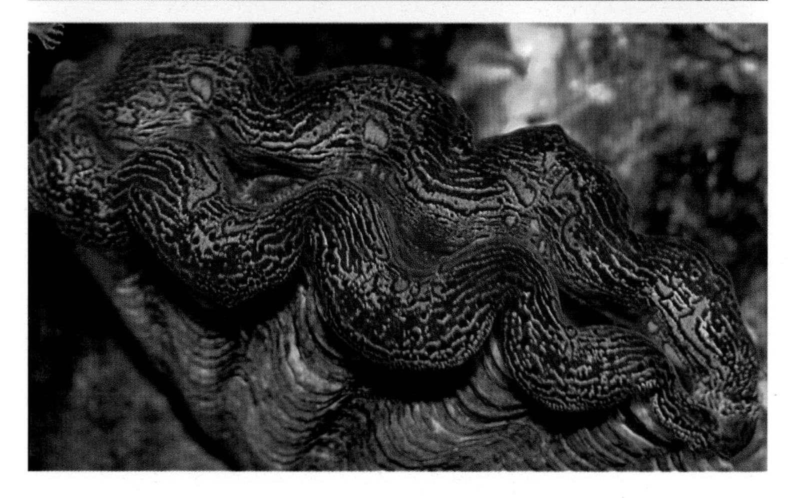

Tridacna crocea. Note eroded scutes on shell. J.C. Delbeek.

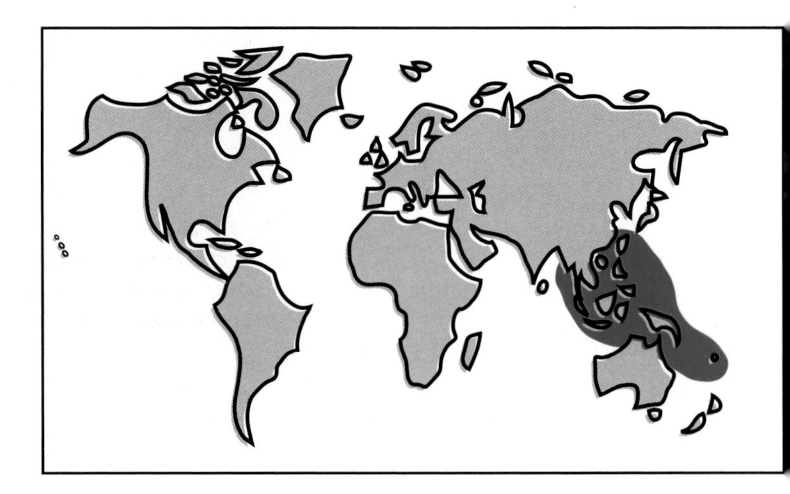

Figure 12.1
Tridacna crocea
Upper and lateral view of shell. Below: Map with geographical distribution of the clam. *After Lucas 1988.*

Distinguishing Characteristics: Large, wide byssus gland opening; normally smooth shell with closely placed scutes restricted to the upper margin; almost symmetrical shape; small size; frequently brightly coloured and; the incurrent aperture has very small, fine tentacles (Lucas, 1988; pers. obs.). Cultured clams will have scutes along the entire body of the shell, as they have not yet ground them down through their burrowing habits. Max. Length: 19 cm (9 in.).

Similar Species: This species is often confused with *T. maxima,* especially by those who rely on colour pattern alone. The shells of these two species, however, are easily distinguished.

Sometimes it is difficult to distinguish between *T. crocea* and *T. maxima* on the basis of mantle colouration alone. This beauty is a good example. It is probably *T. crocea*, but without viewing the shell it is hard to be certain. J.C. Delbeek.

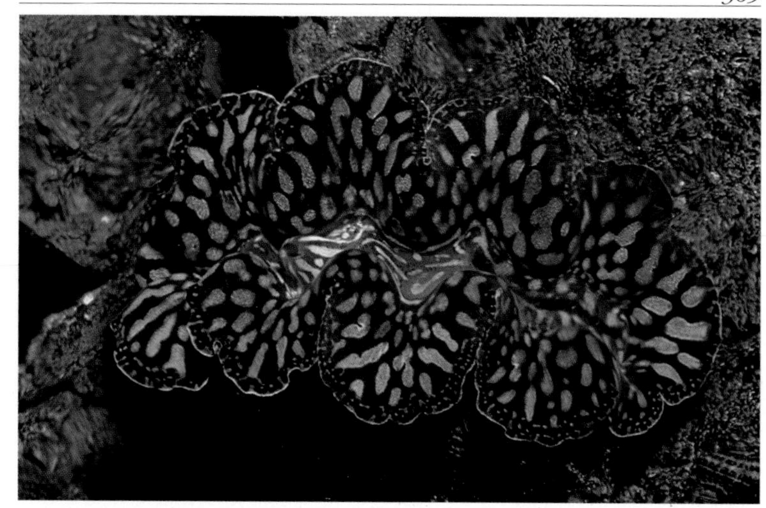

Tridacna maxima usually has a more elongated shell; at times the shell can be 3x longer than it is broad. Although both species have rows of scutes on the outside of the shell, those in *T. maxima* are more pronounced and cover the majority of the shell, while those of *T. crocea* are usually worn away and are generally restricted to the upper margin of the shell. Finally, *T. crocea* has a very long, wide byssus gland opening, extending almost to the edge of the shell.

Natural Habitat: *Tridacna crocea* is commonly found in shallow areas near shore and on the interior reef flat (Crawford and Nash, 1986). As its common name indicates, these clams actively burrow deep into boulders and coral heads by contraction and relaxation

Right: This clam imbedded in a coral head could be either *T. crocea* or *T. maxima*. S.W. Michael.

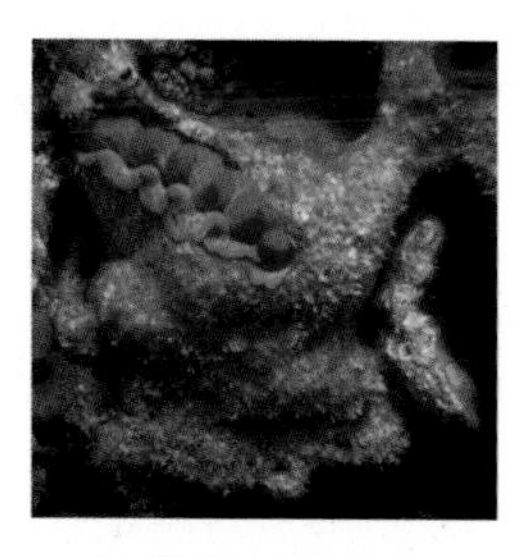

Above: The typical habitat of *T. crocea*, bored into old coral rock in the intertidal zone. Note shell of a dead clam still in the bored hole. J. Sprung.

of their byssal muscles. As a result only the top edges of the shell and mantle are visible. Once embedded they use their byssal threads to hold themselves in position. This species has been found in coral heads on the Great Barrier Reef at densities of up to 200 clams/m^2 (Lucas, 1988).

Aquarium Care: *Tridacna crocea* is probably the most delicate of the tridacnid species available for the home aquarium. This could be due to a combination of factors. Since these clams come from generally shallow waters, intense lighting is necessary to keep them in good condition and to maintain their bright colours. Although we have kept these clams for several years under fluorescent lighting, they were kept fairly close to the surface. As these clams live deeply buried in rock and have strong byssal attachments, collecting them requires extra care. Unfortunately, too often they are not collected properly and many of them arrive damaged. Once the byssus gland has been damaged they usually do not live much longer. This may be due more to subsequent bacterial infections than the actual damage. If you can obtain a specimen that is still attached to a rock, your chances will be much better.

This species is relatively easy to keep, provided a healthy specimen is obtained and enough light is given. Avoid handling or removing this species once it has become attached; they are very sensitive to such handling and can quickly die from the stress. Cultivated specimens of this species are now available and they have proved to be very hardy in aquariums.

Scientific Name: *Tridacna derasa* (Roding, 1819)

Common Names: Smooth Giant Clam or Derasa Clam

Colour: *Tridacna derasa* generally have either a striped pattern of wavy lines or a spotted pattern consisting of various combinations of orange, yellow, black, blue and white. Some specimens can also have brilliant blue or green lines.

Distinguishing Characteristics: Heavy, very plain shell; absence of strong ribbing and scutes; incurrent aperture has pronounced tentacles; loses byssus gland as it grows; very small, narrow byssus gland opening; most have tiny scutes and some have sparser but larger ones and; hinge is usually longer than half the shell length (Lucas, 1988; pers. obs.). Max. Length: 50 cm (20 in.).

Tridacna derasa. J.C. Delbeek.

Similar Species: *Tridacna gigas* can be confused with *T. derasa*, however, *T. gigas* has a different mantle colouration and the top edges of it's shell have triangular projections extending inwards. The mantle of *T. derasa* also extends further over the shell.

An adult *Tridacna derasa* in its natural habitat on the Great Barrier Reef, Australia. In the Coral Sea *T. derasa* are often vivid blue. Elsewhere they are typically shades of golden brown and green with a blue margin on the mantle. J. Sprung.

Natural Habitat: *Tridacna derasa* are common in oceanic environments, particularly in the 4 to 10 m (12 to 33 ft.) range of outer reef edges (Crawford and Nash, 1986). This species loses it's byssus gland fairly early and is often found lying freely on the substrate in lagoons (Yonge, 1975). This species is greatly sought after as a food item and has been hunted extensively throughout its natural range. As a result they are listed as threatened by the International Union for the Conservation of Nature. In protected areas such as the Great Barrier Reef, they can be found in densities of up to 30 clams/hectare (Crawford and Nash, 1986). This species was one of the first, along with *T. gigas*, to be commercially bred. As a result, specimens sold in the aquarium trade today are the product of aquaculture projects and are not wild-caught.

Aquarium Care: This species is the most widely available and hardy of the tridacnid clams. They can be placed almost anywhere in the aquarium and do well under a variety of light intensities. Of course, the more light you can supply, the faster they will grow. These clams can grow extremely quickly in the aquarium and it is not unusual for 6 cm (2.5 in.) individuals to double or triple their size in less than a year, provided they are given abundant calcium (>400 mg/L). It is not unusual for this clam to develop scutes when grown in the aquarium, possibly as a result of the artificial light regime and the effect on expansion of the mantle. Occasional specimens do form scutes clearly as a result of a genetic trait (G. Heslinga, pers. comm.).

Figure 12.2
Tridacna derasa
Upper and lateral view of shell. Below: Map with geographical distribution of the clam. *After Lucas 1988.*

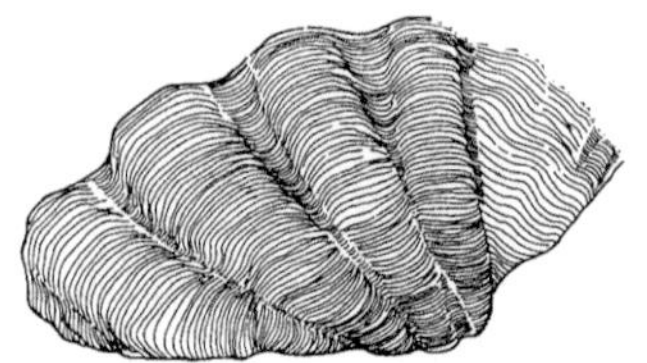

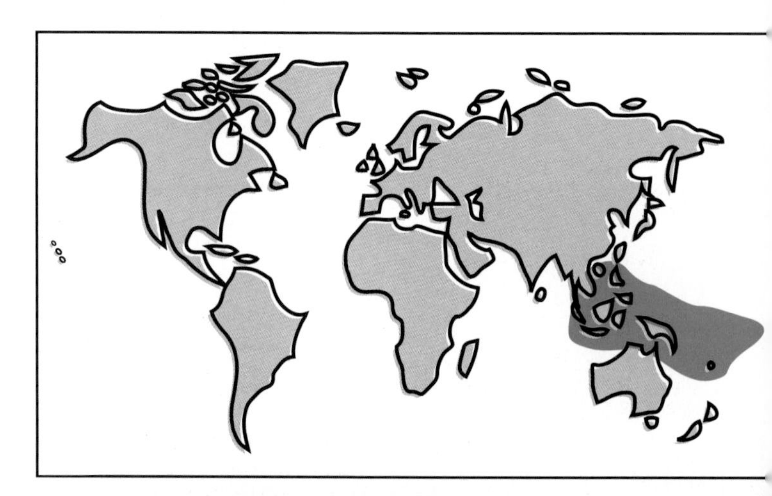

A possible hybrid cross between *T. derasa* and *T. gigas*, as mentioned in the text. The specimen is tank-raised. It is also possible that this is only a localized form of *T. derasa* with features similar to *T. gigas*. Note the clear windows at the center of the mantle, like *T. gigas*, and the unusual small teardrop-shaped clear windows scattered over the mantle.
J.C. Delbeek.

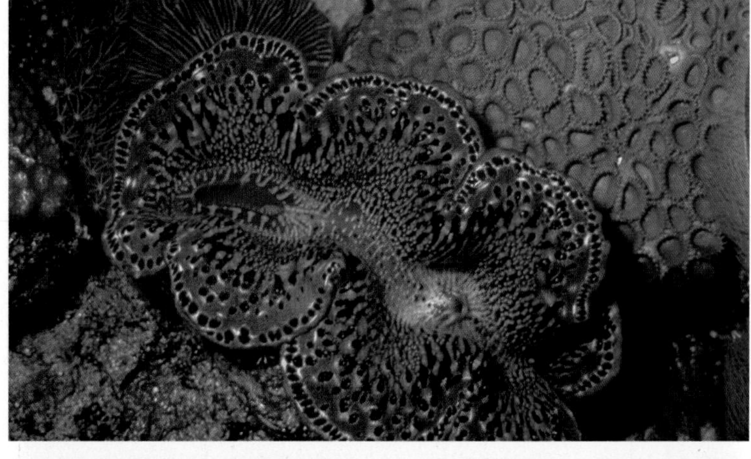
Tridacna maxima. J.C. Delbeek.

Tridacna maxima. J.C. Delbeek.

Tridacna maxima. J.C. Delbeek.

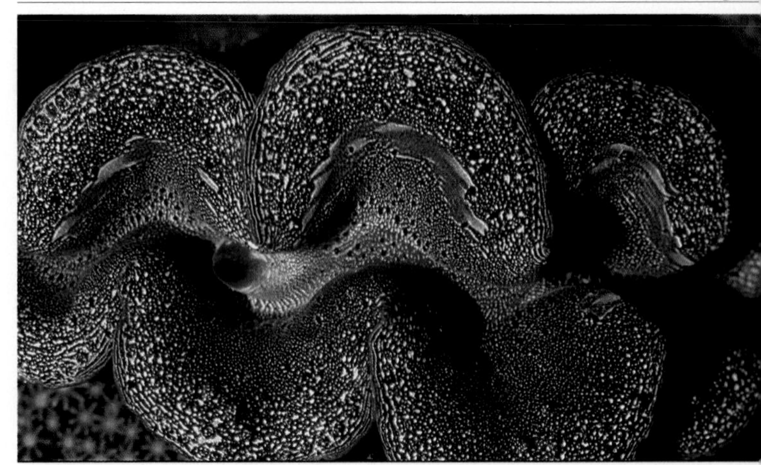

Tridacna maxima. J.C. Delbeek.

The natural habitat of *Tridacna maxima*, on top of the reef in shallow water with plenty of light. This reef has a particularly dense population. S.W. Michael.

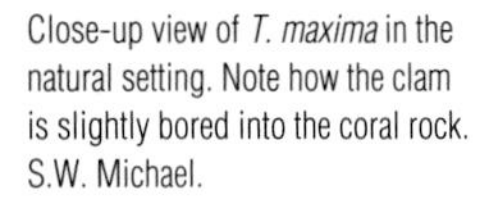

Close-up view of *T. maxima* in the natural setting. Note how the clam is slightly bored into the coral rock. S.W. Michael.

Figure 12.3
Tridacna maxima
Upper and lateral view of shell. Below: Map with geographical distribution of the clam. *After Lucas 1988.*

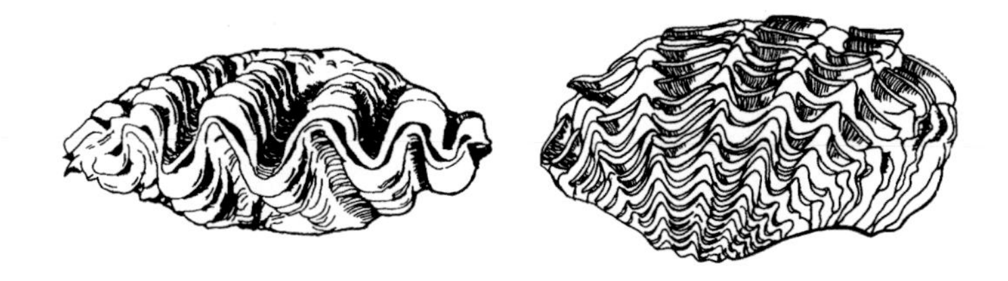

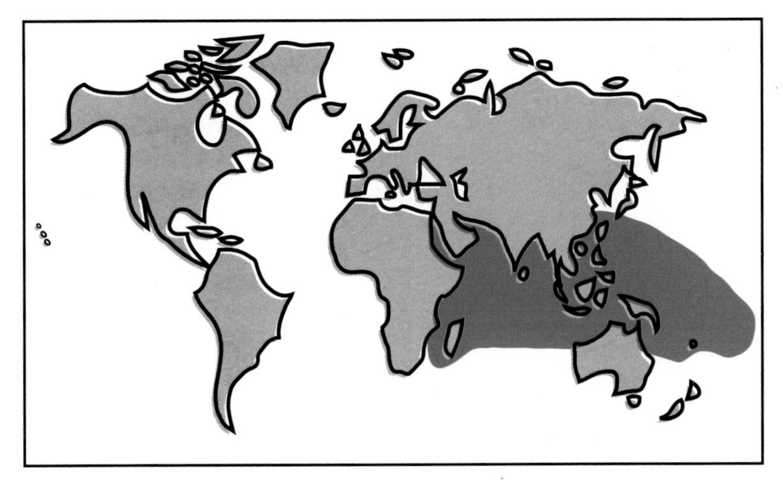

maxima's byssus opening tend also to curl upwards with a chitonous ring surrounding it (Achterkamp, 1987b; pers. obs.).

Natural Habitat: *Tridacna maxima* is one of the most common and widespread giant clam species, being found from the Red Sea and East Africa to Pitcairn Island in the eastern Pacific. They are generally found on reef tops or slopes, partially embedded in the substrate, firmly attached by their byssus filaments. As a result, wild specimens often have various encrusting growths such as corals, coralline algae and sponges on the upper portions of their shell. In some parts of Polynesia these clams can reach densities of up to 60 clams/m^2 and can occupy virtually all of the available surface on some coral patches (Lucas, 1988).

Aquarium Care: This species doesn't require as much light as *T. crocea*, and can therefore be placed a little lower in the aquarium. Once placed on a suitable substrate, a healthy specimen will attach itself in less than a day, so be careful that where you place the clam initially is where you want it to be. This species has proved to be relatively hardy, however, like *T. crocea* they can be delicate. Many of the specimens available in aquarium stores are still wild-caught, but aquacultured specimens are becoming available and the supply can only improve.

Scientific Name: *Tridacna squamosa* (Lamarck, 1819)

Common Names: Fluted or Scaly Clam, Squamosa clam

Colour: The mantle of *T. squamosa* is highly variable in colour with green and blue spotted varieties. However, brown with numerous golden brown or yellow wavy lines is the most common pattern.

Distinguishing Characteristics: Shell almost symmetrical; large well-spaced scutes; hinge half the shell length; incurrent aperture has numerous large, branched, tentacles; shells usually white but can be yellow, orange or pink in colour; mantle extends well over the edges of the shell and; small to moderate byssus gland opening (Lucas, 1988; pers. obs.). Max. Length: 40 cm (16 in.).

Similar Species: This species can be confused with *T. maxima* when small. This is due to the fact that both species have scutes on their shells. However, those of *T. squamosa* are much larger than

A *Tridacna squamosa* extends its exquisite mantle. J. Sprung.

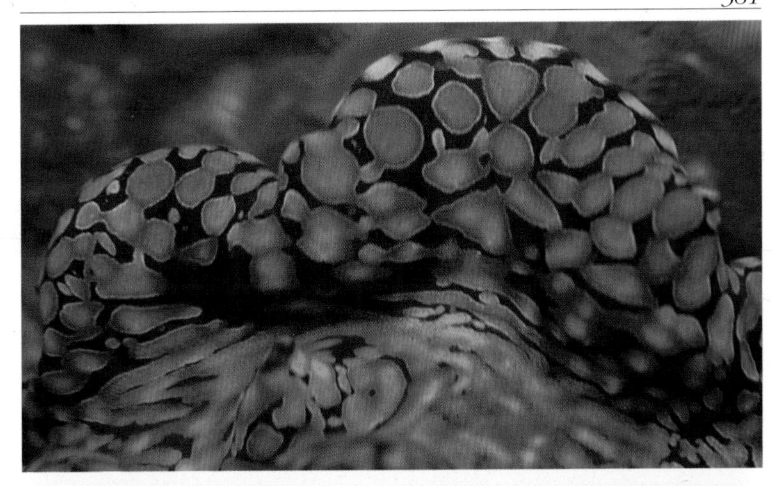

A spotted mantle pattern is common in *Tridacna squamosa*. The blue centers are unusual. J. Sprung.

This astonishing clam is a *Tridacna maxima* with mantle colouration like *T. squamosa*. It may be a rare colour morph only or else a hybrid. J. Sprung.

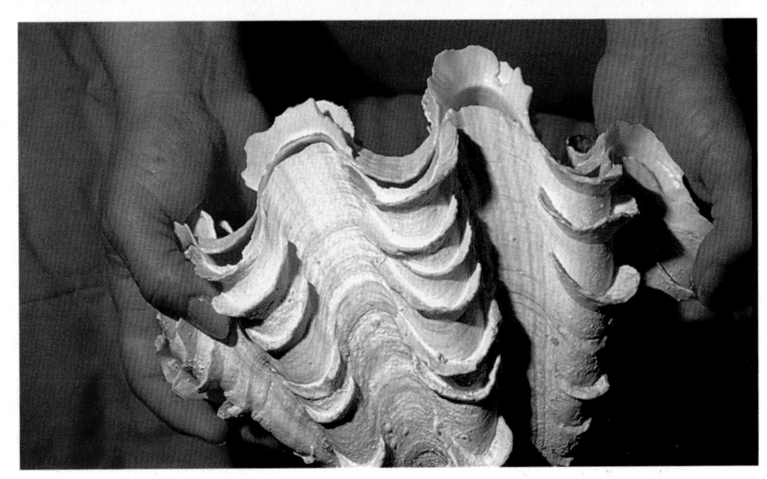

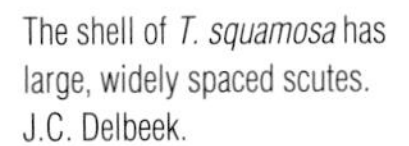

The shell of *T. squamosa* has large, widely spaced scutes. J.C. Delbeek.

Figure 12.4
Tridacna squamosa
Ventral and lateral view of shell. Below: Map with geographical distribution of the clam. *After Lucas 1988.*

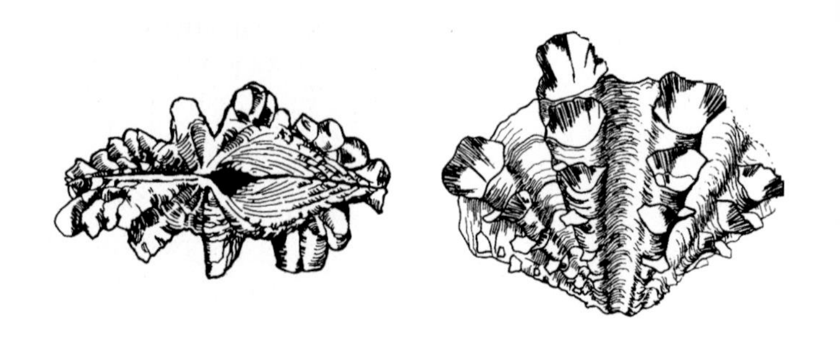

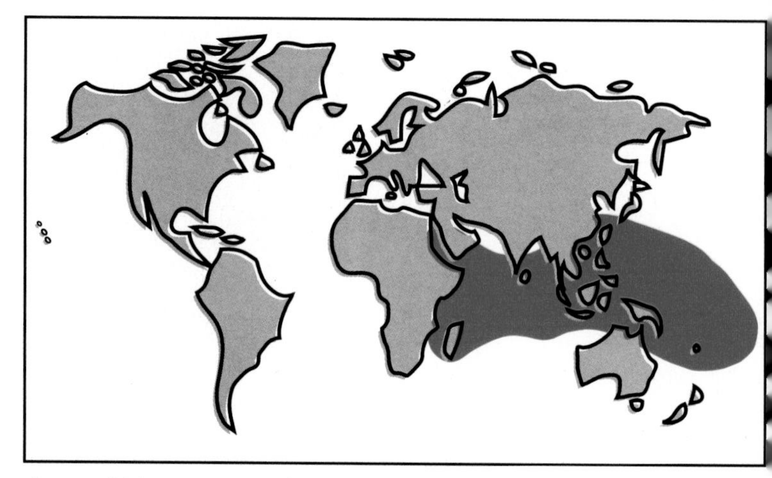

those of *T. maxima* and are not as closely spaced. This species is known to form hybrids with *T. crocea* and *T. maxima.*

Natural Habitat: The Scaly Clam inhabits sheltered environments such as back-reef lagoons, in depths to 15 m (45 ft.) (Crawford and Nash, 1986). They are usually found lying on the substrate, always attached by their byssus threads. Because of their pleasing shape and brightly coloured shell, this species is highly prized in the shell trade and is frequently imported from the Philippines. One wonders whether such novelty imports are as strictly controlled and monitored as the aquarium trade in live clams is.

Aquarium Care: *Tridacna squamosa* has proved to be very hardy with life spans of 10+ years not uncommon. These clams are not as demanding as the others when it comes to light intensity, however, as with any tridacnid clam, the more light you can provide, the better. This species is seldom imported into North America but is still commonly available in some European countries such as Holland. Again, commercial breeding of this species is already underway and small cultured specimens are now available in North America. Small, 3 cm (approx. 1 inch), cultured specimens are more delicate than the other cultured species, and survival is not good. However, larger, hardier farm raised specimens will be made available for sale in the very near future, and they will have nice color (G. Heslinga, pers. comm.).

Large *T. squamosa* should be placed on a firm substrate, preferably the bottom of the tank, providing the light intensity is adequate. They are capable also of producing a strong stream of water out of their excurrent siphon when they close rapidly. This can easily drench any lights and equipment around the aquarium if the specimen is too close to the surface; beware!

Scientific Name: *Tridacna tevoroa* (Lucas, Ledua, and Braley 1990)

Common Names: Devil or Tevoro Clam

Colour: The mantle is predominantly brownish-grey in colour and rugose (having many protuberences). The mantle colour is usually uniform, but it may also have some pattern, with numerous pale spots, including but not limited to the tip of the exhalent siphon and grooves between the mantle protuberences.

Distinguishing Characteristics: Similar to *T. derasa* but has a thinner, sharper edged shell; more prominent guard tentacles on the incurrent siphon; a rugose mantle and; red bands on the shell near the umbo (Lewis and Ledua, 1988). Unlike *T. derasa*, the mantle does not extend far over the edge of the shell. This is a characteristic that *T. tevoroa* shares with *Hippopus* species, in addition to the red bands near the umbo (Lucas, et al., 1991). Max. Length: 55 cm (21 in.).

Similar Species: *T. derasa*; see above for differences.

Natural Habitat: This species was first described by Lewis and Ledua (1988) from specimens found in deep reef areas in very clear oceanic water, 20 m (60 ft.) deep or more, off eastern Fiji. It has since been found in the same type of habitat in Tonga (Lucas, et al., 1991).

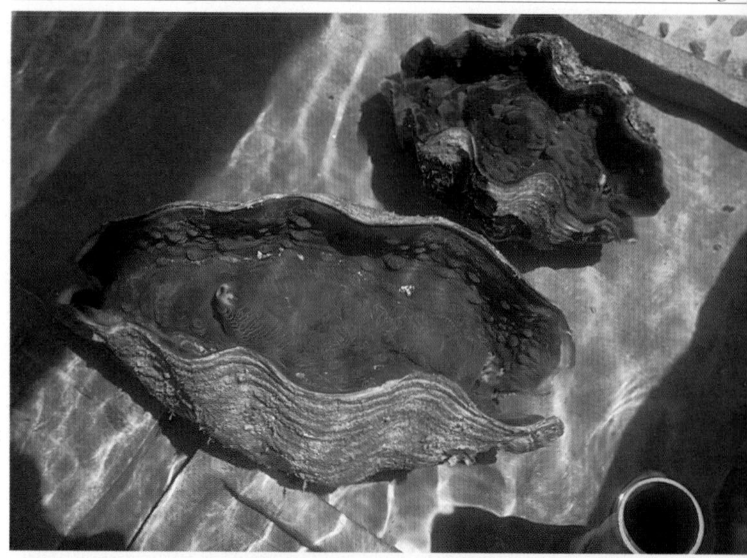

Tridacna tevoroa. The shell is most similar to *T. derasa*. J. Lucas.

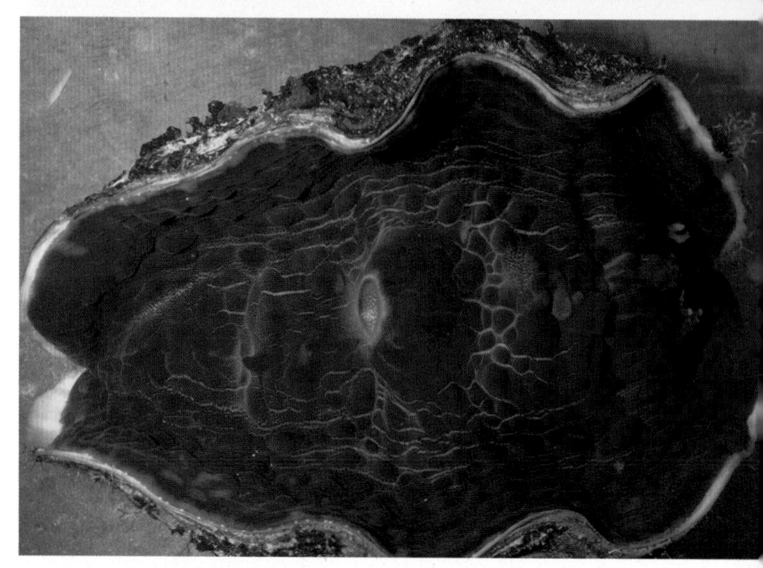

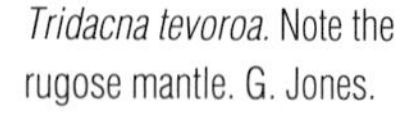

Tridacna tevoroa. Note the rugose mantle. G. Jones.

Tevoro clams have adapted to the low light levels at depth through morphological changes that maximize the light reaching the zooxanthellae. They have dull color due to low density of iridophores (iridocytes) that give bright colours to shallow water giant clam species and reflect back some of the light. Whereas iridophores are found close to the surface in shallow water tridacnids, with most zooxanthellae occuring beneath the iridophore layer, in *T. tevoroa* the zooxanthellae have a shallow distribution along the surface of the mantle (Lucas, et al., 1991).

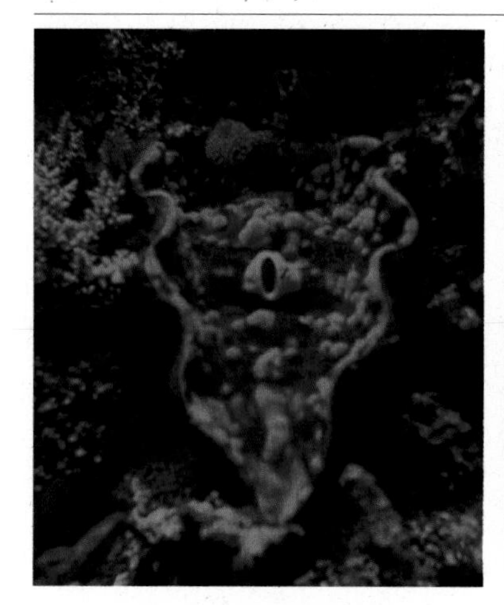

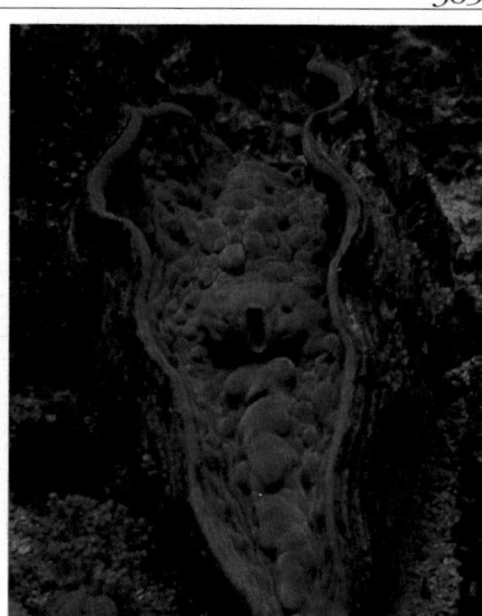

Tridacna tevoroa photographed in its natural habitat. No flash was used, in order to show the blue light characteristic of the deep location where it occurs. This specimen has some pattern on the mantle. R. Braley.

Another *T. tevoroa* in the natural habitat. R. Braley.

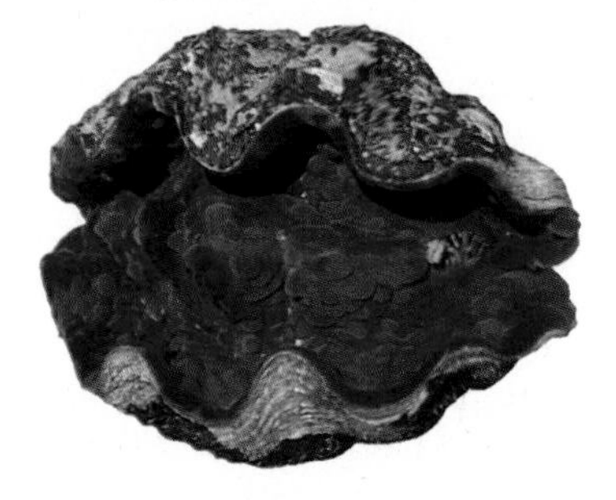

Figure 12.5
Tridacna tevoroa
Upper and ventral view. Note red stripes mostly worn away on the underside of the old shell on the right. R. Braley and J. Lucas.

Below: Map with geographical distribution of the clam. *After Lucas 1988.*

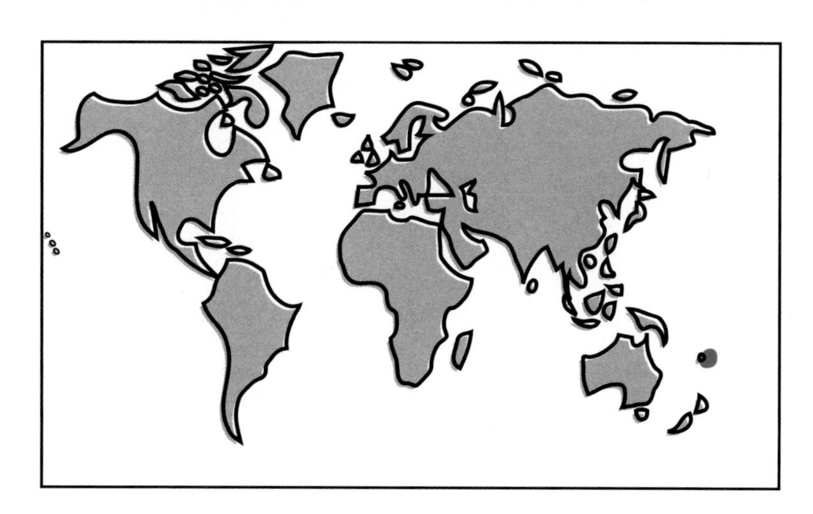

The rugose mantle provides additional surface area for the growth of zooxanthellae, and Tevoro clams also compensate for the lack of mantle expansion over the shell margins by wide gaping of the valves (Lucas, et al., 1991).

Aquarium Care: Preliminary work with this species indicates that it is sensitive to bright light (Lewis and Ledua, 1988). Mortality of the clams when placed in shallow, sunlit aquariums was attributed to light intensity, since they were kept in 3m depth after being collected from 20 m. Tevoro clams may lack sufficient amounts of protective pigments, and therefore may be sensitive to UV light (see Dunlap and Chalker, 1986). They also may lack sufficient enzymes to detoxify excess oxygen produced by increased photosynthesis in bright light (Lucas, et al., 1991) (see Shick and Dykens, 1985).

This species is extremely rare (50:1, *T. derasa:T. tevoroa* in Fiji), a feature that makes it the giant clam species most vulnerable to extinction (Lucas, et al., 1991). It is unlikely that any wild caught specimens would ever be imported for the aquarium trade. It is more likely that captive breeding projects will provide aquarium specimens in the future. They are being bred and reared on a small scale presently by fisheries staff in Tonga (J. Lucas, pers. comm.). Aquarium care should be similar to *T. derasa*, taking into account the possible UV and/or bright light sensitivity. We expect that they would fare best under a combination of standard fluorescent lights, with strong emphasis on blue spectra.

Scientific Name: *Hippopus hippopus* (Linnaeus, 1758)

Common Names: Horse's Hoof, Bear Paw or Strawberry Clam

Colour: Mantle is dull greenish-brown to grey with some faint gold stripes.

Distinguishing Characteristics: Mantle does not extend past the edge of the shell; shells thick and strongly ribbed with reddish blotches; incurrent aperture without tentacles and; very narrow byssal opening bordered with interlocking teeth (Lucas, 1988;). The name, "horse's hoof" refers to the appearance of the shell when the clam is resting on its byssal opening with the valves closed. The broad base of the valves is shaped very much like a horse's foot. Max. Length: 45 cm (18 in.).

these species often produce small polyps around the base that can be broken off with a pointed scissors and planted on the bottom of the tank in sand, or on live rock (see photo, chapter 3).

With a bone scissors, wire cutter, or hammer and chisel, colonies with phaceloid (e.g. *Caulastrea furcata*), or even plocoid development can easily be divided, and these fragments rapidly grow and bud new polyps to form whole colonies. Foliaceous or sheet forming species (e.g. *Pavona* spp.) can be broken like dividing large cookies, and the fragments will heal and continue to grow, even when individual polyps on the edges are torn in half. Coral polyps or fragments must be securely positioned with monofilament line, cement, or other devices to insure survival (see aquascaping, chapter 7, for complete details).

Duncanopsammia axifuga fragment given to Dietrich Stüber, Berlin. J. Sprung.

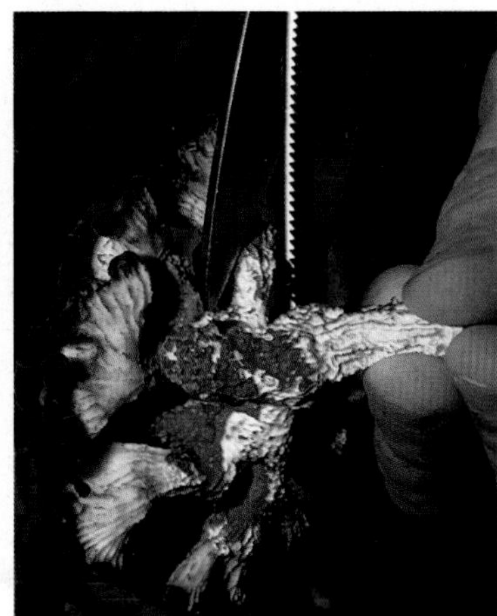

Mike Paletta shows how to cut fragments off of *Caulastrea furcata* using a powerful shears. J.C. Delbeek.

Whole colonies of globe shaped or massive corals (e.g. *Favia*) may be sawed in half or into smaller portions, and each will heal and grow eventually into the characteristic dome or globe shape. Cores of living polyps with the skeleton can also be drilled out of massive coral heads, and the holes left behind heal rapidly, especially if they are filled with cement to the tissue layer. The cores become new colonies genetically identical to the parent and other cores from the same head. With care it is even possible to core individual live polyps.

As we discussed in chapter 3, stony corals exhibit a range of spawning strategies and schedules depending on their geographic location. Cues for sexual reproduction such as temperature, photoperiod, and moonlight also vary among species. Lighting and temperature can be manipulated, and the presence of moonlight can be duplicated with a low wattage blue incandescent lamp. Monthly lunar cycles can be simulated with programmable timers.

For additional information about coral propagation, please review the article written by Michael Paletta in the Winter 1992 Seascope, volume 9, published by Aquarium Systems, Inc.

The Identification and Care of Stony Corals Kept in Aquariums

Subclass Zoantharia
Order Scleractinia

Family Acroporidae (Verrill, 1902)

Scientific Name: *Acropora* spp.

Common Names: Staghorn Coral, *Acropora*

Colour: Usually golden-brown with pale tips that may be white, pink, purple, blue, or green. Whole colonies may also be vividly coloured, with blue, pink, purple, orange, green, and yellow being typical in the Pacific. Caribbean species seem to be only shades of brown to yellow.

Distinguishing Characteristics: *Acropora* species are characterized by rapidly growing terminal polyps at the tips of branches.

Similar Species: The only other species covered by this book that might be confused with *Acropora* species are *Hydnophora rigida, Pocillopora* spp., *Seriatopora hystrix,* and *Stylophora*. The lack of a terminal polyp readily distinguishes these from *Acropora.*

Natural Habitat: With three hundred and sixty eight nominal species (Veron, 1986), one can imagine that there are many niches occupied by *Acropora* species. The reef zone most colonized by *Acropora* is the outermost part, where wave action and currents are strong. Still, many lovely species occur in lagoons and backreef areas, where they may form finely branched, intricate colonies. Most species are abundant in brightly illuminated, clear water zones, but those living in lagoons or coastal reefs may grow in turbid water, and many species can be found in deep water, where they do not receive bright light.

Catalaphyllia jardinei in turtle grass and *Montipora digitata* reef-flat zone, Palau, at 1.5 m (5 ft.) depth. M. Awai.

Aquarium Care: *Catalaphyllia* is one of the hardiest species and one of the most beautiful. Like *Euphyllia* spp., it needs plenty of room. Allow a minimum of 15 cm (6 in.) around the skeleton for tissue expansion. *Catalaphyllia* does not like strong current, but does like flow sufficient to make the tentacles lightly sway, like a breeze over a wheat field. The most common problem with this species is separation of the tissue from the skeleton. This is prevented with 1) adequate calcium levels, 2) regular strontium additions, and 3) adequate light (see troubleshooting section, chapter 10, and reproduction information below). *Catalaphyllia* does well under all types of lighting, but has the best appearance under fluorescent lighting; the use of blue fluorescents brings out it's rich colours. The different colour morphs are all compatible, and may be placed next to each other.

Aquarium Reproduction: This species is a gonochoristic broadcast spawner (Richmond and Hunter, 1990). Emmens (1991) observed gamete release by an apparent male colony. Occasionally new polyps will bud off around the base, as in *Euphyllia* species. Another means of asexual reproduction may occur when a portion of the polyp becomes detached from the skeleton. This mode is similar to both polyp "bail-out" and polyp "ball" strategies. The detached tissue does not reattach to the base. A new skeleton forms inside it that causes it to hang down and pull away from the colony, eventually severing the loose connection with the tissue of the mother polyp. Small dark "balls" have been witnessed inside the tentacles (J. Joos, pers. comm. and pers. obs. by the authors). It is unknown whether these are reproductive structures (i.e. eggs or planulae) or whether they are merely bundles of zooxanthellae accumulating like snowballs in the slight water currents within the tentacle.

Asexual reproduction in *Catalaphyllia jardinei*. A skeleton has formed inside a flap of tissue that had separated from the original skeleton. The weight of the growing skeleton causes the juvenile colony to drag down and separate from the parent. In this photo a scissors is used to artificially sever the attachment. A. Storace.

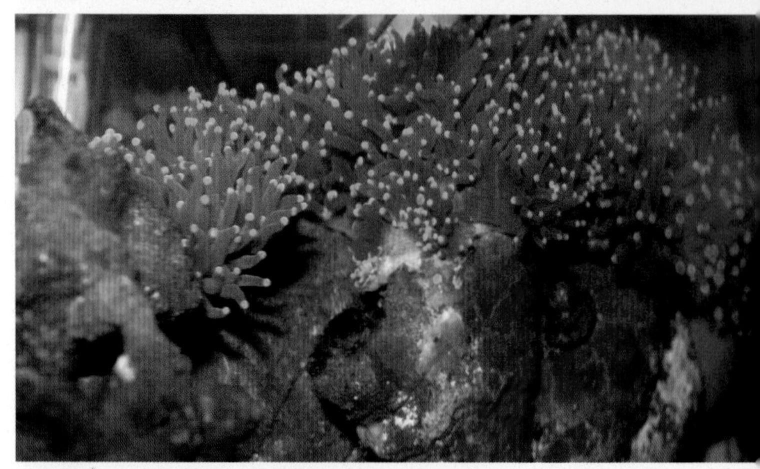

The exposed skeleton after the juvenile coral has been separated. In a healthy coral, tissue and skeletal growth would fill in this gap within a few months. A. Storace.

The baby *Catalaphyllia* moments after separation from its parent. This asexual mode of reproduction produces small numbers of large offspring with high chance of survival, but low dispersal. Sexual reproduction produces many offspring, with high dispersal and low survival. Note that this specimen has developed several separate skeletons inside, like petals of a flower. These could further divide to produce separate corals. A. Storace.

Genus *Euphyllia* (Dana, 1846)

Scientific Name: *Euphyllia ancora* (Veron and Pichon, 1980)

Common Name: Hammer Coral, Anchor Coral, Sausage Coral

Colour: *Euphyllia* species have several different colour morphs. The colour palette includes brown, green, yellow, and gray, and there are numerous shades of these, and combinations of the colours. The tips of the tentacles are usually coloured differently from the rest of the polyp, but solid coloured specimens also occur.

Distinguishing Characteristics: *Euphyllia ancora* is one of the most beautiful corals. The anchor-shaped tips of the tentacles are characteristic. Colonies are phacelo-meandroid.

Similar Species: At least three other species can have anchor shaped tips, *E. cristata*, and two species that seem intermediate between *E. ancora* and *E. glabrescens* on the basis of the skeleton. The name *E. fimbriata* is often used for specimens in which the anchors do not curl inwards, the tips of the tentacles being more "T" shaped. We do not know if *E. fimbriata* is a valid name for another species, or an old name for *E. ancora*.

Natural Habitat: Hammer Coral grows attached to hardbottom in lagoon reefs, in turbid water. Colonies are massive, usually 0.6 m to 1.2 m (2 to 4 ft.) across, composed of meandering walls that form dichotomous interlocking branches. Specimens collected for the aquarium trade with roughly 15 cm (6 in.) skeletons are usually

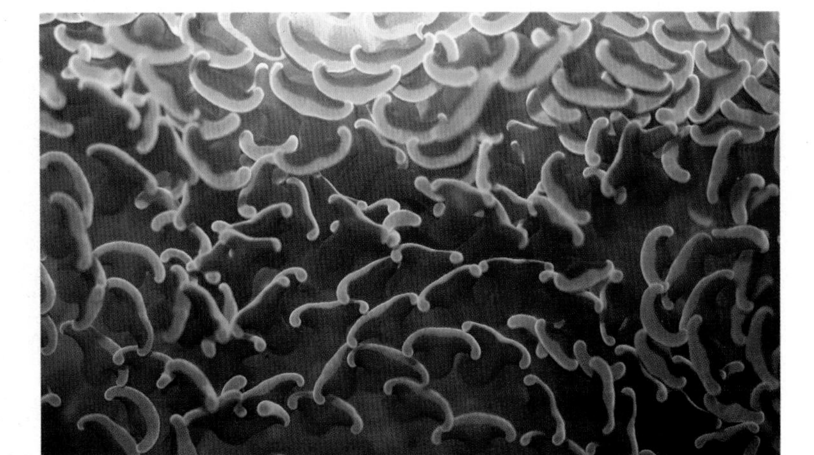

Euphyllia ancora, detail of the tentacles of this most beautiful coral. J. Sprung.

Three colour morphs of *Euphyllia ancora*. J.C. Delbeek.

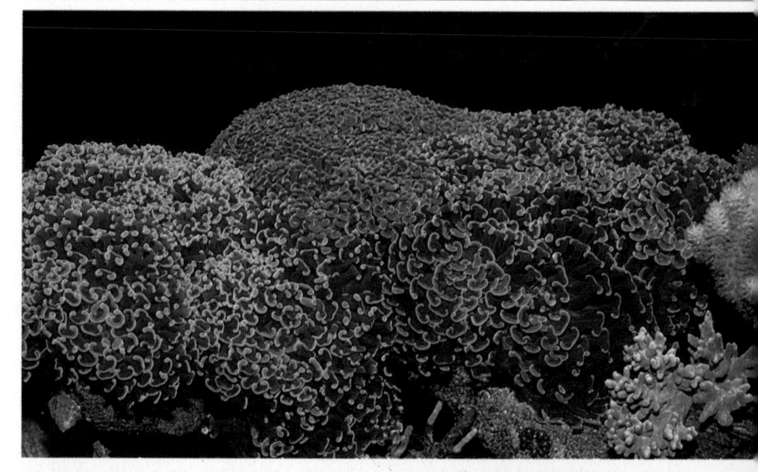

Unidentified *Euphyllia* sp. with thick phaceloid skeleton, 5 cm (2 in.) diameter corallites, septa not strongly exsert, and polyps like *E. ancora*. Another species(?) has the identical skeleton and polyps like *E. divisa*. The whole colonies remain phaceloid, never taking flabellomeandroid shape. J. Sprung.

Unidentified *Euphyllia* sp. This rare form has a thin phaceloid skeleton, 1.5 cm (0.6 in.) diameter corallites, septa not strongly exsert, and polyps like *E. ancora*. J. Sprung

broken off of the main colony with a geological hammer, though sometimes whole, unbroken colonies are collected. Often, the larger colony is composed of growths no longer connected by tissue, as the upward growing polyps divide and shade the base so that the tissue recedes and branches separate. Sometimes pieces are just broken through the polyp tissue. The tissue damage does not harm the main colony, and if the collected specimen is properly handled, it too will heal rapidly. However, such specimens are more prone initially to fatal infections from bacteria and protozoans. It should be noted that the colonies collected for the aquarium trade represent only about one to two years growth. If the main colonies are only periodically harvested and not removed entirely, then the growths taken are renewable.

Aquarium Care: See description for entire genus following the species descriptions.

Scientific Name: *Euphyllia cristata* (Chevalier, 1971)

Common Names: None, or same as *E. divisa.*

Colour: Same as *E. ancora.*

Distinguishing Characteristics: Colonies phaceloid. Corallum is compact. Large septa similar to those of *Plerogyra sinuosa* are visible when the living tissue is contracted. Tentacles are most like those of *E. ancora,* but most of the tips are more sausage-like, not curled. A distinct species has tentacles and appearance nearly identical to *E. cristata,* but the skeleton does not have large septa. It is phaceloid, with septa most like *E. glabrescens,* but the skeletons of whole colonies are much taller than that species. Still another "species" also has a phaceloid skeleton and a polyp indistinguishable from *E. divisa.*

Scientific Name: *Euphyllia divisa* (Veron and Pichon, 1980)

Common Names: Frogspawn Coral, Fine Grape Coral.

Colour: Usually brown with pale tipped tentacles. May be green with paler tentacle tips, brown with green tips, or yellow with pale tipped tentacles.

Distinguishing Characteristics: The mass of lighter coloured tentacle tips gives the impression of a spawn of frog eggs or tiny

Euphyllia divisa. "Frog Spawn Coral" J.C. Delbeek.

A green colour morph of *Euphyllia divisa.* J. Sprung.

grapes, hence the common names. Colonies are phacelo-meandroid. Veron (1986) describes the skeleton as identical to *E. ancora*. This is not often so. Usually the skeleton of *E. divisa* is about twice as broad as that of *E. ancora,* and more robust or denser. In addition, the septa are often larger, more exsert, and more widely spaced than in *E. ancora.* Still, the effects of environmental conditions can produce skeletal forms that are indistinguishable between these species, and there are undescribed species or hybrids that are intermediate in form, both in the skeleton and the live animal.

Scientific Name: *Euphyllia glabrescens* (Chamisso and Eysenhardt, 1821)

Common Names: Torch Coral, Branch Coral.

Colour: Usually brown with white-tipped tentacles. Occasionally with green tentacles or with green tips.

Distinguishing Characteristics: Colonies phaceloid. Tubular tentacles like an anemone or like *Catalaphyllia jardinei*.

Euphyllia glabrescens, "Torch Coral". J.C. Delbeek.

Aquarium Care (genus *Euphyllia*): Once established in the aquarium, all *Euphyllia* species are hardy. *Euphyllia divisa*, the Frogspawn Coral, is the hardiest while *E. glabrescens* is the most sensitive. The rest fall in-between these. All species require a lot of room for expansion. Allow at least 15 cm (6 in.) around the skeleton for expansion to avoid stinging neighboring corals. Sweeper tentacles are regularly formed by *Euphyllia* spp., and these may span 30 cm (1 ft.) or more in search of a targeted neighbor. *Euphyllia* should only be maintained in large aquaria unless one wishes to maintain only *Euphyllia*. In a small aquarium, less than 200 L (50 gal.), most *Euphyllia* species will grow too large within one year. In fact, it would be preferable if collectors would choose smaller specimens only. *Euphyllia* will capture food missed by the fish, but need not be fed directly. Slimy infections, which can rapidly consume tissue, should be treated immediately, according to the directions given in the troubleshooting section, chapter 10. *Euphyllia* can be maintained with almost no current, but they expand better when there is sufficient current to slowly lift and play with the mass of tentacles. If the current is too strong they will not open up.

Most *Euphyllia* species are compatible with each other, which means they can be placed adjacent to each other, allowing the expanded polyps and tentacles to mix. This creates a spectacular display. *Euphyllia glabrescens* is the exception to the compatibility rule. It is not as compatible with other *Euphyllia* species. It can touch them, but fares much better when it is not crowded by other species.

Aquarium Reproduction: *Euphyllia* species often bud new colonies in the aquarium. Normally budding occurs around the base of the coral where tiny new polyps may project out (see chapter 3). Technically this represents formation of new (phaceloid) branches, but these branches form in greater proliferation than can successfully survive in the given space, and as they grow, their skeleton, shaped like an ice-cream cone, is very thin at the point of attachment with the main colony. These cone-shaped polyps easily break off and may attach to live rock or live freely. The other form of budding involves the formation of septa unattached to the main colony. As these septa grow, they become heavy and slowly separate from the main colony, being pulled by gravity. As they fall they drag some tissue with them, which may form into a new polyp (see *Trachyphyllia* and *Catalaphyllia* which have a similar means of budding). *Euphyllia* have separate sexes and hobbyists have documented the release of both sperm and eggs into the water. However, there is at least one report of planula release in the wild (see Fadlallah, 1983), and Veron (1986) states that some *Euphyllia* from equatorial localities may be brooders. He witnessed small "balls" traveling up and down inside the tentacles (J. Veron, pers. comm.), and believes they were brooded larvae. His observation reminds us of the traveling balls seen inside the tentacles of *Catalaphyllia,* which may also be brooded larvae.

Scientific Name: *Nemenzophyllia turbida* (Hodgson and Ross, 1981)

Common Names: Fox Coral, Ridge Coral

Why this coral is called Fox Coral is a complete mystery to us, but the name has been in use in the aquarium trade for years and apparently has stuck. This living coral is truly unique in appearance and biology. It greatly resembles corallimorpharian "mushroom anemones", *Discosoma* sp., except that it forms a

Nemenzophyllia turbida, "Fox Coral". S.W. Michael.

skeleton. The scientific name and classification of this species may change (J. Veron, pers. comm.).

Colour: Usually pale green with cream-coloured stripes. Occasionally just pale brown.

Distinguishing Characteristics: *Nemenzophyllia* has a thin, approximately 0.6 cm (0.25 in.) wide skeleton with flabello-meandroid growth. The arrangement of septa gives the appearance of tiny boxes, and the walls are very fragile and paper thin. The sides of the walls appear much like the sides of *Plerogyra,* to which it may be related (Veron, 1986). When the polyp is expanded it looks like a corallimorpharian. It has no tentacles at all, and a smooth surface. The mouths are numerous along the central portion of the meandering polyp. When fully expanded the polyp becomes ruffly, or "fluffy" looking, and some areas may be so swollen that they appear almost like the inflated ribs on a beach raft. In this expanded state the relation to *Plerogyra* is evident, with a little imagination.

Natural Habitat: Unknown. Based on the appearance and structure, this species should be from areas with very little water movement. One can deduct from this that they may occur either in lagoons protected from heavy surge or tidal currents, or in deep water on outer reefs, likewise protected from strong water motion. Colonies probably grow as large stands, several feet across, and collectors break off branches.

Aquarium Care: Though it is hardy once established, we do not recommended this species for beginners. Most colonies are damaged in collection or shipping, and they succumb easily to white paste or "brown jelly" infections (see troubleshooting section, chapter 10, for details). Even healthy colonies take time to adapt, expanding only about 5 cm (2 in.) across initially. When it is fully adapted to the aquarium, *Nemenzophyllia* expands tremendously, the polyps being up to 10 cm (4 in.) broad, 5 cm in both directions from the central skeletal wall. It prefers the same light and conditions as *Plerogyra* and *Cynarina* i.e. little or no water motion, and bright, indirect light. It does not mind very bright light, which makes it expand even larger. *Nemenzophyllia* does not produce sweeper tentacles, and does not appear to eat large prey. It subsists mostly on the nutrition provided by its zooxanthellae, and perhaps it may trap microscopic prey and bacteria in its mucus, transporting that to the mouth by ciliary action. It probably also consumes dissolved organic and inorganic substances from the water.

Most pieces appear to be broken off larger colonies. Under optimal conditions damaged colonies will heal and grow. Still, the best colonies have no breaks in the wall where there is living tissue, being severed lower down, or being the whole unbroken colony.

Aquarium Reproduction: Presently unknown for aquarium specimens. It is likely that new polyp buds arise around the base as in *Heliofungia, Plerogyra,* and *Euphyllia*. If these appear, they can be clipped off and new colonies can thus be generated.

Scientific Names: *Plerogyra sinuosa* (Edwards and Haime, 1848) *Physogyra lichtensteini* (Quelch, 1884)

Plerogyra sinuosa (left) and *Physogyra lichtensteini* photographed in a green, turbid lagoon at 12 m (40 ft.) depth in Palau. B. Carlson.

Common Names: Bubble Coral, Pearl Coral.

Colour: Usually whitish gray, bluish gray, or brown. Sometimes green, especially the unidentified *Plerogyra* species described below. *Plerogyra sinuosa* often has a "cats-eye" appearance in the bubbles.

Distinguishing Characteristics: Three species of bubble corals are imported from Indonesia, *Plerogyra sinuosa* (Dana, 1846), *Physogyra lichtensteini* (Edwards and Haime, 1851), and an unidentified *Plerogyra* sp. All have bubble-like polyp vesicles, light capturing structures that expand during the day. At night the tentacles expand to capture prey. The most common variety, *Plerogyra sinuosa*, has large, smooth, round bubbles and huge, sharp exsert septa. Colonies are phaceloid to flabello-meandroid.

White morph of *P. sinuosa* with "cat's-eye" appearance on the vesicles. J. Sprung.

Large, bladder-like vesicles of *P. sinuosa*. J.C. Delbeek.

A second *Plerogyra* species is also common, and differs from *P. sinuosa* by having numerous, large "pimples" on the bubbles, and the bubbles are smaller than in *P. sinuosa*, affording a very different appearance. This distinct species is sometimes called "Octobubble Coral" in the aquarium trade. Another distinction of this species is the width and shape of the skeleton. Colonies are primarily meandroid, seldom phaceloid, with the branches being about as wide as in *Euphyllia ancora* or *E. divisa*.

Unidentified *Plerogyra* sp. sometimes called "Octobubble or Pearl Coral". J.C. Delbeek.

Unidentified *Plerogyra* sp. close-up. view. J.C. Delbeek.

The third Bubble Coral, *Physogyra lichtensteini*, is seldom collected. It forms massive, meandroid colonies, which do not have easily collectible side branches or phaceloid columns as in *Plerogyra* species. Generally small, whole colonies are collected. The bubbles are smaller and more numerous than in *Plerogyra sinuosa*, but are also round and smooth.

than *Cynarina,* and the skeleton has smaller, more numerous "teeth". Usually *Scolymia* species do not expand as tremendously as *Cynarina,* but some specimens do. *Scolymia vitiensis* is an occasional import from Indonesia, which can be green, brown or reddish. *Scolymia australis,* see photo, is similar. This red specimen is also extremely unusual. Green or brown are more typical colours.

Another coral, which may be a new species of *Cynarina* (Veron, 1986, reports there are two), *Scolymia,* or possibly a new genus, is an uncommon import from Indonesia. It expands like *Cynarina* and has a skeleton more like *Cynarina* than *Scolymia,* but distinct from *C. lacrymalis.* The tissue is thick and tough and not translucent, more like *Scolymia* than *Cynarina.* It also does not form the large bubble-like swollen sections typical of *C. lacrymalis,* but it does form small sections, particularly at the edge of the expanded polyp, giving it an appearance distinct from *Scolymia.* At night when it expands the tentacles it looks like *C. lacrymalis.* All of the specimens we have seen appear to be free-living, with a cone shaped base. The colouration is like *Scolymia,* brown, green, grey, or red. In our experience this coral always has pale, whitish areas around the mouth that may fluoresce green under blue light. Brilliant red specimens are a rare treasure.

Natural Habitat: *Cynarina lacrymalis* occurs on inshore reefs and hardbottoms in turbid water, usually 5 m (15 ft.) deep or greater. They may be found free-living on muddy sediments, or firmly attached on the hardbottom, nearly always facing straight up, or attached under overhangs. Seldom are the specimens as beautifully expanded as they are in our aquariums. *Scolymia* occur in the same habitat, usually attached on reefs with some vertical profile, often shaded. The unknown coral is usually free-living.

Aquarium Care: This is a very hardy coral that can be recommended to the beginner. It is easy to keep even with standard output fluorescent lights. *Cynarina lacrymalis* prefers low to medium levels of light, and little or no current. At night, when the tentacles come out, it is effective at capturing prey, and the nematocysts which it uses for this purpose are very potent. Be aware that *Cynarina* may capture and eat fish! Feeding is not necessary, but occasional feedings are appreciated, resulting in more rapid growth and larger expansion of the polyp during the day. If you place a *Trachyphyllia* next to the *Cynarina,* you should notice that neither harms the other. *Cynarina lacrymalis* is harmed by mushroom anemones placed next to it, and may be harmed by other corals. It seldom is the winner in a territorial battle; it either is unaffected or it loses. In our experience, *Cynarina lacrymalis* prefers to be located either mid-way down in the tank or on the bottom, where the light may be less intense than at the surface, or indirect. It will open up huge when placed in very still water. *Cynarina* may expand as large as 35 cm (14 in.) in diameter under fluorescent lighting. Care for the similar *Scolymia* species is essentially the same as for *Cynarina* and other mussidae, with one exception. *Scolymia* species are much more sensitive to stings from neighboring corals or anemones of any kind. Use caution when placing them! They quickly suffer injury, and often die from seemingly minor encounters. The mystery coral that resembles both *Cynarina* and *Scolymia* is not so delicate. In fact it is one of the hardiest corals! Its reaction to neighboring corals and anemones is another of the distinguishing characteristics that separate it from *Scolymia* and *C. lacrymalis.* This coral is a "universal" type with respect to burning neighbors. In general it does not harm other corals and other corals do not harm it. It even tolerates corallimorpharians and doesn't bother them. It feeds well, expanding tentacles during the day at the slightest scent of food.

Aquarium Reproduction: Not reported. Colonies with two or three "heads" (polyps) are occasional, indicating the possibility of buds forming around the base, though it is likely that they arise from division of the polyp as it grows. *Scolymia* spp. are hermaphroditic broadcast spawners (Richmond and Hunter, 1990), and it is possible that this characteristic applies to *Cynarina* spp., since they are closely related.

Scientific Name: *Lobophyllia* (de Blainville, 1830)

Common Names: Meat Coral, Brain Coral, Tooth Coral, Modern Meat Coral, Modern Tooth Coral, Root Coral

Colour: Shaded specimens tend to have brilliant colours such as red, orange or green. In turbid water such colours are common even in unshaded specimens. The most brightly illuminated specimens tend to be pale brown or gray, with hints of green or red. Grayish spots or lines are common features.

Distinguishing Characteristics: A Sharp spiny skeleton is the origin of the name Tooth Coral. Thick fleshy polyps that are often red, is presumably the origin of the name Meat Coral. *Lobophyllia*

Favia sp. (?) J. Sprung.

Favites sp. J. Sprung.

Favia sp. (?) J. Sprung.

Favites abdita. S.W. Michael.

Favites sp. (?) J. Yaiullo and F. Greco.

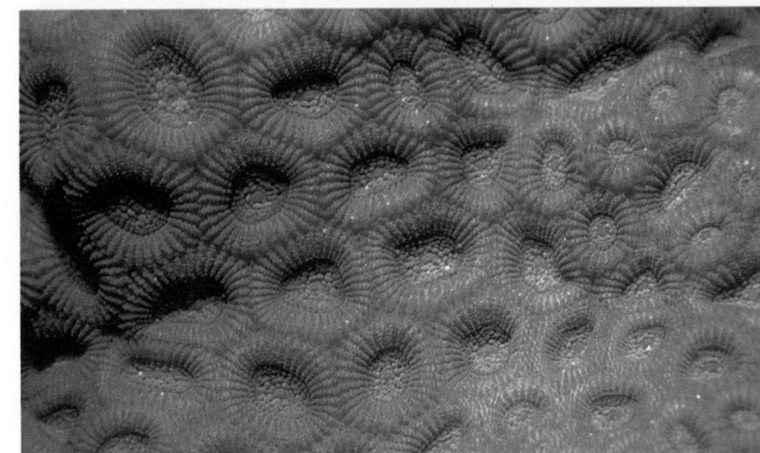

Favites flexuosa. B. Carlson.

Favia sp. (?) J.C. Delbeek.

Favia maxima. The giant expanded polyps in this species resemble those of *Caulastrea.* J.C. Delbeek.

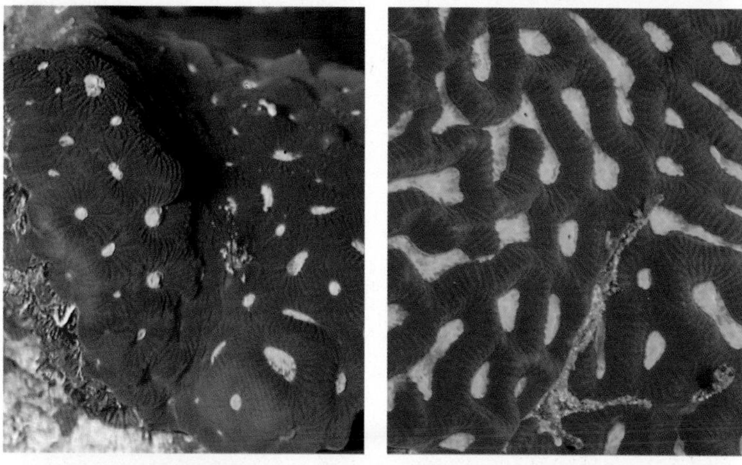

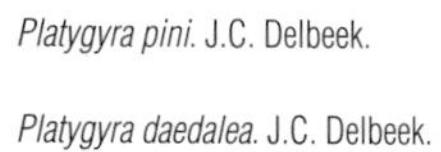

Platygyra pini. J.C. Delbeek.

Platygyra daedalea. J.C. Delbeek.

Leptoria phrygia. J.C. Delbeek.

Oulophyllia crispa. J. Sprung.

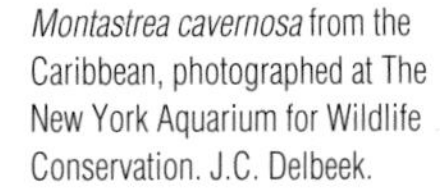

Montastrea cavernosa from the Caribbean, photographed at The New York Aquarium for Wildlife Conservation. J.C. Delbeek.

1986). *Montastrea* spp. have monocentric, plocoid corallites, as in *Favia* spp., but differ by forming daughter polyps through extratentacular budding generally, but not always (Veron, 1986). Indo-Pacific *Montastrea* can be difficult to distinguish from *Favia* species. Fully expanded living colonies of *Favia* may be difficult to distinguish from the other genera since the inflated polyps do not reveal the skeletal distinctions underneath.

Natural Habitat: These are common shallow water inhabitants, particularly *Goniastrea* species, which grow intertidally, but many range into deep water as well.

Aquarium Care: *Favia* species and the other Faviidae described above are easy to maintain in aquaria, and they may grow and attach to the rock within a few months. They feed well on shrimp and blackworms, but need not be fed. Although the tissue does not expand very far off of the skeleton in Faviidae, (2 to 5 cm (1 to 2 in.) only), one must be aware that they can produce amazingly long sweeper tentacles at night to sting neighbors (see chapter 3). They tolerate a wide range of light regimes and appreciate slight currents, which stimulate the polyps to open. Strong illumination enhances growth. Give strong but indirect illumination for best colour development. Placement is facilitated by inserting a plastic screw in the base to form a peg, and inserting the peg in a hole in the rockwork. See aquascaping, chapter 7, for more details about this method.

Aquarium Reproduction: Asexual reproduction is common in aquaria for members of the family faviidae. A complete polyp with a small amount of skeletal material separates from between polyps on the main colony i.e. polyp "ball". This drops off and attaches to

Polyp "ball" satellite colony forming on a faviid coral. J. Yaiullo and F. Greco.

the substrate (see chapter 3). Sexual reproduction should be possible in aquaria since faviids are hermaphrodites (Veron, 1986). They are broadcast spawners, spawning in November on the Great Barrier Reef, and in June and July in the central pacific (Richmond and Hunter, 1990). In chapter 3 we show a polyp of *Favia fragum*, a Caribbean species, which settled on the glass in J. Sprung's aquarium, apparently having arisen from a planula. The polyp has formed a skeleton and divided within just a few months of settlement. Although most faviid corals are hermaphroditic broadcast spawners, some species are brooders, and the planulae of some species have zooxanthellae (Fadlallah, 1983; Veron, 1986).

Family Fungiidae (Dana, 1846)

Scientific Name: *Fungia* spp. (Lamarck, 1801)

Common Names: Mushroom Coral, Plate Coral, Disk Coral

Colour: Brown, green, red, or pink, often with a pink mouth, or pink stripes.

Distinguishing Characteristics: There are many different *Fungia* species, and we do not distinguish them here. Veron (1986) gives a pictorial key to this genus. The different species form round, flat discs or dome-shaped free-living single polyps usually up to several inches (cm) across. Some species can reach diameters of 30 cm (2 ft.) (Veron, 1986).

Natural aggregation of *Fungia* sp., 1.5 m (5 ft.) depth at Kwajalein atoll, Marshall Islands. B. Carlson.

A beautiful red morph of *Fungia* on the reef in New Guinea. S.W. Michael.

Fungia sp. J.C. Delbeek.

Fungia sp. S.W. Michael.

Similar Species: *Cycloseris, Diaseris*. *Fungia* grows much larger, may be elongate and has septa with larger teeth (Veron, 1986). *Herpolitha* is also similar.

Natural Habitat: *Fungia* species live in shallow lagoons and on reef flats, on sand, mud, gravel, or coral rubble. Initially attached to shell, rock, or the parent *Fungia* when small, they break away and become free-living. Those from shallow, brightly lit waters with strong currents tend to be dome-shaped and thicker, while those from deeper or turbid waters with weak currents, tend to be flatter and thinner (Hoeksema and Moka, 1989).

Aquarium Care: *Fungia* should be placed on the bottom with slight or pulsed currents and bright light. They walk around a bit, so allowances should be made for this to prevent them from stumbling into something they might sting or be stung by. Strategic placement of small live rocks can pen them in one area. They will accept food but need not be fed.

Aquarium Reproduction: Formation of anthocauli is a common occurrence, and many offspring have been raised in aquaria, even through the second generation (see chapter 3). New colonies have also arisen in aquaria (D. Stüber, pers. comm.), presumably through asexual formation of planulae, but possibly from sexual reproduction. It is known that the sexes are separate, and that fertilization is external, by release of gametes into the water (Veron, 1986). Sexual reproduction in the aquarium has been observed (B. Carlson and A.J. Nilsen, pers. comms.).

Scientific Name: *Heliofungia actiniformis* (Quoy and Gaimard, 1833)

Common Names: Plate Coral, Sunflower Coral, Disk Coral, Mushroom Coral

Colour: Usually brown with white tips on the tentacles, but also bright green is common, and some colonies have pink tips or even solid pink tentacles. The oral disc has pale stripes.

Distinguishing Characteristics: Looks like an anemone when expanded. Long tubular tentacles. Septa have prominent teeth.

Heliofungia actiniformis. J. Sprung.

Heliofungia actiniformis.
J.C. Delbeek.

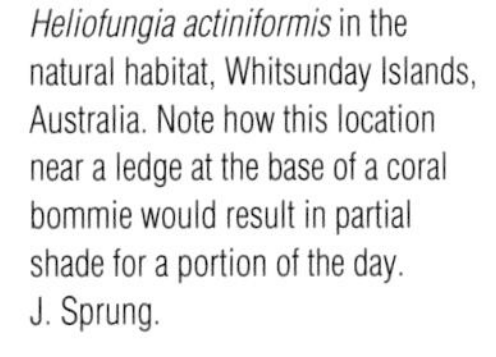

Heliofungia actiniformis in the natural habitat, Whitsunday Islands, Australia. Note how this location near a ledge at the base of a coral bommie would result in partial shade for a portion of the day.
J. Sprung.

Similar Species: None; has much larger tentacles than *Fungia* spp.

Natural Habitat: *Heliofungia* inhabits shallow, calm lagoons, where it can be found among coral rubble, or on sandy or muddy bottoms. They may be found between coral bommies or adjacent to them, where they receive bright light for several hours when the sun is directly overhead, but are partially shaded for the remainder of the day.

Aquarium Care: *Heliofungia* is one of the most frequently imported corals, and is often one of the first corals that beginners try because of its availability. This is unfortunate since it is a delicate coral that should not be recommended to beginners. The puzzling thing about this coral is that it does well in captivity, growing and apparently thriving for years sometimes up until the day it dies. When it dies it usually does so very rapidly, and the fouling tissue may drift and land on other corals killing them in a deadly "domino" fashion.

In the aquarium, *Heliofungia* should be placed on the bottom. It "likes" sand or gravel on the bottom, and bright light. It will walk, so it should be retained by strategic placement of rocks. Allow 15 cm (6 in.) for expansion— it can swell enormously with water. Best expansion occurs with light currents or surge. *Heliofungia* will feed, and should be offered small pieces of shrimp or fish about once per month. *Heliofungia* is easily injured by the stings of other corals, and its habit of wandering complicates this problem. Seemingly minor injuries that should heal sometimes overwhelm this species. It is possible that the lack of some trace substances weakens this coral to lethal attacks of bacteria and protozoans.

Aquarium Reproduction: Budding of new polyps from the underside of this species has been observed by many aquarists. These buds are not anthocauli as seen in the related Fungia and *Herpolitha* species, being more like the polyp buds seen around the bases of *Euphyllia* species. Sometimes newly imported *Heliofungia* have such buds attached. The buds grow and eventually break off at their constricted point of attachment. Under ideal conditions they grow quite rapidly, and the bottom can become a field of *Heliofungia*. *Heliofungia* are reported to be hermaphroditic brooders, releasing planula that settle within 2 days (Fadlallah, 1983). However, Veron (1986) suspects that this genus has distinct sexes like other fungiids, and this has been reported for populations on the Great Barrier Reef, where *Heliofungia* spawns in October and November (Richmond and Hunter, 1990).

Scientific Name: *Herpolitha limax* (Houttuyn, 1772)

Common Names: Tongue Coral, Sea Mole, Slipper Coral

Colour: Usually brown or gray, occasionally green.

Distinguishing Characteristics: It is shaped like a tongue, boomerang, "Y" or "X", with *Fungia*-like septa on its surface and a central groove, called the axial furrow, with numerous mouths, both in the axial furrow and across the corallum surface. The axial

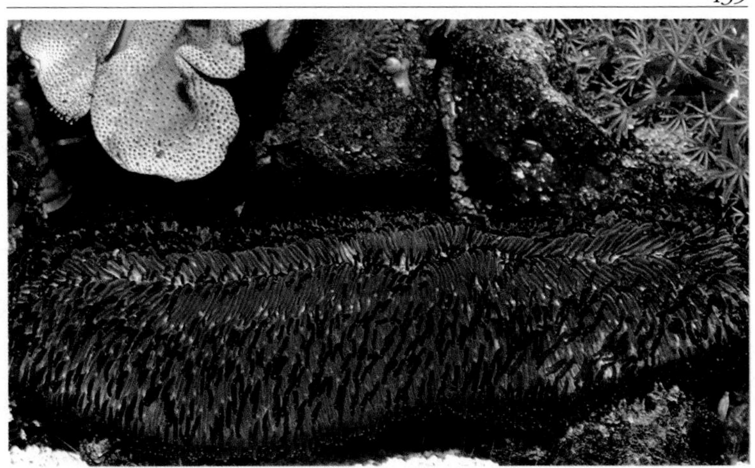

Herpolitha limax. J.C. Delbeek.

furrow extends to the end margins of the corallum. *Herpolitha* grows to be the heaviest of all free-living corals (Veron, 1986), attaining lengths over 1 m!

Similar Species: *Herpolitha weberi* has more sharply pointed ends and sparsely located mouths outside of the furrow. *Polyphyllia talpina* has mouths scattered uniformly over its surface, a less distinct axial furrow, and more abundant tentacles. *Fungia (Ctenactis) echinata* and *F. (Ctenactis) simplex* are quite similar to *Herpolitha limax,* though their large-toothed septa readily distinguish them from it. In addition, *Fungia echinata* has one large mouth in the axial furrow, while *F. simplex* has multiple mouths, like *Herpolitha,* but does not have mouths outside of the axial furrow (Veron, 1986).

Natural Habitat: It occurs on sandy, rubble, or muddy bottoms in lagoons and protected reef slopes, often shaded by adjacent coral bommies. It is capable of "walking" by inflating its tissue and lurching forward.

Aquarium Care: Tongue coral is an interesting and hardy species that can be highly recommended to the novice aquarist. It readily adapts to all light regimes, and swells enormously with water when it is really "happy". In the aquarium, place *Herpolitha* on the bottom, in a light current stream. It likes shade, but be sure that it is not too shaded by corals expanding above it, which may block out too much light. It will probably move around a bit, and its progress can be halted by the strategic placement of small "border" rocks.

Herpolitha limax in the natural setting, 3m (10 ft.) deep on a sandy/muddy bottom in the Whitsunday Islands. J. Sprung.

Aquarium Reproduction: This coral reproduces by forming anthocauli, as in other fungiid species. Hobbyists in Europe have reproduced a few generations of *Herpolitha,* and have traded many offspring. Sexes are probably separate, as in other fungiids.

Scientific Name: *Polyphyllia talpina* (Lamark, 1801)

Common Names: Tongue coral, Sea Mole, Slipper Coral

Colour: Brown, gray, or bright green

Distinguishing Characteristics: *Polyphyllia* is shaped essentially the same as *Herpolitha limax,* but with a much fleshier polyp and numerous long tentacles and mouths all over the "tongue", usually

Polyphyllia talpina. J.C. Delbeek.

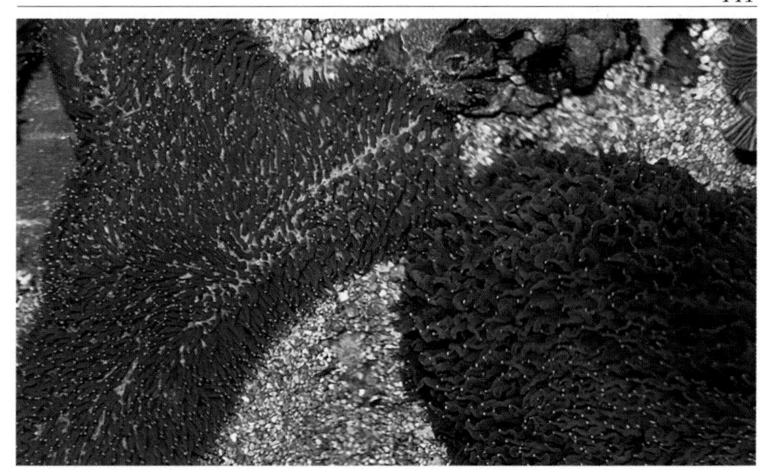

Polyphyllia talpina. Compare the dimensions of the corallum and the density of the tentacles in these two specimens. We suspect that they are distinct species of *Polyphyllia,* but they may just be distinct morphs of *P. talpina.* J. Sprung.

without a distinct central groove. When fully expanded the carpet of tentacles is all that is visible. Three species exist (Veron, 1986); two are common imports from Indonesia. See photo.

Similar Species: *Herpolitha limax*, see previous description.

Natural Habitat: It occupies the same niche as *Herpolitha,* on sandy or rubble bottoms in lagoons, and protected reef slopes.

Aquarium Care: *Polyphyllia talpina* is easy to keep in the aquarium. Since it has many mouths all over the top surface, and a carpet of tentacles capable of capturing small prey, it will catch brine shrimp and other fish foods that drift by, but does not need to be fed at all. Light should be bright, but it adapts to lower light levels. Place it on the bottom with slight currents and it will expand enormously. If the light and/or currents are too strong it will not expand.

Aquarium Reproduction: Unknown. Loose septa may detach with some tissue and form daughter colonies. Like other fungiids, it may reproduce asexually by forming anthocauli or bud off daughter polyps from around the base. Sexes are probably separate, as in other fungiids.

Family Oculinidae (Gray, 1847)

Scientific Name: *Galaxea fascicularis* (Linnaeus, 1767)

Common Names: Crystal Coral, Galaxy Coral, Star Coral, Durian Coral, Brittle Coral

Colour: Polyps are usually brown in colour but polyp tips can be white or green.

Distinguishing Characteristics: *Galaxea* forms large colonies that are lightweight because the skeleton is not very dense. The polyps of *Galaxea* are separate from each other, and look like tubes projecting up about 1.25 cm (0.5 in.) off the flat plain

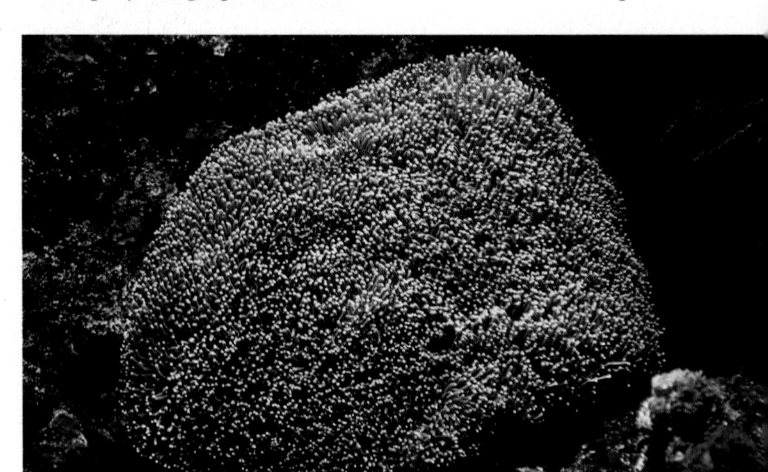

Galaxea fascicularis. J.C. Delbeek.

(coenosteum) connecting them. *Galaxea fascicularis* has corallites up to 5 mm (0.25 in.) diameter or greater, and many septa reaching the center of the corallite (Veron, 1986).

Similar Species: *Galaxea astreata* (Lamarck, 1816) has smaller corallites, up to 4.5 mm diameter max., but usually about 3mm, with only eight to twelve septa reaching the center of the corallite (Veron, 1986).

Natural Habitat: *Galaxea* is a common import from Indonesia, occurring in nearshore reefs and lagoons in shallow, turbid water. J. Sprung observed *Galaxea* growing together with Organ-pipe Coral, (*Tubipora musica*), and sponges in Australia. Their proximity indicated compatibility between this stony coral and

Organ-pipe Coral. This species is very common over a wide range of habitats and may be the dominant coral on inshore, fringing reefs (Veron, 1986). This coral is often the home of numerous obligate commensal pontoniidnid and alpheid shrimp (e.g. *Anapontonia, Platycaris, Ischnopontonia* and *Racilius*), and the aquarist may receive some of these unexpected guests in large heads of *Galaxea fascicularis* (Bruce, 1973).

Aquarium Care: *Galaxea* grows well once established. It is delicate at first mostly because of injury from the method of collection or from transportation. It is especially prone to "brown jelly" infections. Also, if the usual associated sponges are left on the underside of a colony, their death as a result of poor handling can cause an infection that kills the coral. Damaged colonies often exhibit characteristic tissue loss on the coenosteum (between polyps), though the polyps are intact. Under good conditions this tissue rapidly heals and grows back over the exposed skeleton.

While quite beautiful, *Galaxea* is one of the least desirable species because it forms extremely elongate sweeper tentacles with a potent sting. A small fragment of this coral can be interesting for variety, but a large colony will dominate much of the aquarium, particularly if placed in a strong current, which causes the sweeper tentacles to stretch. This coral tolerates low current velocities and medium to high light intensities.

Aquarium Reproduction: *Galaxea* has been propagated in aquaria by fragmentation. It is known to broadcast gametes during mass spawnings in November on the Great Barrier Reef and in summer in the central Pacific and Red Sea (Richmond and Hunter, 1990). It can also brood and release zooxanthellae-bearing planulae daily, all year long (Fadlallah, 1983; Veron, 1986).

Family Poritidae (Gray, 1842)

Scientific Name: *Alveopora* (de Blainville, 1830)

Common Names: (She-loves-me-not) Daisy Coral, Goniopora, Ball Coral, Flowerpot Coral, Yo Stone, Sunflower Coral

Colour: Brown, gray, green, white, or a combination of these.

Distinguishing Characteristics: *Alveopora* spp. are very similar to *Goniopora* spp., but differ in the number of tentacles on each

Alveopora sp. S.W. Michael.

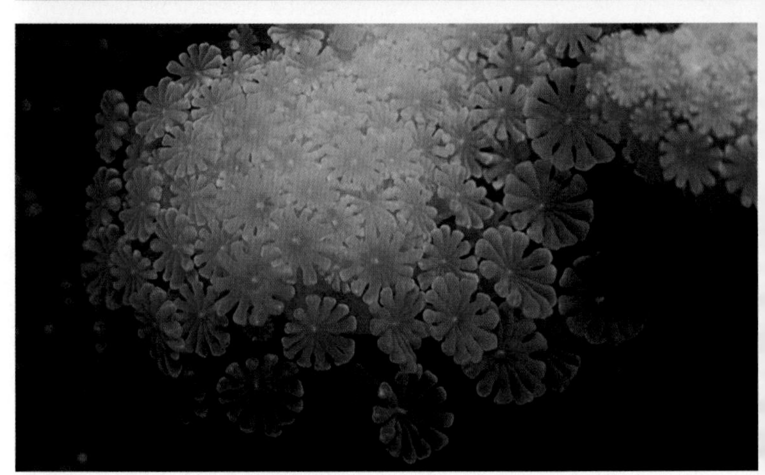

Alveopora sp. J.C. Delbeek.

polyp, 12 for *Alveopora* and 24 for *Goniopora*. The tips of the tentacles in *Alveopora* are typically blunt. The few species available to hobbyists have branched skeletons composed of numerous fingers about 1.25 - 2.5 cm (0.5 to 1 in.) in diameter and 5 - 7.5 cm (2 to 3 in.) long. The tubular polyps expand several inches, making a spectacular display.

Similar Species: *Goniopora* spp.

Natural Habitat: *Alveopora* is not an especially common species, but occurs usually in shallow turbid water in lagoons and coastal reefs with very little wave action, and ranges even into temperate waters. Some species occur in clear water (Veron, 1986).

Aquarium Care: *Alveopora* spp. are rare imports from Indonesia or Singapore which, like the similar *Goniopora,* generally have poor survival in home aquaria. It is likely that the difficulty is related to special trace elements and or dissolved nutrients required by the coral. Once its particular nutritional requirements are discovered, it should be quite easy to maintain. Until that time, it is not recommended for beginners. Fortunately, it is seldom imported. See *Goniopora* for more information.

Alveopora catalai (Wells, 1968) and *A. gigas* (Veron, 1985) occur in turbid or deep water, protected from wave action (Veron, 1986). This suggests that the conditions of medium light levels and little current will afford the best results.

Occasional success with this genus has been reported, and we observed a healthy, three year old colony of *Alveopora gigas* in Bob Goemans' aquarium. It was maintained under fluorescent lighting, with very slight currents. Strong light and more powerful water motion resulted in poor expansion, and excess food added to the aquarium caused the colony to remain closed for several days (B. Goemans, pers. comm.).

Aquarium Reproduction: In branching species fragmentation could be employed. In the Red Sea, *Alveopora daedalea* is an hermaphroditic brooder from fall to winter (Richmond and Hunter, 1990).

Scientific Name: *Goniopora* (de Blainville, 1830)

Common Names: Flowerpot Coral, Daisy Coral, False Brain Coral, Goniopora, Ball Coral, Sunflower Coral, Yo Stone

Colour: Usually brown or gray with green highlights. May be solid green, purple, or coppery coloured. Purple colonies are usually from very shallow water.

Distinguishing Characteristics: *Goniopora* spp. can form free-living globes, encrusting sheets, tall columns, and finger-like branches. There are about forty species. All of them have tubular polyps which extend far off of the skeleton like flowers on tall stems. The polyps have 24 oral tentacles. Common species available include: *Goniopora stokesi* (Edwards and Haime, 1851), *G. lobata* (Edwards and Haime, 1860), *G. tenuidens* (Quelch, 1886), *G. djiboutiensis* (Vaughan, 1907), and an unidentified species (?) that

Goniopora lobata. J.C. Delbeek.

Goniopora stokesi. J. Sprung.

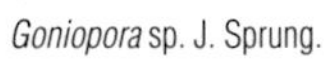

Goniopora sp. J. Sprung.

appears similar to either *G. pandoraensis* (Veron and Pichon, 1982) or *G. fruticosa* (Saville-Kent, 1893). *Goniopora stokesi* forms hemispherical colonies that are usually free-living, but may be attached and encrusting. Asexual reproduction by means of polyp "ball" formation is especially common in *G. stokesi.* In strong currents, the tentacles at the tips of the extremely elongate tubular polyps become long and stringy. *Goniopora lobata* is similar to *G. stokesi,* but the tentacles do not become so stringy in currents, though the polyps are equally long. The skeleton differs as well. *Goniopora stokesi* has a wide columella and ragged walls (Veron, 1986). *Goniopora djiboutiensis* is very similar to G. lobata, but is distinguished by large oral cones on the living polyps (Veron, 1986). *Goniopora tenuidens* has long polyps with short tentacles all of similar length, and is typically shades of green or pink. In the unidentified branched *Goniopora* species the branches are about 2-3 cm (0.8-1.2 in.) in diameter with long polyps, typically brown, sometimes with coppery highlights. The oral disc is lighter in colour.

Similar Species: See *Alveopora.*

Natural Habitat: Given the large number of species, *Goniopora* can be found in a wide range of habitats ranging from shallow, turbid, nearshore waters, to pristine outer reefs. They generally occur where they are protected from strong wave action. The specimens for the aquarium trade are most commonly collected from turbid lagoons.

Aquarium Care: *Goniopora* is one of the most readily available genera, and has been so longer than hobbyists have been keeping reef tanks. Despite their apparent hardiness in nature, their record in aquariums has so far been poor. Actually, their habit is deceptive and alluring. They are beautiful corals, with elongated polyps that tempt the hobbyist like so many daises on stalks. In the aquarium they remain healthy and beautiful for a convincing six to eighteen months, just long enough to assure the hobbyist that warnings about this genus are unfounded. After about a year they often take a slow, downhill plunge that may last another year or more. During this period the polyps don't expand as they did at first. In the worst case (depending on species) the coral will gradually die. Typically it hangs on but just doesn't expand, and is prone to attack by fish, algae, and infection. Occasionally a hobbyist will have success with a *Goniopora* species (years in captivity, no decline, reproduction and substantial growth), and this mystery has lead to many assumptions about the species. The most common assumption is that the methods of

This unidentified (?) branched *Goniopora* sp. from Indonesia resembles both *G. pandoraensis* and *G. fruticosa* (see text). J. Sprung.

Goniopora sp. J.C. Delbeek.

collection and handling are responsible for the lack of success with *Goniopora,* and the occasional success relates to the rare, perfectly collected piece. This is nonsense. If a coral is badly injured by collection or handling, it either dies immediately or recovers. It does not suffer six or more months later. What is happening is best compared to starvation. It is very likely that the missing factor is one or more trace nutrients, and that once this is discovered, and the elements regularly added to the aquarium, *Goniopora* will suddenly be simple to maintain, as in the example of *Acropora* species and the addition of strontium chloride solution. Most species of *Goniopora* do eat food when it is offered, but feeding plankton or other foods is not the complete answer, since many of the success stories have been in the absence of any food additions. The species that fare best in aquariums are *G. stokesi,* and the branched *G. pandoraensis* - like species, the latter having the best record of success.

Goniopora are sensitive to changes in the water quality, and retract their polyps to display their displeasure. In nitrate rich water, it is possible that the algae *Ostreobium* sp. proliferates in the skeleton (Wilkens, 1990), and affects this and other species (see troubleshooting section, chapter 10). It is also possible that some species cannot tolerate accumulation of nitrate or phosphate in the closed system, but this does not explain their general poor record since plenty of colonies have done the year long good display followed by decline in aquaria with very little nitrate and phosphate, and many colonies have remained quite healthy in aquaria with nitrate.

In general, *Goniopora* species prefer strong currents at least for portions of the day, and bright light; The branched *G. fruticosa-pandoraensis* like species tolerates shady conditions. *Goniopora* specimens that are purple, pink, or bluish come from extremely shallow water with high illumination and strong water motion, and should be placed accordingly in the aquarium.

Positioning of colonies can be facilitated by inserting a plastic screw into the base to form a peg. The peg can then be inserted into a hole in the rockwork. See aquascaping, chapter 7, for more detail about this technique.

For now we recommend that the novice avoid this genus, and that importers order fewer numbers of colonies, until such time that the specific requirement(s) of this genus are worked out.

Aquarium Reproduction: Asexual reproduction in *Goniopora stokesi* is by budding, and new satellite colonies (polyp "balls") form among the polyps of the main colony, dropping off when their skeleton becomes heavy enough to tear them free (see chapter 3). This method of reproduction may occur in other *Goniopora* species as well. *Goniopora* species have male and female colonies, and are broadcast spawners (Veron, 1986) in the fall on the Great Barrier Reef, and in the summer in Okinawa (Richmond and Hunter, 1990).

Scientific Name: *Porites* spp. (Link 1807)

Common Names: Finger Coral, Porites

Colour: Brown, yellow, green, or a combination of these. Often pink or purple in shallow, brightly illuminated zones.

Porites sp. J.C. Delbeek.

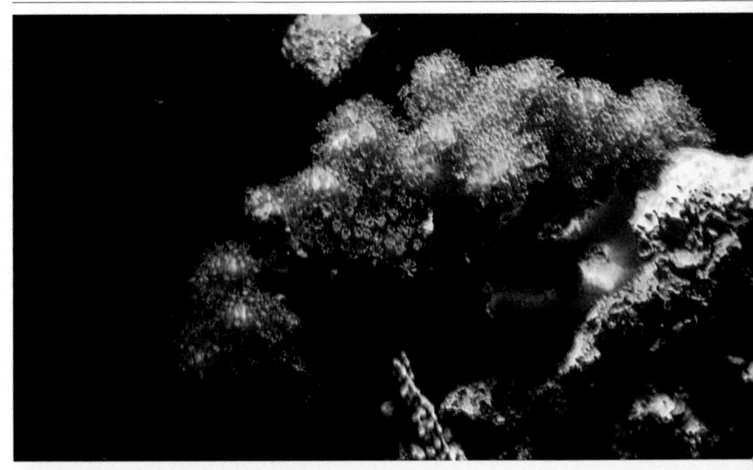

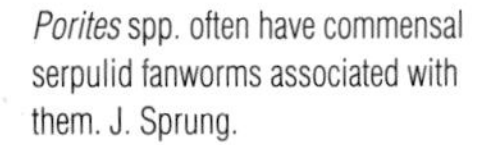

Porites spp. often have commensal serpulid fanworms associated with them. J. Sprung.

Distinguishing Characteristics: Colonies often composed of fingers, but may also be encrusting or dome shaped. Polyps are tiny. *Porites (Synaraea) rus* (Forskal, 1775), which often has fanworms associated with it, is distinctively different in appearance, and resembles *Montipora* species.

Similar Species: Can resemble some species of *Montipora,* but the corallites are usually larger than in *Montipora* and they are filled with septa, while those of *Montipora* have inward-projecting septal teeth (Veron, 1986). *Montipora* is rarely imported for the aquarium trade.

Natural Habitat: All tropical and subtropical reef areas. Also common in lagoons, bays, and grassy areas. Most abundant in areas with strong tidal currents or wave action. Tolerates and seems to prefer extremely bright illumination.

Aquarium Care: Most *Porites* species available to hobbyists are collected incidentally with another target organism. "Christmas tree worm" rocks are typically live heads of *Porites* with the associated boring serpulid worms. Often these heads are so damaged in shipping and handling that they die, or the hobbyist does not offer adequate lighting to grow the coral and it slowly perishes. The worms appear to derive nutrition from the coral mucus, and do not survive long after the coral has died (Wilkens, 1990). With strong illumination and strong currents, it is not only possible, but easy to keep these corals and their commensal worms, and even severely damaged colonies can completely recover as long as a few living polyps survive. New coral tissue rapidly sheets over the dead skeleton, closing in over all exposed areas, and forming thousands of new polyps. Such recovery is only possible when the aquarium is free of filamentous algae and coralline algae are dominant, because filamentous algae readily settle on the exposed skeleton and prevent growth of the coral tissue. *Porites* is not recommended for the novice, though it is hardy when given really strong illumination and strong water movement. *Porites* is one of very few stony corals which, like many soft corals, periodically shed a clear waxy film to remove fouling algae and detritus that settles on its surface. Under low velocity water motion it will shed more frequently than when the polyps are really "blowin' in the wind". The branched, "finger coral" varieties grow very rapidly.

Aquarium Reproduction: Fragmentation of branches, or parts of growing plates in encrusting or tier forming species, is easily accomplished. This genus may reproduce by polyp "bail-out" or asexual formation of planulae, since new colonies have arisen spontaneously in the aquarium (J. Sprung, pers. obs.). It is also possible that these are from brooded planulae and represent true sexual reproduction. Sexes are separate and planulae are brooded, although many species are also broadcast spawners (Fadlallah, 1983; Richmond and Hunter, 1990).

Family Agariciidae

Scientific Name: *Pavona* spp. (Lamarck, 1801)

Common Names: Lettuce Coral, Cactus Coral, Leaf Coral

Several species of *Pavona* have been grown in reef aquaria, getting their start as small colonies attached to live rock with other invertebrates such as mushroom anemones or soft corals. Aquarists with a sharp eye can spot such little treasures and grow nice colonies from them. *Pavona cactus* (Forskal, 1775) and *Pavona decussata* (Dana, 1846) are the most often encountered species.

Colour: Brown with slight greenish hue.

Pavona cactus. J.C. Delbeek.

Pavona sp. J.C. Delbeek.

Distinguishing Characteristics: massive, laminar, or foliaceous colonies, with corallites on both sides of the fronds.

Similar Species: Agaricia (in the Caribbean). Also can easily be confused with *Leptoseris*, which forms foliaceous but generally unifacial colonies.

Natural Habitat: Lagoons and outer reefs. *Pavona cactus* is most abundant in lagoon areas, in shallow, turbid water.

Aquarium Care: *Pavona* are hardy corals and grow rapidly, attaching to the live rock and sending up new branches in just a few months. They fare best when brightly illuminated, but can adapt to lower light levels. They tolerate both strong and weak currents.

Aquarium Reproduction: Propagation is as easy as snapping off portions of the growing blades. Little is known about sexual reproduction in this family, but it appears that colonies have separate sexes (Fadlallah, 1983).

Family Pocilloporidae (Gray, 1842)

Scientific Name: *Pocillopora damicornis* (Linnaeus, 1758)

Common Names: Birdnest Coral, Lace Coral, Seriatopora, "Seriatophora", *S. caliendrum*, Pocillopora, Brush, Cluster, Finger

Colour: Brown with green highlights, especially under blue light. Sometimes pink, or with pink tips.

A large colony of *Pocillopora damicornis* in Berlin, begun from a small fragment. J. Sprung.

Distinguishing Characteristics: *Pocillopora damicornis* is a highly variable species, with numerous growth forms, some dependent on environmental conditions, and some being recognizable subspecies (Veron, 1986). The form typically seen in aquaria forms finely branched hemispherical colonies. Another form has thicker branches and may easily be confused with *Stylophora pistillata*. There are about ten species of *Pocillopora* (Veron, 1986). Several species have thick, heavily calcified branches with bumps called verrucae all over the surface. In *P. damicornis* the verrucae and branches intergrade. Some species may require strong currents for survival in the aquarium. Some specimens (possibly subspecies of *P. damicornis*) show the typical form of *P. verrucosa* or *P. meandrina* when collected, having distinct verrucae on thick, upright branches, but in the aquarium the verrucae can grow into branches that look like *P. damicornis*.

Natural Habitat: *Pocillopora damicornis* has a very wide range, and lives in a variety of environments, including calm turbid bays, wave beaten, clear-water reef fronts and calmer reef slopes. It is obviously an adaptable species, well-suited to aquariums.

Aquarium Care: This species has been propagated for years in European aquariums, beginning in Berlin. This species has been confused with *Seriatopora caliendrum* (Wilkens, 1986), which it does resemble. *P. damicornis* likes bright light and turbulent water motion, which stimulates rapid growth, but it also does well under average lighting, as from standard output fluorescents, and it will expand nicely even in relatively still water. It is not terribly disturbed by rough handling. The polyps retract for only a matter of seconds and expand again immediately after being touched. Low oxygen levels in shipping or toxins in the water may cause portions of a colony to turn white as the polyps "bail-out". These lost polyps may settle in the aquarium and form new colonies, and the bare spot on the original colony rapidly heals. New tissue and polyps grow over the old skeleton in a few days.

Aquarium Reproduction: The skeleton is fragile, and branches are easily broken off. The broken branches readily adhere to live rock, and grow into new colonies. Polyp "bail-out" and asexual formation of planulae are also means of reproduction that have been observed in aquaria. Planulae settle rapidly, already contain zooxanthellae, and a polyp is visible within one day. Four or five more polyps develop around the first one within ten days (J. Sprung, pers. obs.; see chapter 3).

Trachyphyllia geoffroyi, green morph. J. Sprung.

Trachyphyllia geoffroyi, red morph. J. Sprung.

Trachyphyllia geoffroyi, banded morph. J.C. Delbeek.

occurs it may be totally exposed to the light or partially shaded by an adjacent coral head. The brown and green *Trachyphyllia* are generally the most brightly illuminated. The bright red specimens generally are from very turbid water, or they were shaded by an overhang or adjacent coral. They fare best at the bottom of a brightly illuminated tank. The red colour is indicative of green light usage. The use of a green fluorescent bulb will enhance the colour and health of the specimen, though green light makes other specimens appear less colourful.

Distinguishing Characteristics: Colonies are flabello-meandroid with large, fleshy polyps. They are usually bilaterally symmetrical, with one or more mouths (Veron, 1986).

Similar Species: Small *Lobophyllia* may be confused with *Trachyphyllia*. *Lobophyllia* has tougher tissue with rougher texture, spiny septal teeth, and forms larger, taller walled colonies, generally without a conical base. *Scolymia* may also be confused with *Trachyphyllia,* but is not flabello-meandroid, and has spiny septal teeth.

Natural Habitat: *Trachyphyllia geoffroyi* is a common free-living species on shallow mudflats or sandy bottoms around coastal reefs. It grows unattached on the bottom, its ice cream cone-like base holding it in place, and it uses its ability to inflate with water to lift itself up above the sediment to prevent burial (Laboute, 1988).

Aquarium Care: *Trachyphyllia* is hardy, though not as hardy as, for example, *Turbinaria* and *Cynarina*. Iodine (as potassium iodide) is especially critical for the health of *Trachyphyllia*. Symptoms of iodine deficiency are sudden bleaching accompanied by poor tissue expansion, and inability to adapt to a light field in which the specimen had remained healthy previously. At night, oral tentacles are extended, and the coral can be fed small pieces of fresh shrimp, though feeding is not mandatory for this coral. Weekly feedings may result in greater polyp expansion and growth.

Aquarium Reproduction: Asexual reproduction has been noted in the aquarium. First, a portion of skeleton (or one or more septa) form free from the main colony, and their weight drags down on a portion of tissue until it separates from the main colony (de Greef, 1990). *Trachyphyllia* may also bud new polyps around the base as is common in *Euphyllia* and *Plerogyra* species. In nature, this genus most likely sexually reproduces via separate sexes.

Family Dendrophylliidae (Gray, 1847)

Scientific Name: *Duncanopsammia axifuga* (Wells, 1936)

Common Names: None

Colour: Gray, brown, or green.

Distinguishing Characteristics: Large, fleshy polyps held at the end of a dendroid skeleton. Hermatypic.

Similar Species: *Turbinaria heronensis* and *Tubastrea micrantha.* Resembles Elegance coral, *Catalaphyllia jardinei,* when fully expanded, but is most closely related to *Turbinaria.*

Duncanopsammia axifuga.
J.C. Delbeek.

When closed, *D. axifuga* resembles *Tubastrea* and other non-photosynthetic corals such as *Astrangea.* This species may form a link between hermatypic and ahermatypic corals (Veron, 1986).

Natural Habitat: This is a deepwater species restricted to Australia, New Guinea, and eastern Indonesia (Veron, 1986). J. Sprung observed this species occasionally in shallow, turbid water under ledges on coastal reefs along the Whitsunday region of the Great Barrier Reef.

Aquarium Care: *Duncanopsammia axifuga* is a rare species from Australia now being propagated in aquaria. This extremely hardy coral grows rapidly, laying down a dense, heavy skeleton that is difficult to cut. It does well under all types of artificial lighting and, though naturally a shade-loving or deepwater species, it easily adapts to high intensity light. It feeds well on any fish food, and regular (weekly) feedings increase its growth. It expands best under slight currents that make the tentacles sway.

Aquarium Reproduction: A powerful scissors or a hammer and chisel are needed to sever individual polyps or groups of polyps. No other means of reproduction in the aquarium is reported. Sexual reproduction likely involves separate sexes and brooding of planulae as in other Dendrophyllids (see Fadlallah, 1983).

Scientific Name: *Tubastrea* spp. (Lesson, 1829)

Common Names: Orange Cup Coral, Sunflower, Sun, Turret

Colour: Usually orange with yellow polyps with orange mouths. *Tubastrea micrantha* and *T. diaphana* are brown with a greenish fluorescent glow.

Distinguishing Characteristics: *Tubastrea* species have no zooxanthellae, tubular corallites, and typically form fist-sized hemispherical colonies. The most commonly seen species are *T. faulkneri, T. coccinea, and T. micrantha* that forms lovely branching colonies up to several feet tall. These colonies can easily be mistaken for hermatypic corals because of the brown colour and green fluorescence. *Tubastrea micrantha* was formerly called *Dendrophyllia nigrescens*, and the name is still used for it and brown colonies of *T. diaphana*. Both are occasionally imported for the aquarium trade, particularly from Tonga.

Similar Species: *Dendrophyllia*. Distinction is in the septal arrangement in mature polyps (see Veron, 1986).

Natural Habitat: Since *Tubastrea* are ahermatypic, they can grow in areas of little or no light, avoiding competition from the faster growing hermatypic species. Therefore *Tubastrea* are usually found in caves or crevices. *Tubastrea micrantha* can be found growing in the same areas as hermatypic species, provided fast growing *Acropora* spp. are not common (Veron, 1986).

Aquarium Care: *Tubastrea* are common imports from Indonesia and the Caribbean. They are popular because of their bright colour, but they are not especially long lived in most aquaria because of their need for large quantities of food. When they begin

Tubastrea sp., probably *T. faulkneri*, as it appears with the polyps contracted. J.C. Delbeek.

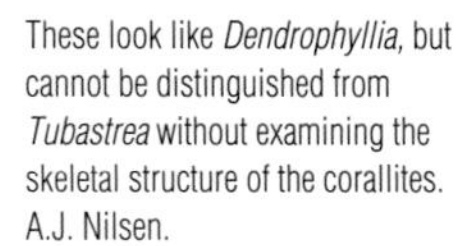
These look like *Dendrophyllia*, but cannot be distinguished from *Tubastrea* without examining the skeletal structure of the corallites. A.J. Nilsen.

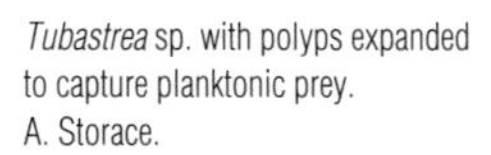
Tubastrea sp. with polyps expanded to capture planktonic prey. A. Storace.

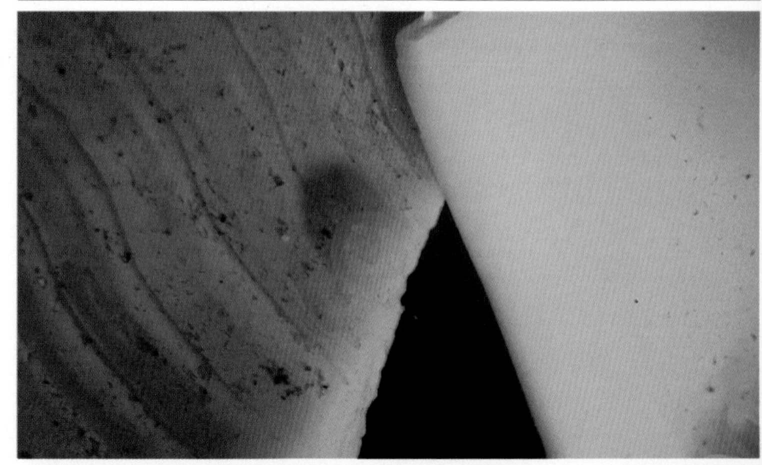

Tubastrea sp. growth series from a planula. This specimen was spawned in a holding tank in the The New York Aquarium for Wildlife Conservation.
J. Yaiullo and F. Greco.

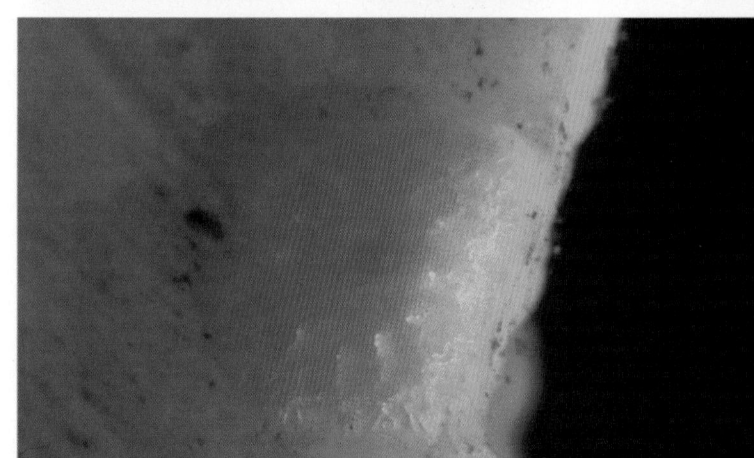

Two weeks later.
J. Yaiullo and F. Greco.

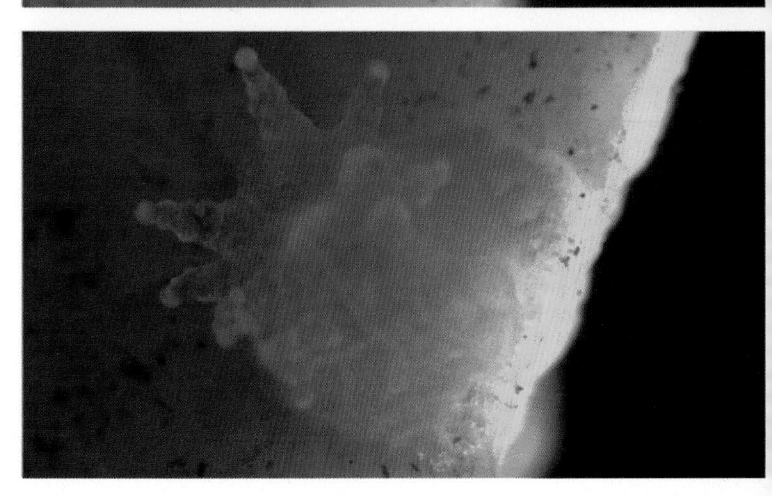

One month later.
J. Yaiullo and F. Greco.

to starve the tissue recedes from the skeleton. Algae can gain a foothold on the exposed skeleton and smother the coral. For this reason we do not recommend them for the beginning hobbyist unless he/she is willing to give this coral the special attention it needs: daily feeding of shrimp tissue, mussel tissue, brine shrimp, worms, or other foods offered to the open polyps by means of a long pipette. Daily additions of live brine shrimp to the aquarium may train this coral to open in the light. When cared for, this coral can be hardy and long lived.

Placement is also critical. *Tubastrea* tolerate light but they don't require it to survive. Therefore it is best to position colonies under ledges, in the shade, either upside-down or right-side up. Current is essential for their health, and stimulates polyp expansion. Ideally the current is strong for part of the time and weak at other times, imitating a tide cycle. The easiest way to position a colony upside-down under a ledge is by drilling a hole in the base and inserting a plastic screw there, for insertion into a hole underneath the ledge. The head of the screw can lock the coral in place, perhaps with the aid of small stones, gum, or epoxy cement to serve as shims. See aquascaping, chapter 7, for more detail about this technique.

Be aware that *Tubastrea* often carries parasitic nudibranchs or the Wentletrap snail, *Epitonium* sp. These will eat the coral's tissue! See chapter 10 for more detail about nudibranchs and snails.

Aquarium Reproduction: *Tubastrea aurea* has produced planulae larvae a number of times in various public and private aquaria, in both North America and Europe (B. Carlson, J. Hemdal, J. Tullock, and J. Yaiullo pers. comms.; Musgrave, 1976; Wilkens, 1976). In most cases the planulae settled within a few days and developed into small colonies. What triggered the release of planulae is not known. There is evidence that planulae can be produced asexually in this coral (Richmond and Hunter, 1990). *Tubastrea faulkneri* on the Great Barrier Reef is a gonochoristic brooder (Richmond and Hunter, 1990).

Genus *Turbinaria*

Scientific Name: *Turbinaria mesenterina* (Lamarck, 1816)

Common Names: Lettuce Coral, Cup Coral, Scroll Coral, "Rugosa"

Colour: Gray, greenish, or pale brown. The growing outer margin is often lighter in colour.

Distinguishing Characteristics: Small corallites are only about 1.5 - 2 cm (0.5 - 0.75 in.) in diameter. This species is very similar to *T. reniformis* (Bernard, 1896), and the two are easily confused. *Turbinaria reniformis* is most often bright yellow, but gray, green or brown colonies resemble *T. mesenterina* so closely that distinguishing them is a job for the coral taxonomist. The polyps are usually more crowded in *T. mesenterina* than in *T. reniformis*, but specimens of either species from shallow water will have crowded polyps and a very highly convoluted shape. The authors suspect that *T. mesenterina* has a shallow water form that can be yellow.

Natural Habitat: This species is a regular import from "Tonga", and rarely from Indonesia where it also occurs but is seldom collected. Shallow, turbid water is where it proliferates. It may also occur in clear water on deep reef slopes. Occasional in tide pools exposed to extremes of temperature, salinity, and light.

Aquarium Care: This species and *T. reniformis* are a little more demanding than *T. peltata*, and are not recommended for the novice. The highly convoluted specimens that are collected in

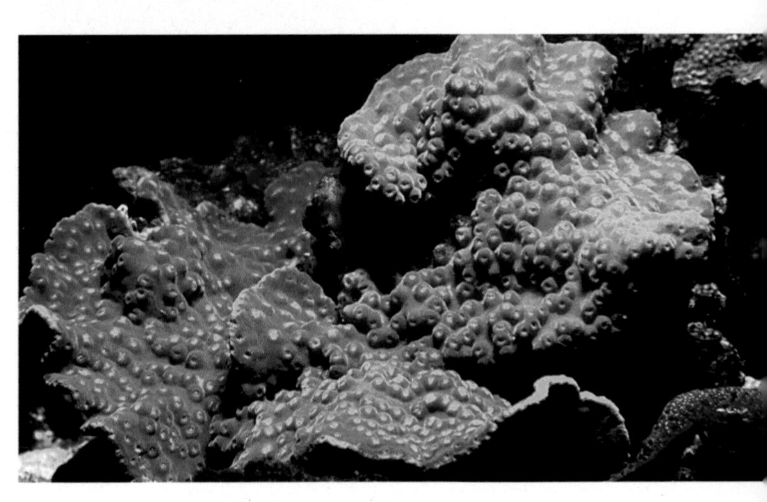

Turbinaria reniformis. J.C. Delbeek.

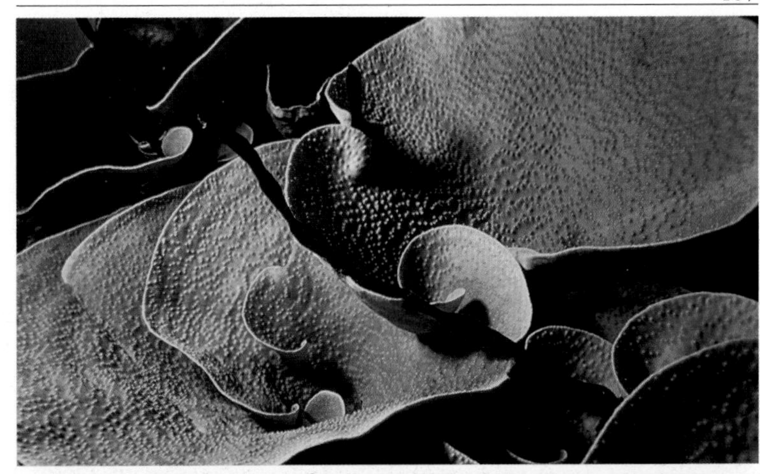

Turbinaria reniformis, Papua New Guinea, at 14 m (45 ft.) depth. B. Carlson.

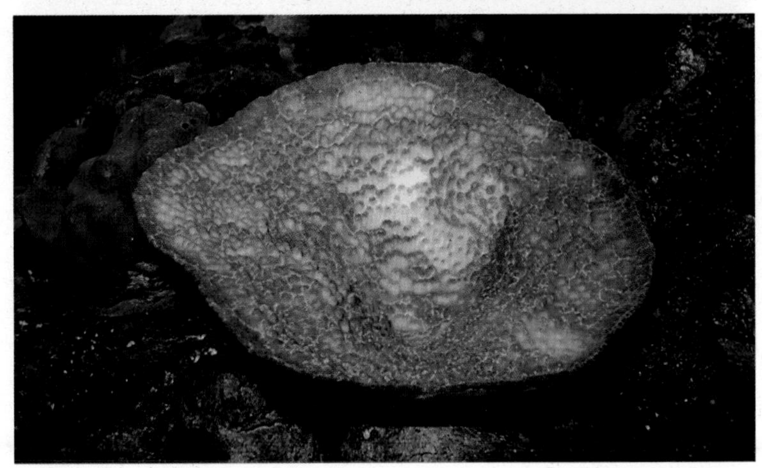

Turbinaria reniformis. This aquarium specimen exhibits bleaching. When the light is not strong enough, yellow colonies turn brown, and the tissue may receed, exposing the coenosteum. Shortage of trace elements, particularly iodide, can also result in bleaching in strongly illuminated specimens. J. Sprung.

shallow water require very strong illumination and water motion. Specimens that are saucer-shaped fare much better, being adapted to shadier conditions with less water flow. The polyps extend mostly at night, but may also extend during the day, though not as prominently as in *T. peltata*. It may fare better with night-time feedings of live brine shrimp nauplii.

Aquarium Reproduction: Not reported. Fragmentation of plates or upright convolutions will produce new colonies. *Turbinaria* species have separate sexes and spawn in autumn when the water becomes cooler. Fertilization is believed to be external (Veron, 1986).

Scientific Name: *Turbinaria peltata* (Esper, 1794)

Common Names: Turban Coral, Cup Coral, Plate Coral, Octopus Coral, Column Coral, Pagoda Coral, Bowl Coral, Vase Coral

Colour: Usually grayish with polyps that are green or brown or a combination of green and brown. Occasionally whole colonies are bright green.

Distinguishing Characteristics: This coral has numerous growth forms, but the common cup, saucer, or plate is most typical. In larger colonies the convolutions at the edge of the plate may grow upward to form erect columns. These are commonly broken off and offered for sale, or the naturally broken pieces around a large

Turbinaria peltata. J.C. Delbeek.

Turbinaria patula is quite similar to *T. peltata*, but it has smaller corallites. J.C. Delbeek.

colony are collected. Wounds heal quickly in this hardy, fast growing coral. The calices are 3-5 mm in diameter, and the polyps may expand over 2.5 cm (1 in.) in diameter, completely obscuring the skeleton in a carpet of tentacles. A similar species, *T. patula* (Dana 1846) is occasionally found. It has more tubular corallites than *T. peltata*, and these incline outward toward the margins (Veron, 1986). Still another species, possibly *T. frondens* (Dana, 1846), has smaller corallites than *T. peltata* and the living polyps, which expand nicely, often have the oral disc of contrasting colour with the tentacles. These two species are also hardy like *T. peltata*, and easy to care for in the aquarium.

Natural Habitat: *Turbinaria* species typically occur in shallow, brightly illuminated water, either turbid inshore, or clear offshore water. *Turbinaria peltata* is most common in inshore water that is very muddy. It is especially common in bays where tidal currents keep fine sediment in suspension, blocking out some of the light. It may be found in tidepools, and tolerates exposure at low tide.

Aquarium Care: *Turbinaria peltata* is very hardy, grows well, and is most desirable because it does not inflate greatly nor send out sweeper tentacles. It fares best under metal halide and actinic combinations, with strong currents at least for portions of the day. It does not need strong currents, but grows best when it receives good water flow. *Turbinaria* may fluoresce bright green under blue fluorescent light, and tolerates dim lighting, under which it will remain healthy, but grow more slowly. It extends its polyps both day and night, often more dramatically at night, which is the best time to offer food. *Turbinaria peltata, T. patula*, and *T. frondens* eagerly take food such as brine shrimp or blackworms, though it is not necessary to feed them at all. Feeding may enhance growth or encourage reproduction.

Aquarium Reproduction: Not reported. Fragments will form new colonies, and will attach to live rock. *Turbinaria* species have separate sexes and spawn in autumn when the water becomes cooler. Fertilization is believed to be external (Veron, 1986).

Fire Corals

Class Hydrozoa
Order Milleporina

Family Milleporidae
Scientific Name: *Millepora* spp. (Linnaeus, 1758)

Common Names: Fire Coral, Ridge Coral

Although fire corals are not true stony corals, we include them here because of their resemblance to them, and similar requirements. Fire corals are hydrozoans, more closely related to jellyfish.

Millepora sp. growing in a reef aquarium. Note that new tissue has encrusted over the adjacent rock. Such growth proceeds very rapidly, and can harm soft corals such as gorgonians. J.C. Delbeek.

Distinguishing characteristics: Colonies form branches, encrusting sheets, upright columns or plates, and may encrust the skeletons of other corals, particularly gorgonians, which can be killed as the fire coral grows. The surface is usually smooth, but can have irregular texture due to the surface of the object being encrusted, or commensal barnacles, or the effects of light and water motion. A few Indo-Pacific species are normally bumpy, forming thick, upright "curtains" with a textured surface. Characteristic of all the species is the appearance of tiny pores on the surface, of two different sizes. The larger pores are called gastropores, the location of gastrozoiids, which consume food. The smaller pores around the gastropores are called dactylopores. Five to seven of these surround a central gastropore, and each contains a dactylozoiid. Dactylozoids are used for prey capture. When extended, they give the colony a fuzzy appearance. They pack a potent sting that can be extremely painful to humans and most fish that come in contact with them. The stings are more severe if they occur on thinner skin, i.e. the underside of the arm. *Millepora alcicornis* from the Caribbean and a few Indo-Pacific and Red Sea species form delicate, anastomosing branches that look like lace. These fast growing hardy species are seldom available to aquarium hobbyists.

Colour: Golden brown to pale yellow. Some species are green.

Similar Species: *Heliopora coerulea*, (Blue Coral), looks like fire coral but has eight-tentacled polyps indicating that it is a soft coral. We will describe this species in volume two. *Porites rus* looks like fire coral but has tiny polyps which are not eight-tentacled, indicating that it is a true stony coral. *Montipora spongodes* also resembles fire coral, but is a true stony coral.

Figure 13.1
Schematic Diagram of *Millepora* colony
After Barnes 1980 and Schuhmacher 1991

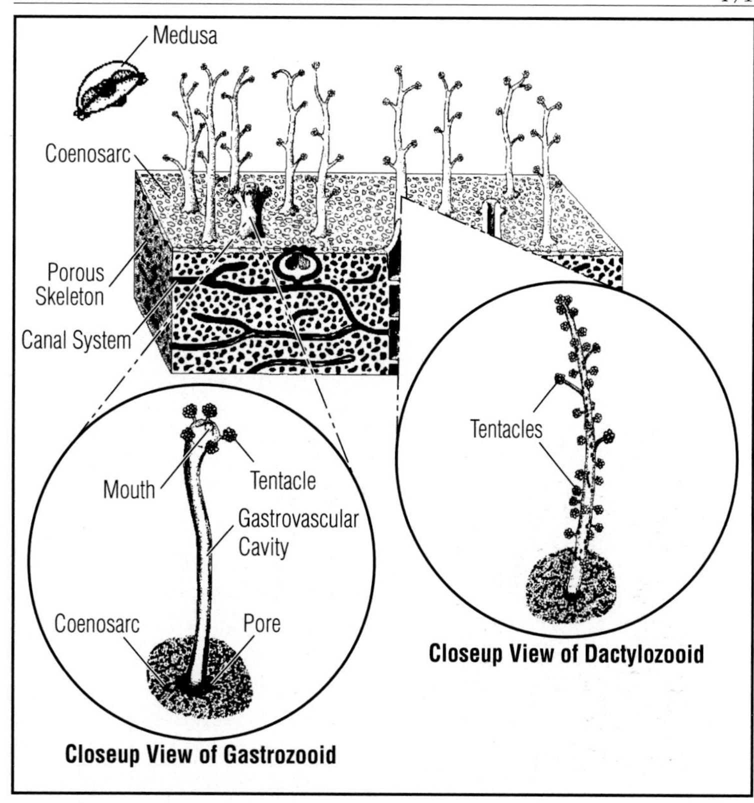

Aquarium Care: *Millepora* adapt well to all light regimes, but grow fastest in the brightest light. Water motion is the most critical factor in stimulating rapid growth, and with *Millepora* the rule for water motion is "the more the merrier". This is a very hardy genus that can be recommended for beginners. Unfortunately it is seldom collected. Growth rate is spectacular, with rapid formation and extension of branches, and encrusting sheets spreading out from the point of attachment with the rock.

Aquarium Reproduction: *Millepora* are easily propagated within the aquarium via cuttings. Sexual reproduction produces tiny free-swimming medusae (Veron 1986).

Chapter Fourteen

Spectacular Reef Aquariums from Around the World

Imagine what it would be like to own a reef aquarium like this! In our travels we have encountered some pretty amazing public and private reef aquariums, and we thought it would be inspirational to give a tour of some of these tanks. Their beauty demonstrates the kind of success that one can attain. With perseverance, patience, and by following the techniques described in this book, one really can reproduce thriving ecosystems like these.

We also want to show some variation in the types of ecosystems that can be created in a reef aquarium. Some of these tanks are primarily dominated by stony corals, some by soft corals, and others are mixed. Some have growth of macroalgae, while others are dominated by coralline algae only. Some have amazing fish populations and some have few fish at all. Each one is still a representation of a section of the reef ecosystem.

There is value in observing the aesthetics of the layout of the reef structure and placement of the corals and clams. The reef aquarium is not just a living ecosystem, it can also be arranged in an aesthetically pleasing manner. Note how the design of the exterior of the aquarium enhances its appearance as well.

We want to thank the people who graciously allowed us to visit with them and photograph their aquariums. It is fantastic that we have met so many people across North America and Europe through our common interest in the beauty and mystery of coral reefs.

Julian is stunned by Klaus Jansen's 8000 L (2,000 gal.) reef aquarium. A.J. Nilsen.

Peter Findeisen's reef aquarium.
J. Sprung.

... full of healthy fish and corals!
A.J. Nilsen.

Aquarium 350x70x60 cm of Mr. Stoop, Moerdijk, Netherlands. L.N. Dekker.

Aquarium 200x105x75 cm of Mr. Leen Dekker, Dordrecht, Netherlands. L.N. Dekker.

Aquarium 120x50x40 cm of Mr. Ansums, Culenborg, Netherlands. L.N. Dekker.

Aquarium 210x60x60 cm of Mr. Vermeer, Aperen, Netherlands. L.N. Dekker.

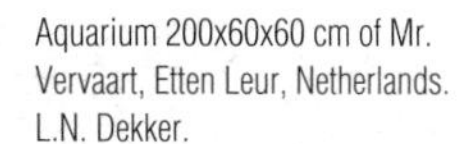

Aquarium 200x60x60 cm of Mr. Vervaart, Etten Leur, Netherlands. L.N. Dekker.

Aquarium 200x60x60 cm of Mr. Leen Dekker, Dordrecht, Netherlands. L.N. Dekker.

Aquarium 220x60x50 cm of Mr. van der Ven, Dordrecht, Netherlands. L.N. Dekker.

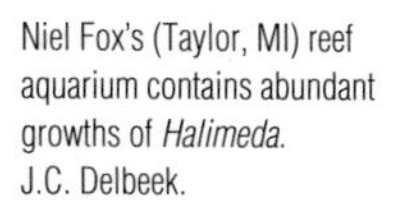

Niel Fox's (Taylor, MI) reef aquarium contains abundant growths of *Halimeda*. J.C. Delbeek.

John Burleson's reef aquarium, Frederick, MD. J. Sprung.

Mick Smith's reef aquarium, Omaha, Nebraska. J.C. Delbeek.

A portion of Steve Tyree's Acropora dominated aquarium. S. Tyree.

A portion of Alf Jacob Nilsen's reef aquarium in Norway. A.J. Nilsen.

Reef aquarium of Mr. Robert James, Ontario, Canada. J.C. Delbeek.

A portion of Dietrich Stüber's reef aquarium in Berlin. J. Sprung.

Part of Julian Sprung's reef aquarium, Miami, Florida. J. Sprung.

Appendix A

Right: Light and water movement table showing acceptable ranges for the different species of stony corals and tridacnid clams. Low numbers correspond to low intensity. High numbers indicate high intensity. Some corals and clams tolerate wider ranges of light intensity and water movement than others. The lighting values also correspond approximately to the light spread drawings on pages 483 and 484, affording some indication of where to place specimens in the aquarium.

Specimen Stony Corals	Light	Water Movement
	2 4 6 8 10	2 4 6 8 10
Family Acroporidae		
Acropora	••••••••	••••••••
Family Caryophylliidae		
Catalaphyllia	•••••••••••	••••••••
Euphyllia	••••••••••	••••••••
Nemenzophyllia	•••••••••••	••••
Plerogyra	••••••••••••••	••••
Physogyra	••••••••••••••	••••••••
Family Mussidae		
Blastomussa	••••••••	••••••••
Cynarina	••••••••	••••••••
Lobophyllia	•••••••••	••••••••••••
Symphyllia	•••••••••••	••••••••••••
Family Faviidae		
Caulastrea	••••••••••••	•••••••••••
Favia	•••••••••••	••••••••••••••
Favites	••••••••••••	••••••••••••••
Goniastrea	••••••••••••	••••••••••••••
Leptoria	•••••••••	••••••••••••••
Montastrea	••••••••••••	••••••••••••
Oulophyllia	•••••••••••	••••••••
Platygyra	••••••••••••	••••••••••••
Family Fungiidae		
Fungia	••••••••••••	•••••••
Heliofungia	•••••••••••	•••••••
Herpolitha	•••••••••••	•••••••
Polyphyllia	•••••••••	•••••••
Family Oculinidae		
Galaxea	•••••••••••	•••••••••••••
Family Poritidae		
Alveopora	•••••••	••••••••
Goniopora	••••••••••••	••••••••••
Porites	••••••••	••••••••••
Family Agariciidae		
Pavona	•••••••••••	••••••••••••••

Stony Corals	Light	Water Movement
	2 4 6 8 10	2 4 6 8 10
Family Pocilloporidae		
Pocillopora damicornis	•••••••	••••••••••••••
Pocillopora spp.	•••••••	•••••••
Seriatopora hystrix	•••••••	••••••••••••••
Stylophora pistillata	••••••••	••••••••••
Family Trachyphylliidae		
Trachyphyllia	••••••••••	•••••••
Family Dendrophylliidae		
Duncanopsammia	•••••••••	••••••
Tubastrea	•••••••	•••••••
Turbinaria peltata	••••••••••	••••••••••
T. reniformis	•••••	••••••••••
T. mesenterina	•••••••	••••••••••
Fire Corals		
Millepora	••••••••	••••••••••••••
Tridacnid Clams		
T. crocea	••••••••	••••••••
T. derasa	•••••••••••	••••••••
T. gigas	•••••••••••	••••••••
T. maxima	••••••••	••••••••••
T. squamosa	•••••••••••	••••••••
T. tevoroa	•••••••	••••••••
Hippopus hippopus	••••••••	••••••••••
H. porcellanus	••••••••	••••••••••

Right: Top view and side view - light spread of metal halide lighting.

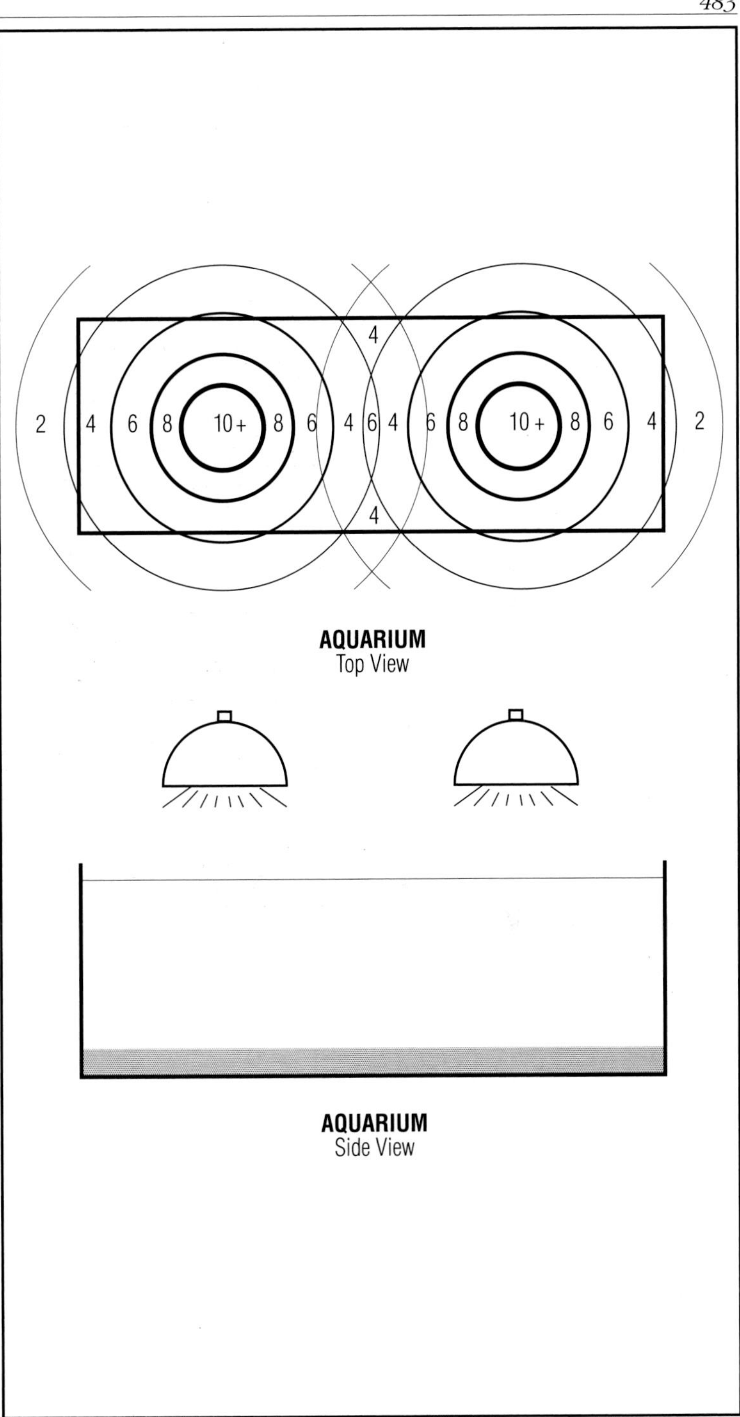

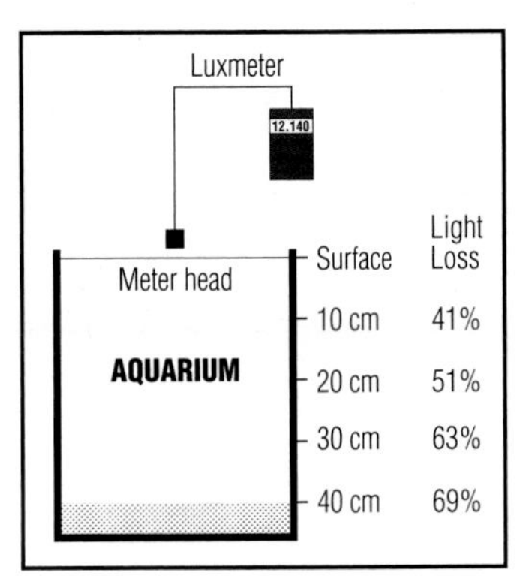

The diagram above provides an estimate of the vertical attenuation of light from a source above the aquarium. The percentages can be applied to give an idea of the approximate values for the light at different depths directly below the center of the bulb. The spread of the light differs between metal halide and fluorescent bulbs (see p. 484), and the intensity lost with depth is different. Reflector design can greatly affect the light spread and penetration.

Right: Top view and side view - light spread of fluoresent lighting.

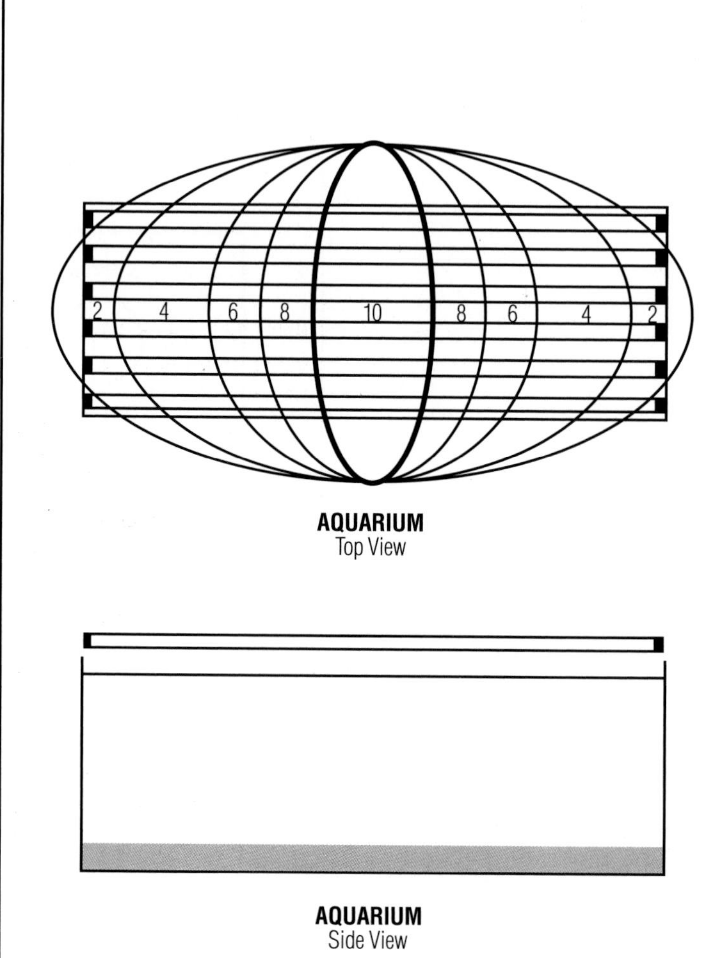

Appendix B General Maintenance Procedures

One might reasonably ask, "How much work is involved in the up-keep of these tanks?" Well, from day to day there might not be a lot that needs to be done, and it is quite possible to leave the tank for extended periods of several weeks or more, depending on the size of your freshwater reservoir to make up for evaporated water. Still, a number of chores need to be done to perpetuate your living ecosystem indefinitely.

Observing the Tank!

Most people don't consider this duty a chore at all, since the pleasure of viewing an aquarium is the primary reason we set one up in the first place. Actually, watching the tank is one of the most important maintenance procedures. The best aquarists have keen observational skills and a kind of empathy for the aquarium and its inhabitants. Noticing early when a specimen has fallen or grown so large that it is stinging or getting stung can make the difference between survival and death for it and other animals in the aquarium. We can't recommend how much time you should spend watching your aquarium, but as long as it doesn't interfere with your work, family, or social life, then observe your reef, often.

Keeping Your Hands Out

While observing the tank is definitely a positive habit, you must resist the temptation to keep re-arranging things and disturbing the environment. The physical contact from your fingers and constant repositioning can prevent growth. Also, oils from your skin, perfumes, etc. can affect the health of the animals, and will impede the performance of your protein skimmer. Still there are many times when you will need to put your hands in, so don't be afraid to do so.

Water Changes

We recommend that about 10 to 25% of the water be changed every few months. See the topic of water changes, chapter 8, for more detail on how and why they are done.

Replacing Evaporated Water

As water evaporates from the system, it must be replenished with freshwater, free of contamination and free of plant stimulating nitrate, phosphates and silicates. The water may simply be poured in daily, or every few days, but much greater environmental stability (and aquarist freedom!) is achieved by installing an automatic water make-up system. Then the frequency of involvement by the aquarist depends on the capacity of the freshwater reservoir and the rate of evaporation of the tank. Automatic water make-up systems can be made with a freshwater reservoir and gravity fed float valve, a float switch and pump arrangement, or a metering pump. Always have a reservoir to draw from. Never use a make-up system attached directly to the water mains of the house, unless you want to find your salt water aquarium suddenly full of freshwater, and your carpet, driveway, or neighbourhood flooded.

Adding Major and Minor Elements and Trace Elements

See chapter 8 for dosage recommendations.

Feeding

How often you feed the fish depends on the number and type of fish, the size of the tank, and its ability to grow algae, crustaceans, and worms on which the fish feed. In small aquariums with a high fish population the food produced within the aquarium may be consumed too rapidly, necessitating food additions. It is ok to feed the tank every day if you have good protein skimming and denitrifying zones. Feeding the aquarium does enhance the growth of corals when the accumulation of nitrate and organic compounds is limited. Don't starve your fish! Many tanks need not be fed at all, and the fish find sufficient algae and small invertebrates on the rocks perpetually. This is especially true of larger aquariums. Even though the fish sometimes need not be fed, feeding them at least few times per week will reduce the incidence of them picking on the corals and tridacnid clams. Leaving the fish without food for a week or two while you are on vacation seldom presents any difficulty for them. Don't worry, they will be fine!

Feeding the corals is not necessary for those species that have zooxanthellae. Non-photosynthetic species such as Tubastrea need to be fed several times per week, preferably daily. You can feed photosynthetic corals occasionally, and they will capture food offered to the fish. See chapters 3 and 13 for additional information on feeding corals.

Cleaning the Glass

Generally it is not necessary to clean the glass unless you expect company to come over and look at your tank, or if you plan to photograph it. The use of herbivorous snails in quantity will keep the glass clean enough for good viewing of the inhabitants, though photography of the tank requires that the glass be bladed free of algae. If the glass becomes so coated with algae that you cannot see through it despite the presence of plenty of herbivorous snails, this is a sign that there is excessive nutrients (phosphate primarily) or trace elements (silicate and iron mostly) in the tank. More protein skimming, the use of purified water for make-up of evaporation, and reduction in feeding and trace element addition will reduce the growth to a point where the snails can further control it. See chapter 9 for more detail on control of algae.

Several tools such as razor blades, credit cards, magnetic algae cleaners with or without blades, and pot scrubber materials are useful for cleaning the glass. See next section for more details.

When there is vigorous growth of coralline algae on the glass, it may be necessary to use a razor blade monthly to prevent the growth from obstructing the view. Be careful that the hard crusts do not scratch the glass when you are cleaning it!

Siphoning

When performing water changes you can siphon out deposits of grey detritus that pile up in locations of little water movement, between rocks and in the sump. This helps to maintain water quality.

Cleaning the Mechanical Filter (if present)

If you use a mechanical filter, it should be cleaned often. See mechanical filtration, chapter 5, for more detail on designs. When filter floss is used, it is best to clean or replace it every few days, to prevent the decay and return to the water of the trapped matter. We don't recommend cartridge type mechanical filters for reef tanks, though they are good (and often very helpful) in fish-only aquariums. They are less likely to be cleaned often, so they become dirty biological filters in a reef aquarium.

Changing Activated Carbon or Other Chemical Filter Media

See chapter 5, chemical filtration, for more detail about the use of activated carbon. If you use activated carbon, it will need to be replaced about every 3 to 6 months. There is a great deal of variability on this requirement, the details of which we explain in chapter 5.

Pruning Algae

See chapter 9 for information on control of undesirable algae.

We refer here not to the undesirable species, but to the pruning of various decorative algae, such as *Caulerpa, Halimeda, Sargassum, Kalymenia, Halymenia,* and *Botryocladia*. *Caulerpa* grows extremely rapidly, and is somewhat undesirable because of its rapid growth rate and ability to injure corals. Nevertheless, *Caulerpa* species are quite beautiful and popular. Managing their growth requires pruning back runners with a scissors about once per week. Otherwise they can take over the entire tank! Herbivorous fishes such as tangs can control the growth of *Caulerpa,* but it is a shaky balance between control and complete consumption as the tangs grow larger. Slower growing algae species such as *Botryocladia, Kalymenia, Halymenia,* and *Sargassum* can be allowed to grow for a month or more before pruning is needed. They grow like houseplants, while *Caulerpa* is more like a weed. Please refer to Littler et al., (1989), an excellent reference, for easy identification of marine algae with colour photographs.

Pruning the Corals, Removing Excess Growth or Big Specimens

It will not take very long either for the corals, anemones, or tridacnid clams to become too large for their position or too large for the tank. Soft corals especially need to be pruned fairly often, depending on the amount of light, trace elements, and available nitrate, all of which enhance their growth. Branches may simply be cut with a sharp scissors (see topic, coral propagation, chapter 13). Stony corals can be pruned similarly, though large single polyps must be moved to allow their continued growth and expansion without harming themselves or neighbouring specimens. Tridacnid clams, particularly *Tridacna gigas, T. derasa,* and *T. squamosa* can easily outgrow a small aquarium in a year or two, depending on the availability of calcium, ammonium, nitrate, and light, which enhance growth. It would be nice if the clams could be propagated by making cuttings, but unfortunately they cannot. When they grow too large they must be moved to a larger aquarium.

Replacing Lamps

Whether you use fluorescent or metal halide lighting, or a combination of them to illuminate your reef tank, the bulbs will need to be replaced at least once per year. Even though they continue to burn bright, after a year of daily use the bulbs have lost intensity and their colour spectrum has shifted. The daily switching on and off ages the bulbs, not just the number of hours they are used. See chapter 6 for additional information.

Cleaning Salt From Bulbs, Light Fixtures

Ideally you should wipe off any salt accumulations on fluorescent bulbs or on shields in the light fixture about once per week, to prevent the salt from reducing the amount of light entering the aquarium. Perform this task when the fixture is off, using a damp cloth.

Cleaning Pumps

About every six months the pumps should be shut off, disconnected from the plumbing, and serviced. When designing your plumbing system, include unions and ball valves to make removal of the pump easy. Servicing magnetic drive pumps simply involves cleaning the impeller with a toothbrush, and cleaning the inside of the impeller chamber to clear out detritus and attached invertebrates (tube worms, sponges). The intake pipe and any strainer grids also need to be cleared.

Cleaning Pipes

The inside surface of the plumbing will become coated by carbohydrates and detritus, or living plants and animals. Sponges and calcareous tube worms commonly coat the insides of pipes and restrict water flow, and clear tubes that receive any light will soon have growths of cyanobacteria or other algae clogging them like plaque in human arteries. In small diameter dark spaces with high water velocity, yellowish carbohydrates form thick gelatinous refractory deposits. A set of "foxtail" brushes can be used periodically to clear the deposits, but some lengths of PVC pipe may remain inaccessible, and that is ok. Flexible hose that can be removed should be cleaned at least once per year, and flexible hose exposed to the light should be cleaned about every three months or more frequently to clear the algae growth and assure continued good circulation. If coralline algae grow in the hose, then the growth of most other algae will be slowed considerably, and it will not need cleaning so frequently. Generally it is best to use dark tape or other shade material to block the light from clear pipes, preventing the growth of algae.

Cleaning Protein Skimmer, Parts, Replacing Air Stones

As with the plumbing, the parts exposed to water in the protein skimmer will become coated with marinelife or refractory organic substances. The riser column just below the cup, or the neck of the cup itself, becomes noticeably coated by deposits of skimmed material within a few days. It should be cleaned at least once per week for ideal operation. The rest of the skimmer becomes coated more slowly. Plan to disassemble your protein skimmer about every four to six months to give it a thorough cleaning, using foxtail brushes, cotton swabs, a toothbrush, and cloth towel to clean off the gelatinous stuff that deposits on the inside. Venturi driven skimmers need special attention for the Venturi, which becomes clogged by salt and dust on the air input orifice, and by organic material in the constricted section that the water is forced through. The Venturi should be cleaned about every three to four months. The impeller on the motor supplying the Venturi should also be cleaned at the same time intervals. In some Venturi skimmer designs, turning the skimmer off daily by means of a timer will flush the Venturi and keep it free of excess salt and dust. Wooden air stones need to be replaced about every two to three months, though sometimes they need to be replaced more frequently. Judge this requirement by the output of the skimmer and the density of bubbles visible in the contact tube if it is made from clear pipe. Calcium build-up on the parts can be dissolved by soaking them in vinegar for a few hours, and then rinsing the parts thoroughly before re-installing them.

Useful Equipment

Routine maintenance can be facilitated by the following instruments that should be kept handy around the aquarium.

A turkey baster - for blowing detritus out from the rocks, picking up small items, washing sediment off specimens, or for spot feeding.

A pipette - for spot feeding fine foods to polyps.

Eyedroppers - for spot feeding, removal of excess mucus or sediment, and for administering trace elements.

Pick-up tool/ tongs- a handy device for picking up fallen objects, especially in deep aquariums.

Scissors - very useful for pruning algae and corals, for picking up small objects, or for cutting monofilament line (see aquascaping).

Magnetic glass cleaner - these make cleaning the glass simple

Toothbrush - for cleaning algae in the corners of the tank, for brushing off filamentous algae.

Old credit card - very handy for cleaning algae off the glass!

Siphon hose and bucket - for performing water changes, removing detritus, and sucking out unwanted introductions, especially *Hermodice carunculata.*

Small containers - It is useful to have some small, capped floating or sinking plastic containers to separate fish or invertebrates temporarily. These can be made from Tupperware™, Rubbermaid™, or similar plastic containers commonly used for carry-out food (especially for hot and sour soup from Chinese restaurants, in our experience). Simply drill or poke holes in the container to allow for adequate circulation of water.

Larger plastic containers - such as pitchers, ice cube bins or other utility containers from the refrigerator, are useful for carrying specimens to and from the display tank, for temporarily holding small specimens when you are decorating the tank, or for gradual acclimation of invertebrates.

Fox tail brushes - for cleaning small diameter pipes, especially for soft tubing.

Wire cutters - For cutting stony coral branches

Finally... A Useful Practice

Keeping a log book may seem like an unnecessary chore for a fun hobby like aquarium keeping, but in practice the insight gained is very rewarding. In a log book you can note all changes and additions to the aquarium. Hindsight is 20:20, but if you don't keep a record, memory may not be so clear. The log will serve you well in your process of learning what maintenance procedures work well for your system, and discovering cause and effect relationships resulting from your tinkering with it. You should note temperature, photoperiod, pH, redox, date and quantities of additions, water change and maintenance dates, and all observations. The information from your log is valuable data for comparison with other captive ecosystems. It is useful for publishing your experience, and demonstrates your professional approach to aquarium keeping.

Appendix C

Suggested Readings, Computer Programs, Videos & Books

Adey, W.H. and K. Loveland. 1991. *Dynamic Aquaria: Building Living Ecosystems.* Academic Press, Inc., 643 pp.

Barnes, R.D. 1980. *Invertebrate Zoology.* W.B. Saunders Co., Toronto, 1089 pp.

Copeland, J.W. and J.S. Lucas 1988. *Giant Clams in Asia and the Pacific.* ACIAR Monograph No.9, Canberra, Australia, 274 pp.

deGraaf, F. 1968. *Handboek Voor Het Tropisch Zeewateraquarium.* A.J.G. Strengholt Boeken, Utrecht, The Netherlands, 362 pp.

Fossa, S. and A. J. Nilsen. 1992a. *Korallenriff-Aquarium.* Band 1. Birgit Schmettkamp Verlag, Bornheim, Germany, 201 pp.

————— and —————. 1992b. *Korallenriff-Aquarium.* Band 2. Birgit Schmettkamp Verlag, Bornheim, Germany, 204 pp.

————— and —————. 1993c. *Korallenriff-Aquarium.* Band 3. Birgit Schmettkamp Verlag, Bornheim, Germany, (333pp.).

Gratzek J.B. and J.R. Matthews 1992. *Aquariology: The Science of Fish Health Management.* Tetra Press, N.J., USA, 330 pp.

Heslinga, G., Watson, T. and T. Isamu. 1990. *Giant Clam Farming.* Pacific Fisheries Development Foundation (NMFS/NOAA), Honolulu, Hawaii, 179p.

Holliday, L. 1989. *Coral Reefs.* Salamander Books Ltd., London, U.K., 204 pp.

Jones, O.A. and R. Endean. 1973. *Biology and Geology of Coral Reefs*: Vol.II, III and IV. Academic Press, New York.

Lebedev, V., Aizatulin, T. and K. Khailov. 1989. *The Living Ocean.* Progress Publishers, Moscow, 327 pp.

Littler, D. S., Littler, M. M., Bucher, K. E., and J. N. Norris. 1989. *Marine Plants of the Caribbean: A Field Guide From Florida To Brazil.* Smithsonian Institution Press, Washington D.C. 263 pages.

Moe, M. Jr. 1989. *Marine Aquarium Reference: Systems and Invertebrates.* Green Turtle Publ., Plantation, FL, 512 pp.

—————— 1992. *Marine Aquarium Handbook: Beginner To Breeder.* Plantation, FL, 318 pp.

Myers, R.F. 1989. *Micronesian Reef Fishes.* Coral Graphics, Guam, U.S.A, 301 pp.

Randall, R.H. and R.F. Myers. 1983. *Guide to the Coastal Resources of Guam* Vol.2: The Corals. University of Guam Marine Laboratory Contribution Number 189. University of Guam Press., 128 pp.

Reader's Digest Book of the Great Barrier Reef. Mead and Beckett Publ., Sydney, Australia, 1984, 394 pp.

Schuhmacher, H. 1991. *Korallenriffe: Verbreitung, Tierwelt, Okologie.* BLV Verlagsgesellschaft mbH Munchen Wein Zurich, 275 pp.

Spotte, S. 1979. *Seawater Aquariums: The Captive Environment.* Wiley-Interscience, John Wiley and Sons, New York, 413pp.

—————— 1992. *Captive Seawater Fishes: Science and Technology.* Interscience, John Wiley and Sons, New York, 942 pp.

Tullock, J.H. 1991. *The Reef Tank Owner's Manual.* Aardvark Press, Bridgeport, CT, USA, 272 pp.

Veron, J.E.N. 1986. *Corals of Australia and the Indo-Pacific.* Angus and Robertson, Publ., North Ryde, Australia, 644 pp.

Vine, P. 1986. *Red Sea Invertebrates.* IMMEL Publishing, London, England, 224 pp.

Wilkens, P. 1973. *The Saltwater Aquarium for Tropical Marine Invertebrates.* Engelbert Pfriem Verlag, Wuppertal, Germany, 216 pp.

—————— 1981. *Niedere Tiere in Tropischen Seewasseraquarium.* Engelbert Pfriem Verlag, Wuppertal, Germany, 455 pp.

———— 1990. *Invertebrates: Stone and False Corals, Colonial Anemones.* Engelbert Pfriem Verlag, Wuppertal, Germany, 136 pp.

———— and J. Birkholz. 1986. *Invertebrates - Tube-, Soft- and Branching Corals.* Engelbert Pfriem Verlag, Wuppertal, Germany, 134 pp.

Wood, E.M. 1983. *Reef Corals of the World: Biology and Field Guide.* T.F.H. Publications, Inc., Ltd., Neptune, NJ, U.S.A., 256 pp.

Periodicals

Aquafauna. Inter-Club D'Aquariophilie et D'Ichthyologie Francophone. Rue Deltour 25, B-4431 Loncin, Belgium.

Aquamar: Peces y Aquarios Tropicales y Marinos. Retablo 5, Alcorcon, Madrid, Spain.

Aquarama. Revue bimestrielle d'aquariophilie et de terrariophilie 24, Rue De Verdun, 67000 Strasbourg, France.

Aquarium Fish Magazine, P.O. Box 53351 Boulder, CO, USA, 80322-3351.

Aquarium Frontiers Quarterly Journal, P.O. Box 190957 Miami Beach, FL. 33119.

Belgische Bond voor Aquarium- en Terrariumkunde V.Z.W., Leliestraat 1, 8800 Roseselare, Belgium.

Corals Reefs, Journal of the International Society for Reef Studies, Springer Verlag, Journal Production Department, Postfach 10 52 80, D-6900 Heidelberg, Germany.

Das Aquarium, Albrecht Philler Verlag GmbH, Postfach 2860, D-4950 Minden, Germany.

DATZ (Die Aquarien- und Terrarien-Zeitschrift), Verlag Eugen Ulmer, Postfach 70 05 61, D-7000 Stuttgart 70, Germany.

Freshwater and Marine Aquarium, P.O. Box 487, Sierra Madre, CA, USA, 91025.

Het Zee-Aquarium, Postbus 45, 7680 AA, Vroomschoop, Holland.

Revue Francaise D'Aquariologie Herpetologie. 34 Rue Sainte Catherine, F-5400, Nancy, France.

SeaScope (Free to aquarists), Aquarium Systems Inc., 8141 Tyler Blvd., Mentor, OH, 44060, USA.

Sea Frontiers P.O. Box 498 Mount Morris, Illinois 61054, USA.

Tropical Fish Hobbyist, P.O. Box 427, Neptune, NJ, USA, 07754-9989.

Computer Programs

Aquadoc: Diagnostic Computer Program, professional and hobby versions. Red sea fish pHarm ltd., Free Trade Industrial Zone P.O. Box 4050 Eilat 88000, Israel.

AquaPro 2.0: Software Guide to Successful Aquariums. Coral Desert Systems, P.O. Box 15125, Scottsdale, AZ, USA, 85267-5125

Professor Fish: Disease Diagnostic Program. Oddbirds, 220 Willoughby Ave., Brooklyn, NY, USA, 11205

Sandpoint Aquarium Manager: Data Logging and Analysis Software. Sandpoint Aquarium Products, 1365-B, Interior St., Eugene, OR, USA. 97402

Computer Networks

Compuserve, Fishnet and Aquadata Forums
Contains message boards and file libraries filled with information for marine aquarists. To obtain an introductory membership for Fishnet call 1-800-848-8199 and ask for kit #164.

Internet
International computer network accessed by universities, businesses and individuals.

Prodigy, Pets Forum

Videos

An Introduction to the Hobby of Reef Keeping. The Creation of a Captive Ecosystem by Julian Sprung. 45 min. color VHS. Two Little Fishies, Inc., 4016 El Prado Blvd., Coconut Grove, FL, USA 33133

Coralife® *The Reef Aquarium. Basic Principles & Setup.* 35 min. color VHS. Energy Savers Unlimited, Inc., Harbor City, CA, USA 90710

Appendix D

North American Aquarium Societies

Brooklyn Aquarium Society: P.O. Box 290610, Brooklyn, NY, 11229-0011. Dues: $15. Monthly newsletter, "Aquatica".

Bucks County Aquarium Society: 7 Fayette Dr., Yardley, PA, 19067, USA. Dues: $15, monthly newsletter, the "Buckette".

Chesapeake Marine Aquaria Society (CMAS) c/o James Skinner P.O. Box 224 Olney, MD, 20830, USA .

Chicagoland Marine Aquarium Society (CMAS) c/o Dennis Gallagher, 1455 Nottingham Lane, Hoffman Estates, Illinois, 60195, USA.

Cleveland Saltwater Enthusiasts Association 20897 Fairpark Dr., Fairview Park, OH 44126, USA. Dues $7 individual, $10 family, Monthly newsletter

Dallas/Fort Worth Aquaculturist's Association c/o Bob Coco, 2010 Klondike Dr,. Grand Prarie, TX, 75050, USA.

Desert Aquarist Society (DAS) c/o Terry Gillpin 2061 S. Sunburst Dr., Tucson, AZ, 85748, USA.

Desert Marine Society (DMS) c/o Jeff Lodge P.O. Box 15125 Scottsdale, AZ, 85267, USA.

Florida Marine Aquarium Society 3280 South Miami Ave. Miami, FL, 33129, USA. Monthly newsletter, "Under The Surface"

Gulf Shores Marine Aquarium Society Inc., P.O. Box 6954 Lake Charles, LA, 70606, USA.

Long Island Aquarium Society P.O. Box 5993 Hauppauge, NY, 11788, USA.

Louisville Marine Aquarium Society - L.M.A.S. 2020 Woodbourne Ave. Louisville, KY, 40205, USA.

Marinelife Aquarium Society of Michigan - c/o Kathy Mykolajenko 1090 Dye Krest Dr. Flint, MI ,48532 ,USA. Dues: $10, monthly newsletter.

1989. *Bull. de l'Institut Oceanographique, Monaco, No. special 5.*:187-193.
Jeffrey, S.W. and F.T. Haxo. 1968. Photosynthetic pigments of dinoflagellates (zooxanthellae) from corals and clams. *Bio. Bull. 135*:149-165.

———— and K. Shibata. 1969. Some spectral characteristics of chlorophyll c from *Tridacna crocea* zooxanthellae. *Bio. Bull. 136*:54-62.

Jenkins, M.C. and W.M. Kemp. 1984. The coupling of nitrification and denitrification in two estuarine sediments. *Limnol. Oceanogr. 29*:609- 619.

Jerlov, N.G. 1950. Ultraviolet radiation in the sea. *Nature 166*:111.

Johannes, R.E. 1963. A poison-secreting nudibranch. *Veliger 5*:104-105.

————, Coles, S.L. and N.T. Kuenzel. 1970. The role of zooplankton in the nutrition of some scleractinian corals. *Limnol. Oceanogr. 15*:579-586.

Jokiel, P.L. 1980. Solar ultraviolet radiation and coral reef epifauna. *Science 207*:1069-1071.

———— and R.H. York. 1982. Solar ultraviolet photobiology of the reef coral *Pocillopora damicornis. Bull. Mar. Sci. 32*:301-315.

Jones, D.S., Williams, D.F. and C.S. Romanek. 1986. Life history of symbiont-bearing giant clams from stable isotope profiles. *Science 231.*

Kanel, J. de and J.W. Morse. 1978. The chemistry of orthophosphate uptake from seawater onto calcite and aragonite. *Geochem. Cosmochim. Acta 42*:1135-1340.

Kaplan, E.H. 1982. *A Field Guide to Coral Reefs of the Caribbean and Florida.* Houghton Mifflin Co., Boston.

Keith, R.E. 1980. Protein skimmers in the marine aquarium. *Freshwater and Marine Aquarium 3(9)*:20-21.

Kinsey, D.W. 1983. Short-term indicators of gross material flux in Coral Reefs - how far have we come and how much further can we go? In: Baker, J.T., Carter, R.M., Sammarco, P.W. and Stark, K.P. Eds. *Proceedings: Inaugural Great Barrier Reef Conference, Townsville.* J.C.U. Press. pgs:333-339.

———— and P.J. Davies. 1979. Effects of elevated nitrogen and phosphorus on coral reef growth. *Limnol. Oceanogr. 24*:935-940.

Kinzie, R.A. III, Jokiel, P.L. and R. York. 1984. Effects of light of altered spectral composition on coral zooxanthellae associations and on zooxanthellae in vitro. *Mar. Biol. 78*:239-248

Klostermann, A. F. 1991. The calcium question. *Freshwater and Marine Aquarium 14(7)*:104,108,110.

Knudsen, J.W. 1967. Trapezia and Tetralia (Decapoda, Brachyura, Xanthidae) as obligate ectoparasites of pocilloporid and acroporid corals. Pac. Sci. 21:51-57.

Kuhlmann, D. 1985. *Living Coral Reefs of the World.* Arco Publ.

Laboute, P. 1988. The presence of scleractinian corals and their means of adapting to a muddy environment; the "Gail Bank". *Proc. 6th Int. Coral Symp. vol.3*:107-111.

Lang, J. 1973. Interspecific aggression by scleractinian corals. 2: Why the race is not only to the swift. *Bull. Mar. Sci. 23*:260-279.

Lange, J. and R. Kaiser. 1991. *Niedere Tiere: Tropischer und Kalter Meere im Aquarium.* Eugen Ulmer Gmbh and C., Stuttgart, Germany.

Larsen, J. and ——. Moestrup. 1989. *Guide To Toxic And Potentially Toxic Marine Algae.* The Fish Inspection Service, Ministry of Fisheries, Copenhagen, Denmark.

Lesser, M.P. and J.M. Shick. 1989. Effects of irradiance and ultraviolet radiation on photoadaptation in the zooxanthellae of *Aiptasia pallida*: primary production, photoinhibition, and enzymic defenses against oxygen toxicity. *Mar. Biol. 102*:243-255

—————, Stochaj, W.R., Tapley, D.W. and J.M. Shick. 1990. Bleaching in coral reef anthozoans: effects of irradiance, ultraviolet radiation, and temperature on the activities of protective enzymes against active oxygen. *Coral Reefs 8*:225-232

Lewis, A.D. and E. Ledua. 1988. A possible new species of *Tridacna* (Tridacnidae: Mollusca) from Fiji. In: Copeland, J.W. and J.S. Lucas Eds. *Giant Clams in Asia and the Pacific.* ACIAR Monograph No.9, 274 p., Canberra, Australia.

Littler, D.S., Littler, M.M., Bucher, K.E., and J.N. Norris. 1989. *Marine Plants of the Caribbean, A Field Guide From Florida To Brazil.* Smithsonian Institution Press, Washington D.C. 263 pages.

Livingston, H.D. and G. Thompson. 1971. Trace element concentrations in some modern corals. *Limnol. Oceanogr. 16*: 786-796.

Lobban, , C.S., Harrison, P.J. and M. Duncan. 1985. *The Physiological Ecology of Seaweeds.* Cambridge University Press.

Lomax, K.M. 1976. *Nitrification with Waste Pretreatment on a Closed Cycle Catfish Culture System.* Unpublished Ph.D. thesis. Dept. of Agricultural Engineering, U. of Maryland, College Park.

Lucas, J.S. 1988. Giant clams: Description and Life History. In: Copeland, J.W. and J.S. Lucas Eds. *Giant Clams in Asia and the Pacific.* ACIAR Monograph No.9, 274 p., Canberra, Australia.

———————, Ledua, E. and R.D. Braley. 1991. *Tridacna tevoroa* Lucas, Ledua, and Braley: A recently-described species of giant clam (bivalvia; tridacnidae) from Fiji and Tonga. *The Nautilus 105(3)*: 92-103.

McConnaughey, B.H. 1978. *Introduction to Marine Biology.* 3rd Ed., The C.V. Mosby Company, St. Louis, Missouri, U.S.A.

Meyer, J.L. and E.T. Schultz. 1985. Tissue condition and growth rate of corals associated with schooling fish. *Limnol. Oceanogr. 30*:157- 166.

Meyer, K. 1991. Survey and analysis of natural and artificial aquarium seawater. Proceedings of the American Association of Zoological Parks and Aquariums nat. conf.

Miller, J.A. 1980. Reef alive. *Sci. News 118*: 250-252.

Moe, M. Jr. 1989. (revised 1992) *Marine Aquarium Reference: Systems and Invertebrates.* Green Turtle Publ., Plantation, FL, 512 pp.

——————— 1982. (revised 1992) *Marine Aquarium Handbook: Beginner To Breeder.* Green Turtle Publ., Plantation, FL, 318 pp.

Mohan, P.J. 1990. Ultraviolet light in the marine reef aquarium. *Freshwater and Marine Aquarium 13(1)*: 4-6,156-160.

Muscatine, L. 1973. Nutrition of corals. In: Jones, O.A. and R. Endean Eds. *Biology and Geology of Coral Reefs*: Vol.II. Academic Press, New York.

————— and C. F. D'Elia. 1978. The uptake, retention, and release of ammonium by reef corals. *Limnol. Oceanogr. 23*: 725-734.

————— and J.W. Porter. 1977. Reef corals: Mutualistic symbioses adapted to nutrient-poor environments. *BioScience 27*: 454- 460.

Musgrave, G. 1976. Breeding and raising live coral. *Marine Hobbyist News 4(8)*: 1, 7.

Myers, R.F. 1989. *Micronesian Reef Fishes.* Coral Graphics, Guam, U.S.A, 301 pp.

Nilsen, A. J. 1990. The successful coral reef aquarium: part 3. *Freshwater and Marine Aquarium 13(11)*: 32

————— 1991. Coral reef vs. reef aquarium: Part 2. *Aquarium Fish Magazine 4(1)*: 18-26.

Norton, J.H., Shepherd, M.A., Long, H.M. and W.K. Fitt. 1992. The zooxanthellal tubular system of the giant clam. *Biol. Bull. 183*: 503-506.

———————, ———————, and H.C. Prior. 1993a. Intracellular bacteria associated with winter mortality in juvenile giant clams, *Tridacna gigas.* J. Invert. *Pathol. 62(2)*: 204-206.

——————, ——————, Abdonnaguit, M.R. and S. Lindsey. 1993b. Mortalities in the giant clam, *Hippopus hippopus*, associated with Rickettsiales-like organism. J. Invert. *Pathol. 62(2)*: 207.

Nooyen, J.W. 1990. De truc met de trechter: Een hulpmiddel om grote borstlewormen te vangen. *Het Zee-Aquarium 40(10)*: 205-206.

Ohta, K. 1979. Chemical Studies on Biologically Active Substances in Seaweeds. Proc. Int. *Seaweed Symp., 9*, 401-411.

Ommen, van, J. 1992. Licht boven het zeeaquarium. *Het Zee-Aquarium 42(3)*: 59-63.

Paletta, M. 1989. Suggestions for reef maintenance. *SeaScope Vol.6, Summer 1989.*

—————— 1990. Coral aggression in reef aquaria. *SeaScope Vol.7, (Winter)*: 1-2.

—————— 1993. Enemies in the invertebrate aquarium. *Aquarium Fish Magazine 5(7)*: 18-26.

Patterson, M.R., Sebens, K.P. and R.R. Olson. 1991. In situ measurements of flow effects on primary production and dark respiration in reef corals. Limnol. *Oceanogr. 36*: 936-948.

Patton, D. 1974. Community structure among the animals inhabiting the coral *Pocillopora damicornis* at Heron Island, Australia. In *Symbiosis in the Sea* (W.B. Vernberg, ed.) pp.219-243. Univ. of S. Carolina, Columbia, S. Carolina.

Pingitore, N.E., Tanger, Y. and A. Kwarteng. 1989. Barium variation in *Acropora palmata* and *Montastrea annularis. Coral Reefs 8*: 31-36.

Perron, F.E., Heslinga, G.A., and J.O. Fagolimul. 1985. The gastropod *Cymatium muricinum*, a predator on juvenile tridacnid clams. *Aquaculture 48*: 211-221.

Randall, R.H. and R.F. Myers. 1983. *Guide to the Coastal Resources of Guam Vol.2: The Corals.* University of Guam Marine Laboratory Contribution Number 189. University of Guam Press.

Richmond, R.H. and C.L. Hunter. 1990. Reproduction and recruitment of corals: comparisons among the Caribbean, the tropical Pacific and the Red Sea. *Mar. Ecol. Progr. Ser. 60*: 185-203.

Riseley, R.A. (1971). *Tropical Marine Aquaria: The Natural System.* Allen and Unwin, London.

Rosewater, J. 1965. The family Tridacnidae in the Indo-Pacific. *Indo-Pacific Mollusca 1(6)*: 347-396.

Rougerie, F. and B. Wauthy. 1993. The endo-upwelling concept: from geothermal convection to reef construction. *Coral Reefs 12*: 19-30.

Rudman, W.B. 1984. Molluscs. In: *Reader's Digest Book of the Great Barrier Reef.* Mead and Beckett Publ., Sydney, Australia.

Ruetzler, K., and D.L. Santavy. 1983. The black band disease of Atlantic reef corals. I. Description of the cyanophyte pathogen. PSZNI: *Marine Ecology 4*: 301-319.

Ruiter, D.N. 1987a. Parasieten op lagere dieren/1. *Het Zee-Aquarium 37(1)*: 11-12.

—————— 1987b. Slakken die kolonie-anemonen eten. *Het Zee-Aquarium 37(3)*: 59.

Sammarco, P.W. 1982. Polyp bail-out: an escape response to environmental stress a new means of reproduction in corals. *Mar. Ecol. Prog. Ser. 10*:57-65.

——————, Coll, J.C., La Barre, S. and B. Willis. 1983. Competitive strategies of soft corals (Coelenterata: Octocorallia): allelopathic effects on selected scleractinian corals. *Coral Reefs 2*: 173-178.

Schiller, C. and G.J. Herndl. 1989. Evidence of enhanced microbial activity in the interstitial space of branched corals: possible implications for coral metabolism. *Coral Reefs 7*: 179-184.

Schlichter, D. and H.W. Fricke. 1986. Light harvesting by wavelength transformation in a symbiotic coral of the Red Sea twilight zone. *Mar. Biol. 91*: 403-407.

Schuhmacher, H. 1991. *Korallenriffe: Verbreitung, Tierwelt, Okologie.* BLV Verlagsgesellschaft mbH Munchen Wein Zurich, 275 pp.

SeaScope 1991. Tridacnid clam culture. *SeaScope Vol. 8*: 1,3,4.

Sebens, K.P. and A.S. Johnson 1991. Effects of water movement on prey capture and distribution of reef corals. *Hydrobiologia 226*: 91-101.

—————— and J.S. Miles. 1988. Sweeper tentacles in a gorgonian octocoral: morphological modifications for interference competition. *Biol. Bull. 175*: 378-387.

Segedi, R.M. 1976. Trace elements in closed systems. *Marine Aquarist 6(6)*: 5-12.

Sepers, A.B.J. 1977. The utilization of dissolved organic compounds in aquatic environments. *Hydrobiologia 52*: 39-54.

Sheppard, C.R.C. 1979. Interspecific aggression between reef corals with reference to their distribution. *Mar. Ecol. Prog. Ser. 1*: 237-247.

Shick, J.M. and J.A. Dykens. 1985. Oxygen detoxification in algal-invertebrate symbioses from the Great Barrier Reef. *Oecologia 66*: 33-41.

Shinn, E.A. 1989. What is really killing the corals? *Sea Frontiers* (March-April): 72-81.

Siddall, S.E. 1977. Some design ideas. *Mar. Aquar. Vol.8 No.5*: 5-57.

Simkiss, K. 1964. Phosphates and crystal poisons of calcification. *Biol. Rev. 39*: 487-505.

Smit, G. 1986. Marine aquariums. Part One: Is it time for a change? *Freshwater and Marine Aquarium 9(1)*: 35.

Sorokin, Y. 1973. On the feeding of some scleractinian corals with bacteria and dissolved organic matter. Limnol. *Oceanogr. 18*: 380-385.

——————— 1992. Phosphorus metabolism in coral reef communities: exchange between the water column and bottom biotopes. *Hydrobiologia 242*: 105-114.

Spotte, S. 1979. *Seawater Aquariums: The Captive Environment.* Wiley-Interscience, John Wiley and Sons, New York, 413 pp.

Sprung, J.F. 1988. Captive Reefs. *Tropical Fish hobbyist, October 1988*:72-84.

——————— 1990. Reef Notes. *Freshwater and Marine Aquarium 13(10)*: 71.

——————— 1993. Reef Notes. *Freshwater and Marine Aquarium 16(6)*: 155.

Steen, R.G. and L. Muscatine. 1984. Daily budgets of photosynthetically fixed carbon in symbiotic zoanthids. *Biol. Bull. 167*: 477-487.

Stephens, G.C. 1962. Uptake of organic material by aquatic invertebrates. I. Uptake of glucose by the solitary coral *Fungia scutaria. Biol. Bull. 123*: 648-657.

Stoddart, D.R. 1973. Coral Reefs of the Indian Ocean. In: Jones, O.A. and R. Endean Eds. *Biology and Geology of Coral Reefs: Vol.I.* Academic Press, New York.

Sutton, D.C. and R. Garrick. 1993. Bacterial disease of cultured giant clam (Tridacna gigas) larvae. *Diseases of Aquatic Organisms 16(1)*: 47-54.

Stuber, D. 1980. De vermeerdering van een *Seriatopora* - soort in het aquarium. *Het Zee-Aquarium 40 (10)*: 207-210.

Swart, P. 1980. The effects of seawater chemistry on the growth of some scleractinian corals. In: Tardent, P. and R. Tardent Eds. Developmental and Cellular Biology of Coelenterates. Elsevier/North-Holland Biomedical Press.

Szmant-Froehlich A. 1983. Functional aspects of nutrient cycling in coral reefs. Symp Ser Undersea research, *NOAA Progr. 1*: 133-139.

Teh, Y.F. 1974. Keeping Live Coral. *Mar. Aquarist 5(1)*: 19-25.

Thiel, A. 1988. *The Marine Fish and Invert Aquarium.* Aardvark Press, Bridgeport, CT.

——————— 1989. *Advanced Reef Keeping I.* Aardvark Press, Bridgeport, CT.

Thomason, J.C. and B.E. Brown. 1986. The cnidom: an index of aggressive proficiency in scleractinian corals. *Coral Reefs 5*: 93-101.

Thurman, H.B. and H.H. Webber. 1984. *Marine Biology.* Scott, Foresman and Co., IL, U.S.A.

Trainor, F.R. 1978. *Introductory Phycology.* New York: Joihn Wiley & Sons. 525pp.

Trench, R.K. 1979. The cell biology of plant-animal symbiosis. *Annu. Rev. Plant Physiol. 30*: 485-532.

——————— and R.J. Blank. 1987. *Symbiodinium microadriaticum, S. goreauii, S. kawagutii* and *S. pilosum:* gymnodinoid dinoflagellate symbionts of marine invertebrates. *J. Phycol. 23*: 469- 481.

———————, D.S. Wethey and J.W. Porter. 1981. Observations on the symbiosis with zooxanthellae among the Tridacnidae (Mollusca, Bivalvia). *Biol. Bull. 161*: 180-198.

Tullock, J.H. 1991. *The Reef Tank Owner's Manual.* Aardvark Press, Bridgeport, CT, 274 pp.

Veron, J.E.N. 1986. *Corals of Australia and the Indo-Pacific.* Angus and Robertson, Publ., North Ryde, Australia, 644 pp.

Vine, P. 1986. *Red Sea Invertebrates.* IMMEL Publishing, London, England, 224 pp.

Wafar, M., Wafar, S. and J.J. David. 1990. Nitrification in corals. *Limnol. Oceanogr. 35*: 725-730.

Webb, K.L., DuPaul, W.D., Wiebe, W., Sottile, W., and R.E. Johannes. 1975. Enewetak (Eniwetok) atoll: Aspects of the nitrogen cycle. *Limnol. Oceanogr. 20(2)*: 198-210.

Wiebe, W.J. 1988. Coral reef energetics. In: Pomeroy, L.R. and J.J. Alberts Eds. *Concepts of Ecosystem Ecology: A Comparative Review.* Ecological studies: Analysis and Synthesis. Vol. 67., Springer- Verlag.

———————, Johannes, R.E. and K.L. Webb. 1975. Nitrogen fixation in a coral reef. *Science 188*: 257-259.

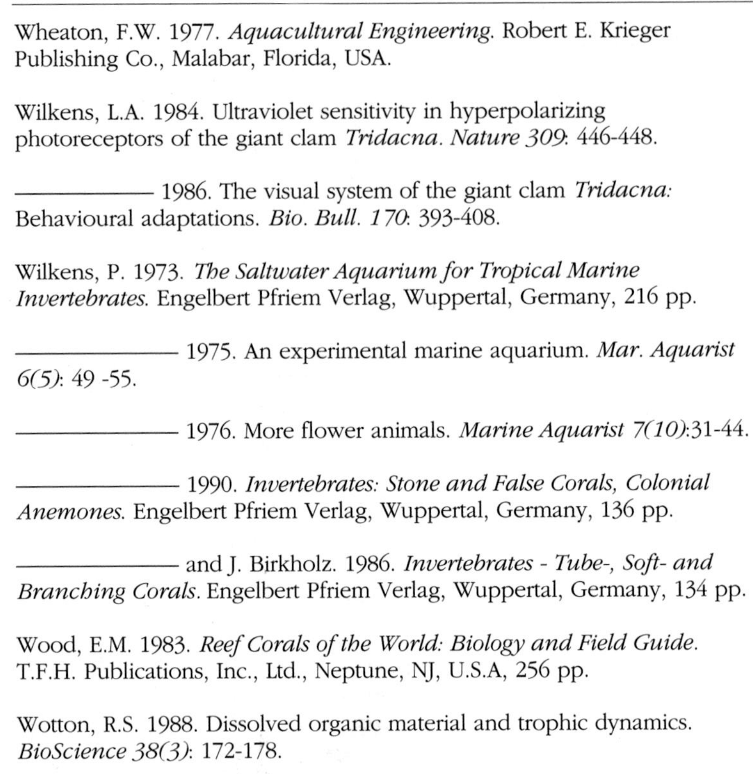

Wheaton, F.W. 1977. *Aquacultural Engineering*. Robert E. Krieger Publishing Co., Malabar, Florida, USA.

Wilkens, L.A. 1984. Ultraviolet sensitivity in hyperpolarizing photoreceptors of the giant clam *Tridacna*. *Nature 309*: 446-448.

——————— 1986. The visual system of the giant clam *Tridacna:* Behavioural adaptations. *Bio. Bull. 170*: 393-408.

Wilkens, P. 1973. *The Saltwater Aquarium for Tropical Marine Invertebrates.* Engelbert Pfriem Verlag, Wuppertal, Germany, 216 pp.

——————— 1975. An experimental marine aquarium. *Mar. Aquarist 6(5)*: 49 -55.

——————— 1976. More flower animals. *Marine Aquarist 7(10)*:31-44.

——————— 1990. *Invertebrates: Stone and False Corals, Colonial Anemones.* Engelbert Pfriem Verlag, Wuppertal, Germany, 136 pp.

——————— and J. Birkholz. 1986. *Invertebrates - Tube-, Soft- and Branching Corals.* Engelbert Pfriem Verlag, Wuppertal, Germany, 134 pp.

Wood, E.M. 1983. *Reef Corals of the World: Biology and Field Guide.* T.F.H. Publications, Inc., Ltd., Neptune, NJ, U.S.A, 256 pp.

Wotton, R.S. 1988. Dissolved organic material and trophic dynamics. *BioScience 38(3)*: 172-178.

Yonge, C.M. 1975. Giant clams. *Sci. Am. 232*: 96-105.

the nitrifying bacteria are grown. The prefilter should be cleaned often.

proboscis - a part of the body that extends from the mouth and aids in feeding.

protein skimmer - a very useful filter that strips nitrogen rich proteins, as well as fatty acids and other surface active organic compounds out of the water by using tiny bubbles to collect them as a stable foam.

redox - reduction-oxidation reactions can be measured with an electrode to afford a numerical value of water quality, though the number value is not an indication of other water quality parameters such as specific gravity or nitrate level. Redox measurement is often combined with controlling equipment for the administration of ozone, an oxidant.

respiration - a cellular process by which oxygen and glucose are used in the production of energy and where CO2 is released as a waste gas.

RNA - ribonucleic acid. Macromolecule found in the nucleus and cytoplasm of cells. Responsible for decoding of DNA and for protein assembly.

S-320 - series of pigments found in aquatic organisms that absorb or reflect harmful wavelengths of ultraviolet light.

salinity - a measure of the quantity of dissolved solids in water. Used to measure the strength of seawater solution. The unit of measure is parts per thousand (ppt.). Full strength seawater has a salinity of about 35 ppt.

Scleractinia - stony or hard corals.

seeded - this term means that enough time has passed to allow the growth of colonies of nitrifying bacteria on rocks or filter media. It is used interchangeably with the term, "cured" with respect to live rock.

septa - radiating vertical plates, lying inside the corallite wall.

septa-costae - radial elements of the corallite, form the septa and the costae.

septal teeth - lobes or spines along the margins of septa.

sexual reproduction - reproduction that involves the fusion of male and female gametes.

shelf reef - reefs that occur on the continental shelf.

siphon - inhalant and exhalant. Used by clams to allow for gas exchange and to expel wastes.

soft coral - Octocorallia. Those corals that have eight tentacles on each polyp. Many different forms exist. They may be soft, leathery, or may even produce a hard skeleton.

specific gravity - the ratio of the density of a given liquid to that of pure water, used to measure the relative salinity of seawater compared to distilled water. Distilled water has a specific gravity of 1.000 while seawater ranges from 1.022 to 1.030.

strontium - a metal cation. Has the same valence as calcium. Corals incorporate it along with calcium in the building of their skeletons. Normal seawater concentrations range from 8-10 mg/L.

sump - the reservoir below the "dry" section in a trickle or "wet/dry" filter, or merely a reservoir below or behind the tank when no trickle filter is used. The water level in the sump varies as water evaporates from the system because the surface skimmer is fixed in position and thus maintains a constant water level in the aquarium. A level switch may be incorporated in the sump to control an automatic water make up system.

surface active - property of certain compounds that makes them "stick" to the air-water interface at the surface of a body of water, or at the surface of bubbles.

surface skimmer - typically, the water drawn into the prefilter comes off of the surface of the tank. This way slimy proteins, which can form a slick on the water's surface, may be continuously drawn off of the surface and trapped in a mechanical "prefilter" or in a protein skimmer. This name is often used interchangeably with prefilter.

surface tension - with respect to water: Molecules at the interface between water and air experience unequal attraction by the air and the water. The measure of this unequal attraction is called surface tension. Surface tension increases with increasing salinity, and decreases with increasing temperature. The size of the bubbles in a protein skimmer is a result of surface tension.

sweeper polyp - elongated polyps that can act in an aggressive manner by stinging neighbouring corals and other sessile invertebrates.

sweeper tentacle - elongated tentacles of polyps that have increased numbers of nematocysts and can be used in aggressive encounters with neighbouring invertebrates.

tentacles - finger-like (feather-like in soft corals) structures that encircle mouth of a polyp. Used for prey capture, defense, gas exchange, reproduction and light absorption.

terpenoid - organic compounds produced by soft corals for defense against predation and for aggressive colonization of new substrate.

trace element - elements such as barium, lithium, iron, iodine, and molybdenum that naturally occur in seawater in very low concentrations (< 1.0 ppm).

trochophore - free-swimming, first planktonic stage of mollusc larvae;

top-shaped with a girdle of cilia.

turbidity - reduced clarity in water. Usually caused by suspended organic or inorganic particles.

turf algae - fast growing filamentous and gelatinous algae that form turfs on the living rocks. These efficiently strip nitrogen and phosphate out of the water. Herbivores should be used to curb their growth inside the aquarium.

ultraviolet light - high energy, short wavelength light between 200 and 400 nm. Three types A, B and C. UV C, 200 - 280 nm, is germicidal. UV B, 280 - 320 nm causes sunburn. UV A, 320 - 400 nm is the least harmful, but can still cause damage to tissues that are not protected by UV absorbing pigments.

undesirable algae - any rapidly growing algae that may cover the rocks and glass and may smother invertebrates. Usually refers to filamentous green algae, brown sheeting diatoms and dinoflagellates, or red sheeting cyanobacteria. Controlled by limiting nutrients (nitrogen, phosphate, silicate), and by herbivores.

veliger - second larval planktonic stage of molluscs where the foot, shell, and other structures first make their appearance.

xanthophyll - accessory pigment found in chloroplasts of plants. Capture light and transport electrons to chlorophyll.

zoanthids - small anemone-like anthozoans with no skeleton; solitary or colonial e.g. *Palythoa, Parazoanthus* and *Zoanthus.*

zonation - a series of regions or zones that contain characteristic organisms.

zooplankton - animals that drift in the water column. Most are microscopic. Some are larval forms of larger organisms.

zooxanthellae - these are the tiny plants called dinoflagellates that live symbiotically with corals, tridacnid clams, and some sponges, providing food to their host and in return getting the nitrogen, phosphorous and carbon dioxide they need for growth.
Symbiodinium spp.

zygote - the first stage of development after the gametes fuse.

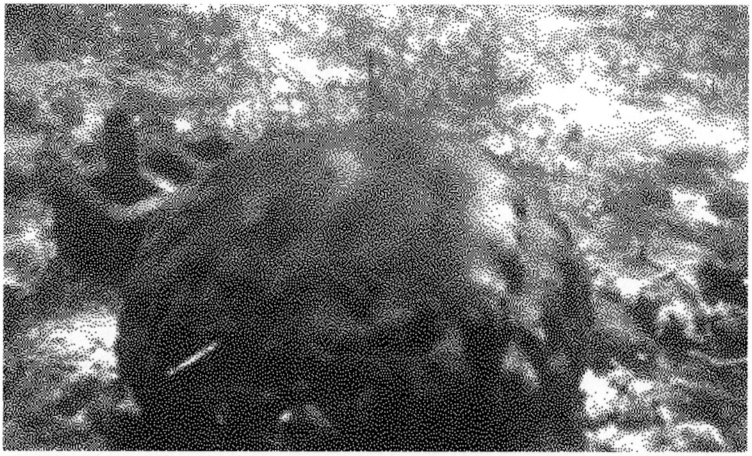

Photo 1
A *Montastrea annularis* head at Grecian Rocks in 1961. The cinder block behind it gives an idea of its size. E. Shinn.

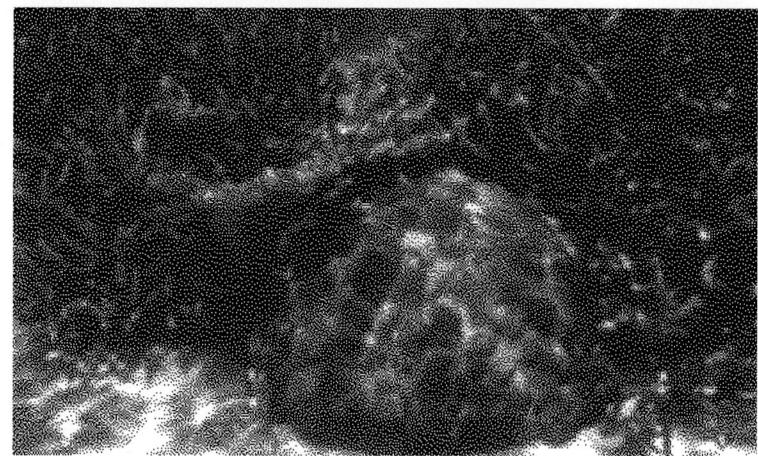

Photo 2
In 1971 the same colony is being overgrown by *Acropora cervicornis* and *A. palmata*.
E. Shinn.

The Cycle of Growth and Death on a Coral Reef

This series of photographs taken by Eugene Shinn at a site selected for growth studies at Grecian Rocks Reef off Key Largo, Florida, documents growth and death of corals in a small area over a period of 27 years (see Shinn, 1989. In addition to the slow demise of the *Montastrea annularis* in the center of the photos, what is striking about the series is that it also demonstrates the rapid rate at which live rock can be produced by fast-growing *Acropora* species. The algae-encrusted loose, dead branches in the photo taken in 1988 are typical of live rock harvested for aquariums. In this example it is produced on a biological time scale, not a geological one. Note that the accumulation of sand and branches has smothered or buried much of this *Montastrea,* though living healthy tissue remains where it can still receive adequate light and water flow. If a storm cleared away the rock and sand, the *Montastrea* could send sheets of living tissue back over the bare skeleton, and other species of coral would likely colonize it too, given adequate herbivory. Much of the loose tangle of dead branches will be broken down into gravel and sand by boring

Photo 3
In 1976 the *A. palmata* is dead, though its skeleton is still visible below the hand-held light, upper left. The *A. cervicornis* continues to overshadow the *Montastrea* head. E. Shinn.

In 1977 a severe cold front that produced snow as far south as Miami damaged temperature sensitive *Acropora* stands. A photo (not shown) taken by Shinn in 1978 reveals the *A. cervicornis* had largely been killed between 1976 and 1978.

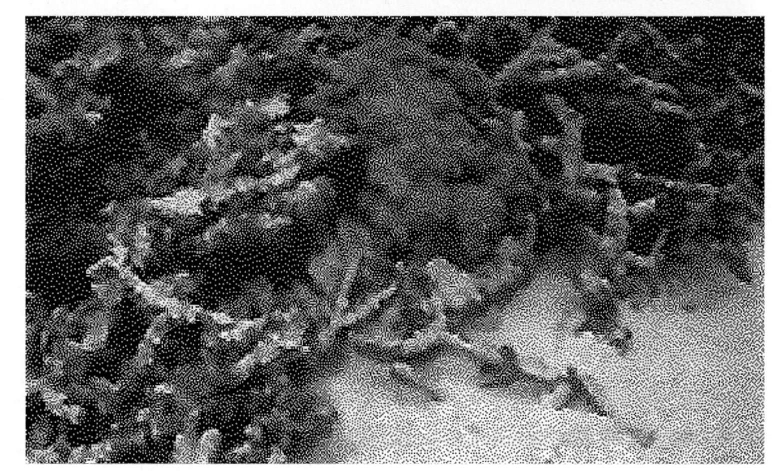

Photo 4
By 1988 few live branches of *A. cervicornis* remain. The dead branches are piled like algae-covered bones. Though healthy, the *Montastrea* has only about one quarter the amount of live tissue it had in 1961. E. Shinn.

organisms, herbivores, and surge. The way reefs survive against the pounding pulverizing surf is by growing, and dying. As corals grow upward and outward, portions must die. New growth covers the old dead growth.

Slow-growing head corals can be very long lived, large colonies often being several hundred years old, sometimes more than one thousand years old. Stony corals are essentially immortal creatures, but change in sea level, natural disasters, predation, and competition for space, (not to mention the impact of people), eventually kills individual colonies. Fragmentation increases the chances that the tissue will survive, while planulae allow dispersal essential to survival of the species. Large colonies or stands of *Acropora* are generally only a decade or decades old, though the dead skeletons beneath them may span hundreds or thousands of years. By growing faster, they make up for the fact that their tissue and skeletons are less able than more durable head corals to withstand environmental extremes.

What *is* that?

A. Serpulid polychaete worms. These filter feeding worms build calcareous tubes found on live rocks and attached to the walls of the aquarium and sump. They multiply prolifically aquariums.

B. Terrebellid "spaghetti" worm. Many different species can be found assoiciated with live rocks and sand substrates. Typically they live in a chitonous tube surrounded by attached shell fragments. The long tentacles are often mistaken for worms. They are detritivores, particulate food running along the tentacles as if on a conveyor belt. They gather food from all over the bottom substrate, and deposit piles near the entrance to their tubes.

C. Spionid worms. Similar to terrebellids, but smaller generally, with only two tentacles. These are common on live rocks and corals, and some live in the sand. They form chitonous tubes with or without sand attached (top of tubes and tentacles shown). Detritivores.

D. Vermitid snails. Sessile snails that feed by means of a mucous net that traps bacteria and particulate matter. Often mistaken for worms. They form hard calcareous tubes and multiply prolifically.

E. Sipunculan "peanut" worm. These harmless, 2-4 cm tough worms bore holes in live rock and coral skeletons. Deposit feeders.

F. Tunicate. Individual or colonial animals that filter bacteria, phytoplankton, and dissolved organic compounds from the water. Require specific trace elements.

G. Barnacle. A sessile filter feeding crustacean that kicks its feet out to catch passing particulate food.

H. "Sessile" Ctenophore. A relative of pelagic comb jellyfish. This tiny (1 cm) or less soft creature puts out two fine 20 cm long tentacles to snare particulate food. The delicate beauty of these tentacles is a marvel. They collapse into a tangle and then instantaneously unravel into neat perfection. Occasionally found on *Caulerpa* algae and on *Sarcophyton* soft corals. They crawl like a flatworm and occasionally swim with a flapping motion.

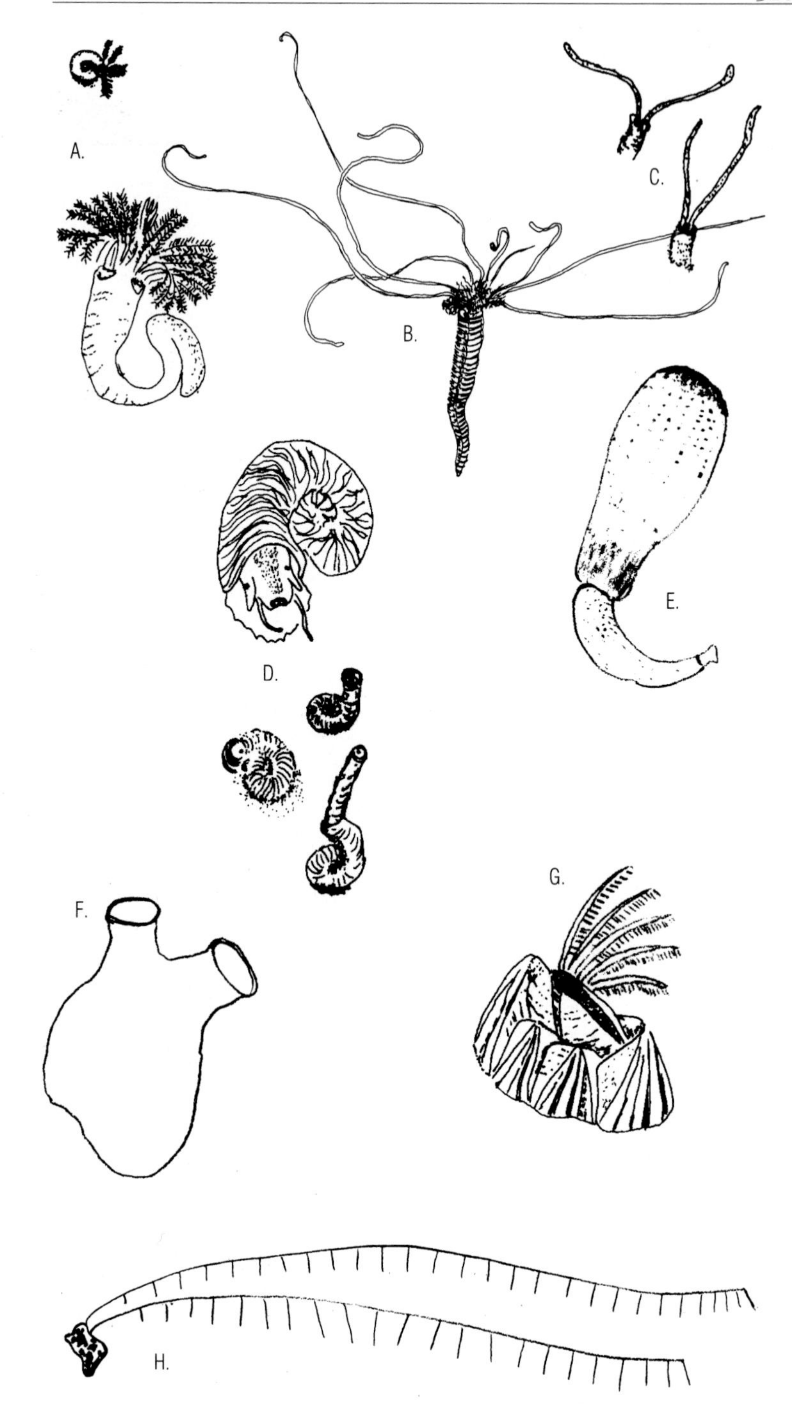

I. Chiton. A primitive mollusk characterized by eight calcareous plates imbedded in a tough integument. Most species are herbivorous.

J. Iosopod. Bug-like crustaceans with dorso-ventrally flattened bodies. Most are oval shaped, but some are elongated and tubular. Most are harmless scavengers, but some species are parasitic. They crawl over the substrate, burrow in sand, and some species swim with amazing speed! Some marine and terrestrial isopods can roll into a ball like armadillos do.

K. Amphipod. Small shrimp-like crustaceans (most 1 cm or less) with (usually) laterally compressed bodies. Harmless scavengers. Many are herbivorous, and play an important role in the control of algae. Most active at night when the lights are off. They can be seen scurrying all over the rocks and bottom substrate.

L. Copepod. Tiny (1 mm or less, generally) crustaceans that swim with a jerky motion. Often have pair of egg sacs hanging off sides. Found crawling over the substrate and walls of the aquarium, and swimming in the water. Feed on algae, detritus, and decaying matter. Bloom dramatically when something has died and is decomposing.

M. Caprellid "Skeleton" shrimp. A specialized amphipod common among algae and bryozoans on live rock. Crawl like inch worms. Feed on plankton and particulate matter.

N. Pycnogonid "Sea spider". Aberrant marine arthropods. These strange creatures often appear to be all legs and no body. Found crawling among algae, hydroids, and bryozoans on live rock. Sometimes swim in the water column. Feed with an anterior proboscis. Commensal on some invertebrates. Some species are predators of bryozoans or hydroids.

See Barnes 1980 for detailed descriptions.

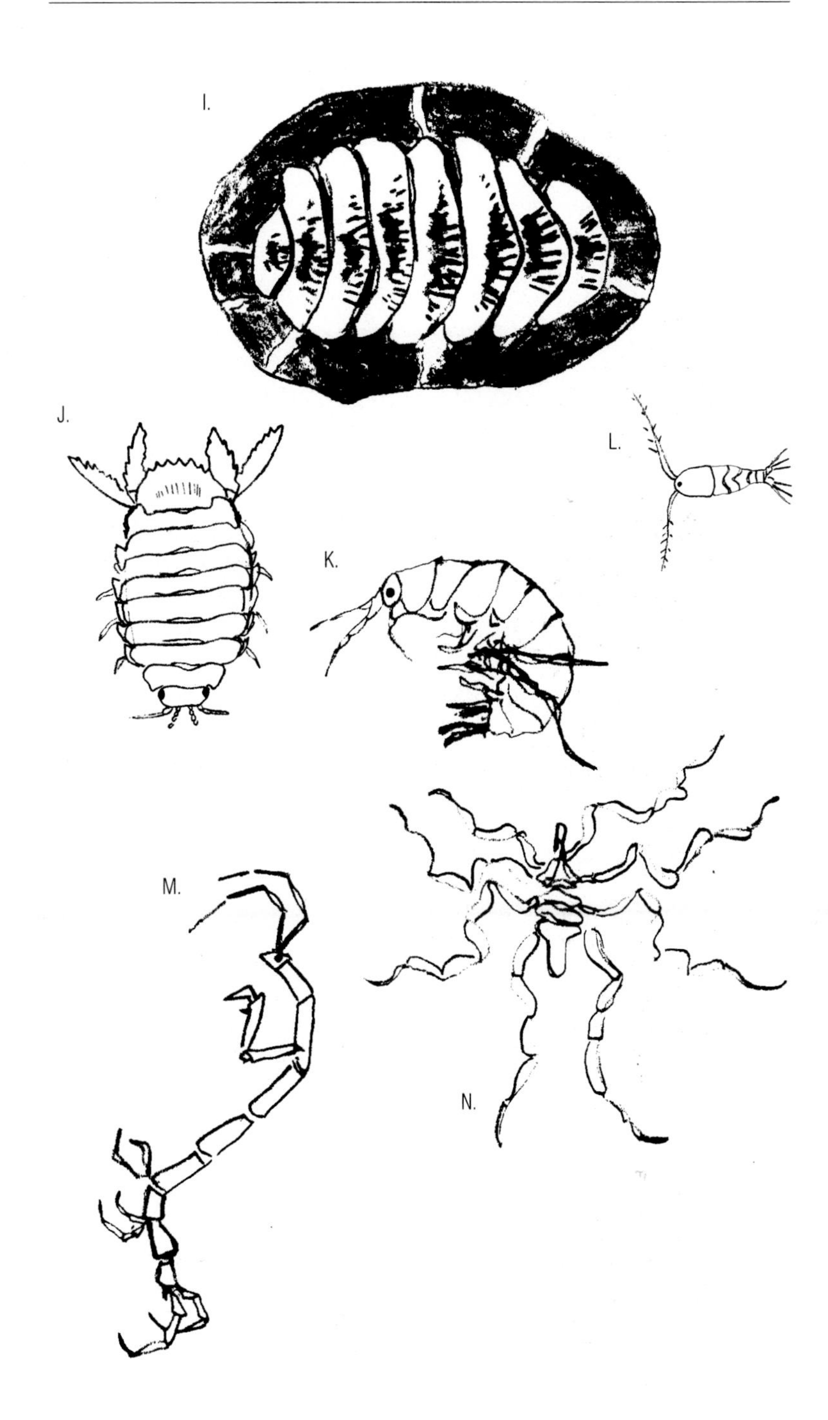

We protect what we love.

Index
